GW01112209

Understanding Change

Also by Andreas Wimmer

FACING ETHNIC CONFLICTS. TOWARDS A NEW REALISM (*co-edited*)
DIE KOMPLEXE GESELLSCHAFT. EINE THEORIENKRITIK AM BEISPIEL DES INDIANISCHEN BAUERNTUMS
NATIONALIST EXCLUSION AND ETHNIC CONFLICTS. SHADOWS OF MODERNITY
TRANSFORMATIONEN. SOZIALER WANDEL IM INDIANISCHEN MITTELAMERIKA

Also by Reinhart Kössler

IN SEARCH OF SURVIVAL AND DIGNITY: Two Traditional Communities in Southern Namibia Under South African Rule (2005)
ENTWICKLUNG (1998)
POSTKOLONIALE STAATEN (1994)
DESPOTIE IN DER MODERNE (1993)
ARBEITSKULTUR IM INDUSTRIALISIERUNGSPROZESS (1990)

*For Yehuda
All good wishes
Andreas*

Understanding Change

Models, Methodologies, and Metaphors

Edited by

Andreas Wimmer

and

Reinhart Kössler

palgrave
macmillan

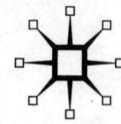

Editorial matter and selection © Andreas Wimmer and Reinhart Kössler 2006
All chapters © their authors 2006

All rights reserved. No reproduction, copy or transmission of this publication may be made without written permission.

No paragraph of this publication may be reproduced, copied or transmitted save with written permission or in accordance with the provisions of the Copyright, Designs and Patents Act 1988, or under the terms of any licence permitting limited copying issued by the Copyright Licensing Agency, 90 Tottenham Court Road, London W1T 4LP.

Any person who does any unauthorized act in relation to this publication may be liable to criminal prosecution and civil claims for damages.

The authors have asserted their rights to be identified as the authors of this work in accordance with the Copyright, Designs and Patents Act 1988.

First published in 2006 by
PALGRAVE MACMILLAN
Houndmills, Basingstoke, Hampshire RG21 6XS and
175 Fifth Avenue, New York, N.Y. 10010
Companies and representatives throughout the world.

PALGRAVE MACMILLAN is the global academic imprint of the Palgrave Macmillan division of St. Martin's Press, LLC and of Palgrave Macmillan Ltd. Macmillan® is a registered trademark in the United States, United Kingdom and other countries. Palgrave is a registered trademark in the European Union and other countries.

ISBN-13: 978–1–4039–3941–8 hardback
ISBN-10: 1–4039–3941–1 hardback

This book is printed on paper suitable for recycling and made from fully managed and sustained forest sources.

A catalogue record for this book is available from the British Library.

Library of Congress Cataloging-in-Publication Data

    Understanding change : models, methodologies, and metaphors / edited by Andreas Wimmer and Reinhart Kössler.
      p. cm.
    Includes bibliographical references and index.
    ISBN 1–4039–3941–1 (cloth)
      1. Social sciences – Philosophy. 2. Change. 3. Social change. 4. Social sciences – Methodology. 5. Science – Methodology. 6. Interdisciplinary approach to knowledge I. Wimmer, Andreas. II. Kössler, Reinhart.

H61.U55 2005
303.4—dc22   2005043284

10 9 8 7 6 5 4 3 2 1
15 14 13 12 11 10 09 08 07 06

Printed and bound in Great Britain by
Antony Rowe Ltd, Chippenham and Eastbourne

Contents

List of Tables and Figures   vii

Notes on Contributors   ix

1   Models, Methodologies, and Metaphors on the Move   1
    Andreas Wimmer

Part I   Chaos and Order in Climate Change

2   Climate Change: Complexity, Chaos and Order   37
    Paul Higgins

3   Chaos in Social Systems: Assessment and Relevance   51
    L. Douglas Kiel

4   Economics, Chaos and Environmental Complexity   59
    Hans-Walter Lorenz

Part II   Genetic Variation in Evolution

5   The Topology of the Possible   67
    Walter Fontana

6   Neutrality as a Paradigm of Change   85
    Rudolf Stichweh

7   Using Evolutionary Analogies in Social Science:
    Two Case Studies   89
    Edmund Chattoe

Part III   Economics of Continuity: Path Dependency

8   The Grip of History and the Scope for Novelty: Some
    Results and Open Questions on Path Dependence
    in Economic Processes   99
    Carolina Castaldi and Giovanni Dosi

9   Analyzing Path Dependence: Lessons from the Social Sciences   129
    James Mahoney

10   Path Dependence and Historical Contingency in Biology   140
    Eörs Szathmáry

Part IV Institutional Inertia

11 The New Institutional Economics: Can It Deliver for Change and Development? 161
Jeffrey B. Nugent

12 Institutions, Politics and Culture: A Case for 'Old' Institutionalism in the Study of Historical Change 177
John Harriss

13 Exporting Metaphors, Concepts and Methods from the Natural Sciences to the Social Sciences and *vice versa* 187
Raghavendra Gadagkar

Part V The Multilinear Modernization of Societies

14 Multiple Modernities in the Framework of a Comparative Evolutionary Perspective 199
Samuel N. Eisenstadt

15 On Modernity and Wellbeing 219
Oded Stark

16 Multiplicity in Non-Linear Systems 222
Somdatta Sinha

Part VI Constellations of Contingency: Political History

17 Historical-Institutionalism in Political Science and the Problem of Change 237
Ellen M. Immergut

18 Social Science and History: How Predictable Is Political Behavior? 260
Roger D. Congleton

19 Reconstructing Change in Historical Systems: Are There Commonalties Between Evolutionary Biology and the Humanities? 270
Joel Cracraft

20 History, Uncertainty, and Disciplinary Difference: Concluding Observations by a Social Scientist 285
Reinhart Kössler

Index 303

List of Tables and Figures

Tables

10.1	Examples of social learning	147
10.2	The major transitions in evolution	150
15.1		220
15.2		220

Figures

1.1	A basic transition probability matrix	10
1.2	Contingency	11
1.3	Transformation	12
1.4	History I: event chains	13
1.5	History II: path dependency	14
2.1	Schematic bifurcation diagram for thermohaline circulation as measured by North Atlantic Deep Water (NADW) production in Sverdrups	42
2.2	Equilibrium results of a simple climate model under different forcing scenarios	43
2.3	Model response of the coupled atmosphere–biosphere system to vegetation perturbations for the Sahel and starting from a vegetation distribution for West Africa close to today's	44
2.4	Sensitivity of modelled biome distributions to initial vegetation	45
5.1	The folding of RNA sequences into shapes as a proxy of a genotype–phenotype map	68
5.2	RNA shape	70
5.3	Epistasis	72
5.4	Sequence space for sequences of length 4 over the binary alphabet {0,1}	73
5.5	Neutral networks and shape space topology	75
5.6	A: An example of a discontinuous shape transformation RNA; B: Punctuation in evolving RNA populations	78
8.1	The Fujiyama single-peaked fitness landscape	104
8.2	A fitness landscape with several local maxima peaks (Schwefel's function)	105
8.3	A non-linear transition function that implies multiple steady states	110

9.1	Illustration of contingency in self-reinforcing sequence	134
10.1	The formose 'reaction', which is, in fact, a complex network of autocatalytic sugar formation. (a) The 'spontaneous generation' of the autocatalytic seed is a very slow process; and (b) the autocatalytic core of the network. Each circle represents a group with one carbon atom	144
10.2	DNA methylation as a chromatin marking system	146
10.3	Memes and Lamarckian inheritance. (a) The Weissmanist segregation of soma and germ line; (b) transfer of memes passes through the performance level, which is mostly absent in the molecular world; and (c) in most cases a meme becomes multiplied by the interactions of two memes	148
10.4	Schematic evolution of ECP and EDN proteins	150
10.5	The radiation of Mexican salamanders	154
16.1	Processes involved in the evolution of a complex system	224
16.2	Structure and dynamical behaviour: (a) logistic; and (b) exponential maps	226
16.3	Structure and dynamics of Lorenz system	227
16.4	Dynamics of (a) logistic; and (b) exponential maps under external perturbation	228
16.5	Dynamics of H and P when external perturbation is applied to (a) H, (b) P, and (c) to both H and P	229
16.6	Paradigms of evolutionary change	230
17.1	First chamber real and counterfactual results, 1911–94	250
18.1	How predictable?	263
19.1	A simplified causal nexus for the origin of a species	278

Notes on Contributors

Carolina Castaldi is currently a Robert Solow Post-doctoral Fellow of Saint-Gobain Center for Economic Studies, France and she is affiliated with the Eindhoven Center for Innovation Studies, the Netherlands. She obtained her Phd in Economics and Management from Sant'Anna School of Advanced Studies, Pisa, Italy. Her research interests include theoretical and empirical issues in economic growth and evolutionary economics.

Edmund Chattoe is Nuffield Foundation New Career Research Development Fellow and Research Fellow of Nuffield College, Oxford, UK. His research interests are in the application of computer simulation and evolutionary models to social behaviour. His current research involves simulating the labour market to understand ethnic disadvantage. His previous research has used simulation and qualitative research to understand pensioner money management, innovation diffusion amongst farmers and evolution of firm strategies under oligopoly.

Roger D. Congleton has been a member of the Economics faculty at George Mason University, USA and a Senior Research at the Center for Study of Public Choice since 1989. His research explores the effects of institutions, information, and culture on political and economic outcomes. His most recent books focus on constitutional design and evolution within modern democracies. They include *Politics by Principle, Not Interest: Towards Nondiscriminatory Democracy* (with James M. Buchanan, 1998) and *Improving Democracy through Constitutional Reform, Some Swedish Lessons* (2003).

Joel Cracraft is Curator in Charge at the Department of Ornithology, American Museum of Natural History and Adjunct Professor at Columbia University and at the City University of New York, USA. His most recent books include *Assembling the Tree of Life* (co-edited with Michael Donogue, 2004), *The Living Planet in Crisis* (co-edited with Francesca Grifo, 1999), and *Philogenetic Analysis of DNA Sequences* (with Michael Miyamoto, 1991). Besides pursuing his more specific interests in studying the evolution of birds, he works on systematic and biogregraphic theory and methods as well as on the diversification and the evolution of biotas.

Giovanni Dosi is Professor of Economics at the Sant'Anna School of Advanced Studies, Italy. He serves as the Continental Europe Editor of *Industrial and Corporate Change*, and is author and editor of several works in the areas of the economics of innovation, industrial economics, evolutionary theory, and organisational studies, including *Technical Change and Industrial*

Transformation (1984), *Technical Change and Economic Theory* (with C. Freeman, R. Nelson, G. Silverberg and L. Soete, 1988), *The Nature and Dynamics of Organizational Capabilities* (co-edited with R. Nelson and S. Winter, 2000), and *Innovation, Organization and Economic Dynamics* (2000).

Samuel N. Eisenstadt is Professor Emeritus at the Hebrew University of Jerusalem, Israel. He is member of many academies and received honorary doctoral degrees of the Universities of Tel Aviv, Helsinki, Harvard, Duke, Hebrew Union College and Central European University. Prizes and awards include the International Balzan Prize, the McIver Award of the American Sociological Association, the Israel Prize, the Rothschild Prize in Social Sciences, the Max Planck Research Award, the Amalfi Prize for Sociology and Social Sciences, the Ambassador of Cultural Dialogue Award of the Polish Asia Pacific Council. Major publications are *Political Systems of Empires; Power, Trust and Meaning* (1995); *Japanese Civilization; Paradoxes of Democracy, Fragility, Continuity and Change; Fundamentalism, Sectarianism and Revolutions; Die Vielfalt der Moderne; Comparative Civilizations and Multiple Modernities; Explorations in Jewish Historical Experience.*

Walter Fontana was formerly at the Institute for Theoretical Chemistry and Molecular Structural Biology of the University of Vienna and at the Santa Fe Institute in Santa Fe. He is now Professor in the Department of Systems Biology at the Harvard Medical School, USA. He was a member of the Program on Theoretical Biology of the Institute for Advanced Study in Princeton and a research scholar at the International Institute for Applied Systems Analysis (IIASA) in Laxenburg near Vienna. Walter Fontana is a computational systems biologist whose interests revolve around the molecular mechanisms and system architectures that generate and maintain phenotype from genotype during development and in evolution. The themes he pursues include evolvability, robustness, aging, and 'biolanguages' that represent molecular information processing.

Raghavendra Gadagkar is Professor at the Centre for Ecological Sciences of the Indian Institute of Science Bangalore, Vice-President of the Indian National Science Academy and Non-resident Permanent Fellow of the Wissenschaftskolleg zu Berlin. Gadagkar heads an active research group engaged in the study of insect societies. The goal of Gadagkar's research is to understand the resolution of conflict and the evolution of cooperation in animal societies. A non-technical exposition of this field of research may be found in his book *Survival Strategies – Cooperation and Conflict in Animal Societies* (1997) and a more detailed account of his work in *The Social Biology of* Ropalidia marginata: *Toward Understanding the Evolution of Eusociality* (2001).

John Harriss is Professor of Development Studies at the London School of Economics, UK and a former Director of the Development Studies Institute at the LSE. He was previously Dean of the School of Development Studies at the University of East Anglia. An anthropologist, with long standing interests in the politics and political economy of India, he is the author (with Stuart Corbridge) of *Reinventing India: Liberalization, Hindu Nationalism and Popular Democracy* (2000). Recent research deals with the idea of social capital and with local politics and democratisation, on which see *Depoliticizing Development: the World Bank and Social Capital* (2002); and *Politicising Democracy* (with Kristian Stokke and Olle Tornquist, 2004).

Paul Higgins is a Visiting Research Fellow at UC Berkeley, USA, a fellow of the National Science Foundation, and co-founder of scienceinpolicy.org. He received his PhD from Stanford University in 2003 and was a Fellow in the Department of Energy's Global Change Education Program from 1999 to 2003. His research examines human caused climate change (global warming) and the role of biological systems in global change. Specific interests include: abrupt climate changes that may accompany human perturbations of the climate system; the influence of climate on the distribution of biological communities and biodiversity; climate feedbacks that arise from changes in the interaction between the land surface and the atmosphere; and developing approaches that decrease conflict between human activities and conservation.

Ellen M. Immergut is Professor of Comparative Politics in the Department of Social Sciences at the Humboldt University of Berlin, Germany. She did her graduate work at Harvard University, was Assistant and Ford Career Development Associate Professor in the Department of Political Science at the Massachusetts Institute of Technology, Visiting Professor at the Instituto Juan March in Madrid, then Professor of Political Theory at the University of Konstanz. She is author of the book *Health Politics* (1992), as well as various articles on the new institutionalism, institutional design and health policy. Her current work focuses on the crisis of governance models in negotiated economics, pension politics and the politics of constitutions.

L. Douglas Kiel is Professor of Public Administration at the University of Texas at Dallas, USA. He is the author of *Managing Chaos and Complexity in Government: A New Paradigm for Change, Innovation and Organizational Renewal* (1994), an editor of *Chaos Theory in the Social Sciences: Foundations and Applications* (1996) and a theme editor for the UNESCO sponsored, *Encyclopedia of Life Support Systems* (2003). His more than 50 publications are cited in over 70 different academic journals ranging across fields as diverse as business management, psychology, nuclear science and music. He is currently conducting research exploring the inhibitions to change produced by conditions of organizational stress.

Reinhart Kössler is Adjunct Professor in the Department of Sociology, University of Münster, Germany. In 2000–02, he has been a Senior Research Fellow at the Centre of Development Research in Bonn. His fields of interest include sociology of development, political sociology and a regional main interest in Southern Africa. He recently published *In Search of Survival and Dignity: Two Traditional communities in Southern Namibia Under South African Rule* (2005) and *Entwicklung* (1998).

Hans-Walter Lorenz is Professor for Macro-Economics at the University of Jena, Germany, and Dean of its Faculty of Economics. His English book publications include *Business Cycle Theory* (with G. Gabisch, 1987/89) and *Nonlinear Dynamical Economics and Chaotic Motion* (1993, Japanese edition 2001). His research focuses on nonlinear dynamics, and especially nonlinear economic models, chaos, and applications to business cycle theory. His is also interested in business cycle and growth theory and the adaptation of expectations with bounded rationality.

James Mahoney is Associate Professor of Sociology at Brown University, USA. His first book *The Legacies of Liberalism: Path Dependence and Political Regimes in Central America* (2001) received the Barrington Moore Jr Prize. With Dietrich Rueschemeyer, he co-edited *Comparative Historical Analysis in the Social Sciences* (2003), which received the Giovanni Sartori award. His current research deals with long-run development and the legacy of colonialism in Spanish America.

Jeffrey B. Nugent is Professor of Economics at USC in Los Angeles, USA where he teaches Development Economics. He has applied a variety of analytical techniques to a wide variety of development issues to countries from all parts of the developing world. Recent interests include: trade, foreign investment, household investments, income distribution, new institutional economics and political economy of development. His books include: *Economic Integration in Central America* (with Pan Yotopoulos, 1976), *New Institutional Economics and Development* (with Mustapha Nabli, 1989), *Fulfilling the Export Potential of Small and Medium Firms* (with Albert Berry and Brian Levy, 1999).

Somdatta Sinha is a Senior Scientist and Group Leader at the Centre for Cellular and Molecular Biology in Hyderabad, India. Trained in physics, she applies physical principles to describe biological organisation. She is interested in formulating a common framework for the understanding of the evolution of generic properties of complex systems that encompass both the living and nonliving world. She has published many original research articles in the interdisciplinary area of theoretical biology in physics and biology journals. She is a fellow of the scientific academies in India. She has visited and lectured in many universities and research institutions around the world, and spent extended periods of time at the Center for Advanced Studies Berlin, the Institute of Theoretical Physics in Santa Barbara, the Santa Fe Institute, the National Institutes of Health in Bethesda, and Centre for Mathematical Biology of Oxford University.

Oded Stark is a Professor of Economics at the Center for Development Research, University of Bonn, a University Professor and Chair in Economic and Regional Policy at the University of Klagenfurt, an Honorary University Professor of Economics at the University of Vienna, a Distinguished Professor of Economics at Warsaw University, and the Research Director of ESCE Economic and Social Research Center, Cologne and Eisenstadt. He served as a Professor of Economics (Chair in Development Economics) at the University of Oslo, and prior to that as a Professor of Population and Economics and as the Director of the Migration and Development Program at Harvard University. He has written on development economics, labor economics, population economics, international economics, urban economics, and the theory of the firm. He is the author of the critically acclaimed books *The Migration of Labor* (1991 and 1993), and *Altruism and Beyond, An Economic Analysis of Transfers and Exchanges Within Families and Groups* (1995 and 1999), and is the co-editor of the *Handbook of Population and Family Economics* (in *Handbooks in Economics*, 1997 and 2004).

Rudolf Stichweh was researcher at the Max Planck Institute for the Study of Societies, the Maison des Sciences de l'Homme and the Max Planck Institute for European Legal History. From 1994 to 2003 he was professor for sociological theory at the University of Bielefeld. Since 2003 he has been Professor for Sociological Theory at the University of Lucerne. His main publications include *Zur Entstehung des modernen Systems wissenschaftlicher Disziplinen* (1984), *Der frühmoderne Staat und die europäische Universität* (1991), *Wissenschaft, Universität, Professionen* (1994), and *Die Weltgesellschaft* (2000). Rudolf Stichweh's current research interests are sociological theory and systems theory, world society, sociology of the stranger, sociology of science and universities, sociocultural evolution, and historical sociology.

Eörs Szathmáry is a theoretical evolutionary biologist who is Professor at the Department of Plant Taxonomy and Ecology of Eotvos University, Hungary and a permanent fellow of the Collegium Budapest (Institute for Advanced Study). He has been a Soros fellow at the University of Sussex, a research fellow at the National Institute for Medical Research in London, a fellow of the Wissenschaftskolleg zu Berlin (Institute for Advanced Study), a guest professor at the University of Zurich, and a guest of the College de France. Major books include: *The Major Transitions in Evolution* (1995, with J. Maynard Smith), *The Origins of Life* (1999, with J. Maynard Smith). Current research interests are origins of life, Martian astrobiology, and the origins of natural language.

Andreas Wimmer is Professor in the Department of Sociology, UCLA, USA. He previously directed the Center for Development Research of the University of Bonn and the Swiss Forum for Migration Research at the University of Neuchtel (Switzerland). His research aims at understanding social change in a comparative and historical framework, starting with his study of the transformations of indigenous communities in Mexico and

Guatemala. His interests later turned to the macro-level, where he examined how political modernization has changed the role of ethnic and national differences. His publications include *Transformationen* (1995), *Nationalist Exclusion and Ethnic Conflicts* (2002), and the edited volume *Facing Ethnic Conflicts* (2004).

1
Models, Methodologies, and Metaphors on the Move[1]

Andreas Wimmer

The plan of the book

Most of our contemporaries would agree that we live in a time of rapid and deep-going change. Globalization, the end of certainty, and post-modernity are three prominent catch-words describing our current condition. Many are concerned about declining political steering capacities, run-away financial markets, global warming, the biotechnological and micro-electronic revolutions, to name just a few particularly prominent issues. While it is hard not to be impressed by the impact of these various processes unfolding before our eyes, we may be well advised to distrust our perceptions. After all, it belongs to the most salient, if not defining characteristics of modern societies that each generation witnesses a fundamental transformation and an upheaval unprecedented in dynamic and impact – a phenomenon that Fowles (1974) has aptly described as 'chronocentrism'.

Is it just another inescapable illusion to perceive a fundamental and unprecedented change in the way the sciences describe and understand phenomena of change? I believe there is enough ground to believe that we are not victims of a chronocentric distortion when making such a claim. All the major disciplines have moved – some earlier than others – beyond older teleological views, which saw change unfolding along a pre-defined path from stage to stage until it reached a known end point: homo sapiens sapiens, the modern society, a free market economy in equilibrium, etc. Today, processes rather than stages have moved to the centre of attention. Notions of equilibrium, reversibility, and determinacy have been displaced by disequilibria, irreversibility, and contingency (cf. Prigogine 1997).

This book reviews some of these innovations in the natural sciences, economics, and the social sciences. Six paradigms have been particularly influential in bringing about this pan-disciplinary paradigm shift: chaos theory and evolutionary theory in the natural sciences; path dependency and new institutionalism in economics; new modernization theory and neo-historical approaches in the social sciences. They all belong, as I will show in the following section, to a larger group of post-mechanistic models of change that share four fundamental properties. They contain elements of non-linearity: pathways of change depend on initial conditions, or a system may behave chaotically during certain periods. They are at least partially probabilistic and describe certain aspects or phases of change in a non-deterministic language. They foresee different possible trajectories of change and thus are multilinear in nature. And they postulate an irreversible process where past conditions determine possible changes in the future in a way that make a return to earlier states impossible.

Many of these paradigms and their core models have originated in one disciplinary field and then been applied to other areas of research, sometimes in a rigorous fashion, sometimes in more loosely metaphorical terms, thus 'migrating' across disciplinary boundaries. This volume discusses the experiences with such concept migration. It will not lead us, perhaps an unnecessary caveat, to a new meta-theory for explaining change, such as envisioned by the Gulbenkian Commission headed by Immanuel Wallerstein (1996). Nor are the editors inspired by what some have termed the 'Santa Fe *Zeitgeist'* that is, the search for common properties of all complex evolving systems (see the *Sante Fe Institute Studies in the Sciences of Complexity*, published by Addison-Wesley). We believe, as Reinhart Kössler will argue in more detail in his conclusion, that there are too many fundamental differences between natural and human systems to make this latest quest to find the hidden construction principles of the world more viable than its various predecessors.

More modestly and certainly less metaphysically inspired, we intend to document and at the same time foster the dialogue among members of a family of similar approaches. Rather than fusion or absorption into a meta-theory, we believe that selective borrowing and mutual learning are the adequate strategies for improving our understandings of change in the various branches of the scientific enterprise. The book is planned accordingly. Each paradigm will be introduced by a scholar from the disciplinary field it originated from and then commented upon by representatives of the other disciplinary fields to which the paradigm has already been – or has the potential of being – applied to.

In this introduction, I should first like to briefly introduce the six paradigms and then offer a preliminary analysis of their commonalities and differences, including an admittedly speculative attempt at describing these in the language of stochastic matrices. The third section will explore the role of concept migration in more detail, offering a typology as well as a discussion of the difficulties and opportunities for innovation that the cross-disciplinary exchange of models, metaphors, and methodologies provides. The final section, to which the efficient reader may jump after having finished the first, will review the individual chapters. I begin with an overview of our six paradigms.

Chaos and order in climate change

Research on climate change addresses one of today's most pressing and broadly advertised issues, and perhaps represents one of the best funded and most transnationally integrated research enterprises. Beyond this obvious policy relevance, understanding climate change forms a specific intellectual challenge, both theoretical and empirical, given the sheer complexity and scale of the issues. This has posed formidable difficulties for modeling: Not only is it hardly possible to know all the relevant factors but also the integration of the various sub-processes into an overarching model poses difficulties, as the parameters proliferate in ever more complex equations. The fact that many sub-models contain important probabilistic elements does not make the task of explanation and prediction easier.

A climate system may have multiple stable states and therefore may respond to a temporary perturbation by moving to a new equilibrium – but it may also contain feedbacks that re-establish a equilibrium state. Chaos theory has proved to be an interesting tool to analyse complex patterns of change with non-linear properties such as for example bifurcations. Research on climate change thus offers an important starting point to question received notions of structure and change in a variety of scholarly fields. It is especially interesting for economists and social scientists because its object is large scale and complex and represents, as do societies and economies, an empirical entity that cannot be subjected to experimental manipulation.

Genetic variation in evolution

Evolution represents, since over a century, one of the major paradigms for studying change in the natural and social sciences. While the conceptual triad of variation, selection and inheritance (retention) has become commonplace since the days of Darwin, important features of evolutionary biology have been frequently overlooked. A striking example is the combination of chance and determinacy in evolutionary models, that has been somewhat obscured in what is called the modern synthesis of Darwinism stressing the gradual accumulation of mutations leading to the appearance of ever fitter species (cf. Gould 2002). This teleological perspective survives in fields that have borrowed evolutionary concepts from biology. Recent advances within the natural sciences, in particular biology, using up-to-date technology for research on the cellular and the molecular levels, but also in paleontology, once again have thrown the original features into sharp relief.

Perhaps the most exciting strand of this new research focuses on 'development', i.e. how genetic structures relate to phenotype, or more precisely, how genetic variation translates into shifts in phenotypical design. It turns out that 'chance' in the production of phenotypic variation is a much more patterned process than isotropically random. Genetic variation drifts non-deterministically along extended, phenotypically neutral pathways across genetic space until it 'hits' clearly identifiable points where it causes a change in phenotype as well. Thus, in contrast to the modern synthesis of Darwinism, the direction of evolutionary change is shaped as much by the pathways of possibilities generated by genetic variations as by external selective pressures producing adaptive change. The three chapters by Fontana, Stichweh and Chattoe (Chapters 5–7) will explore whether this molecular model holds promises for economics and the social sciences as well.

Economics of continuity: path dependency

Path dependency and the theorem of increasing returns have challenged some well established notions of mainstream economy. In the meantime it has been adopted rather enthusiastically by social science disciplines such as sociology and political science. The basic idea, originally formulated by Brian Arthur (Arthur 1994), may be summarized as follows: Contrary to what classical economics predicts, a growing company may not face decreasing returns with every additional product sold, but increasing returns. The reasons are manifold and include technical, social and psychological factors: a product may be combined in an optimal way with already established products; people may need the product in order to communicate with each other; or it may be too costly to learn how to handle a different design.

It depends on initial conditions, whether such externalities do indeed lead to increasing returns and, consequently, to non-equilibrium situations such as monopolies of the Microsoft type. Thus, there is a contingency element introduced into economic thinking: Small differences in initial conditions can set future economic development (of firms, of countries) onto different paths which later are only abandoned at overwhelming costs. The most celebrated case of path dependency has been the QWERTY set-up of the typewriter keyboard in the Anglo-Saxon world, which has never been abandoned although ergonomically more efficient layouts have been proposed (David 1985). Path dependency models have now been used in a wide variety of fields. They play a prominent role, to give two examples, in studies of the post-communist transition to market economies or in the process of democratization in developing countries.

Institutional inertia

The starting point of New Institutional Economics was to consider how rational man relates to institutions, thus going beyond the basically 'institution free' market models of neo-classical economics. At the beginning, the main puzzle to solve was how non-economic institutions such as property laws could emerge from the interaction of economic decision makers. In Coase's path-breaking answer to this question, they would agree on property laws if this reduces transaction costs for negotiating disputes and thus benefits all participants in a market independent of the properties they hold

(Coase 1990). In a later stage, the influence of existing institutions on the individual decision making process was analysed as well (North 1994) and institutions were conceived as products of real-world historical processes (David 1994), thus moving away from the idealized concept of a pre-historical original state from which institutions would emerge. At the same time, the meaning of institutions broadened considerably to include all types of rules, including informal ones, and consolidated routines.

Neo-institutional economics is ideally suited to map out the various trajectories of economic development since these may be preconditioned and continuously influenced by different institutional settings. Similar economic stimuli (such as market reforms) may thus lead to different economic developments, depending on the institutional set up. New institutionalism thus converged on a notion of irreversibility similar to the concept of path dependency (ibid.). It has stimulated research in political science (e.g. Thelen 1999) and sociology (Mahoney 2000 as well as in this volume), which have reformulated much older versions of 'institutionalisms' in parallel, but also in opposition to the economic strand of thinking.

The multilinear modernization of societies

The classical sociological theory of modernization envisaged a largely uniform process through which societies around the world would evolve, passing through a number of more or less predetermined stages at different speeds. The final stage was best represented by Western societies, and the US was usually taken as the apogee of modernity. The unilinearism and the teleology of these models have been criticized for decades. Against this backdrop, a series of new approaches have been developed that analyze the multiplicity of modernization paths – beginning with Julian Steward's 'multilinear evolution' (Steward 1955), to Collier and Collier's (1991) 'critical junctures', Wolfgang Zapf's 'crossroad theory' (Zapf 1996), and Shmuel Eisenstadt's 'multiple modernities' (in this volume). These different accounts vary in how they explain the mechanisms of 'branching off' into the different paths. In general, however a combination of cultural and political factors is evoked: different cultural and institutional backgrounds will produce varying reactions to modernization impulses, e.g. triggered by economic growth; and depending on the specific relations of power between social groups at critical junctures in history, a different reform path will be followed. In their emphasis of the importance of initial conditions and of institutional and cultural rules that reduce the horizon of possible social transformations, these approaches parallel the more formalized theories of path dependency and neo-institutionalism in economics.

Constellations of contingency: political history

Thinking about the significance of events for processes of change has for long been the exclusive domain of history. Traditionally, history saw the unfolding of events as a strictly deterministic process: Each event 'causes' later events to happen in a complex, idiosyncratic, yet fully deterministic way: the fog that obscured the battlefields of Austerlitz is of a different causal nature than Napoleon's brilliant strategic decisions. Both together, and a host of other events, determined the outcome of the battle. The task of the historian was to find the crucial events and to understand, through interpretation and extrapolation, how exactly they impacted on each other. Contrafactual reasoning, such as Blaise Pascal's famous dictum that 'Had Cleopatra's nose been shorter, the whole face of the world would have been different', was seen as irrelevant since Cleopatra's nose had exactly the form it purportedly did (Ferguson 1997).

In the past decade, the social sciences have re-approached history and adopted event chains as a basic explanatory model of change. There are several related strands of this 'historical turn' in the social sciences (McDonald 1996). Some have elaborated the concept of 'event' as a theoretical term encompassing the notions of sequentiality, contingency, and causal heterogeneity (e.g. Sewell 1996). In the sociology of the life course, much attention has been given to the 'turning points' of a biography, where the logic of a socially determined pathway of development is suspended and singular historical forces reshape an individual's life (Abbott 2001, ch. 8). Others in sociology, political science and history have attempted to formalize traditional historical analysis and to determine the causal importance of a particular event chain by rehabilitating contra-factual analysis (Fearon 1991; Ferguson 1997; Immergut, in this volume; Hawthorn 1991; Tetlock and Belkin 1996). Still others have reached for game theory or other tools such as event structure analysis or sequential models to understand the relevant enchainment of individual decisions and events (Abbott 2001). Finally, a group of authors from economics offered to reconcile rational choice models with the analysis of singular historical trajectories in what they termed 'analytical narratives' (Bates *et al.* 1998).

Commonalities and differences

The six paradigms have been chosen because they are all based on post-mechanistic models of change. I hasten to elaborate and justify using the notoriously chronocentric adjective 'post'. According to one definition,

> mechanisms are regular in that they ... work in the same way under the same conditions. The regularity is exhibited in the typical way that the mechanism runs from beginning to end; what makes it regular is the productive continuity between stages. Complete descriptions of mechanisms exhibit productive continuity without gaps from the set up to the termination conditions, that is, each stage gives rise to the next. (Darden 2002: 356)

Many older models for analysing change described the world as composed of such machine-like mechanisms, defined by linear relationships between its parts. Cybernetic models, time series or event history approaches are examples from the social sciences and economics. If the behavior of these machine-like objects were not fully covered by the model, it was attributed to a lack of information, lack of specification of certain functions, or noise and external perturbances. Scientific progress, the credo that usually pairs well with mechanistic thinking, would bring us asymptotically close to a full understanding of the machine's functioning and a better prediction of its behavior. More precisely, mechanistic models of change may be characterized by the following four properties.

First, most models described change as the transition from one steady state to another, for example as a process driven by feedback mechanisms. The idea of systemic stability was very prominent in the functionalist tradition of the social sciences and in neo-classical economics. Societies were described in analogy to a body in a healthy state; economies appeared as perfectly balanced mathematical equilibriums modeled after equations in physics. Calls for a processual approach to understand how change actually occurred, appeared in the fifties and again in the eighties and nineties (e.g. by anthropologists Barth 1995; Firth 1992) but were largely left unanswered.

Secondly, change was seen as linear and continuous, leading from low values on a specific dimension of change to higher ones. In economics, development was modeled as a continuous process of capital accumulation and infrastructure development by early growth theorists such as Rostow (1991). Similarily, the Darwinist–geneticist synthesis of the fifties and onwards saw evolution as a continuous move, driven by selection pressures on the individual organism, towards species ever better adapted to their environment. The idea of multiple equilibria at the same level of systemic complexity was not yet well developed in economics, nor in evolutionary biology (where multi-level selection had not yet been accepted) or the social sciences (where 'Western' culture and society still counted as the model for everybody else to follow).

Third, the end point of the transition curve was known to the researcher: the models had a teleological character. In biology, it was taken for granted that evolution would necessarily lead to the higher levels of complexity of contemporary species, an idea widely copied by the social sciences in the 20th century. Fourth, change was described in many disciplines (neither in evolutionary biology, to be sure, nor in the historically minded social sciences) as a reversible process. If the behavior of a system is governed by linear relationships between its component parts, a process may be reversed to an anterior stage by lowering the value of one variable, leading to adjustments in the other variables that perfectly mirror the initial transformation, thus eventually arriving at the original state. Time, according to Einstein and also quantum theory, was an illusion (cf. Prigogine 1997). The same held true for neo-classical economics, where equilibrium can be reached in a history free space from different starting points situated in the past, present or future.

The six paradigms that will be discussed in this book go beyond such mechanistic understandings of change. They all emphasize non-linearity, partial determination, branching effects, and irreversibility, albeit to different degrees and with varying importance for the overall theoretical argument. Here is a brief summary of these four elements:

1. *Non-linearity.* In many of the paradigms presented here, a continuous change of the value of one variable may lead to discontinuous behavior of the entire system. Chaos and bifurcations are the most obvious exemplars of such non-linear behavior; they will be discussed with reference to climate change. Non-linearity is also found, albeit in a different form, in path dependency models, where changes are self-reinforcing and transition functions may expose a non-linear pattern. In climate change and path dependency models, in new modernization theory and in neo-institutional economics, small (or in some models even arbitrary) changes in initial conditions may produce different reactions to external stimuli and alternate equilibria. In evolutionary models of selection, based on population genetics and ecology, the main dependent variable is the frequency of genes whose change is often described by a nonlinear dynamical system.

2. *Partial determination.* Most paradigms include probabilistic elements and describe zones of partial determination or even of non-determination. The patterned, but aleatory moves in genotypical space in micro-biological analysis of development, the sensitivity to arbitrarily chosen initial conditions and first actors' choices in path dependency models, and the event driven trajectories in neo-historical approaches are the most obvious examples of such non-deterministic properties.

3. *Branching effects.* Non-linearity and partial determination imply that the final outcome will depend on the pathway of transition chosen. The multi-linearity that results from such branching effects is a common characteristic of most models that will be discussed in this book. It is obvious in path dependency, multiple modernities, and in event chains that may "branch out" at those events that could as well not have happened (remember Cleopatra). Branching effects can also be seen in the genotypical variations that follow a certain pathway of mutation which in turn determines the future possibilities for phenotypical change.

4. *Irreversibility.* Non-linearity and path dependency produce irreversible trajectories in many of our six paradigms of change. The economics of path dependency, climate change as a result of irreversible sub-processes such as desertification, and the sequential analysis of event chains stress irreversibility in the most obvious ways, but it can also be found in evolutionary theory (with some exceptions, as the patient reader will discover) and neo-institutional economics.

Contingency, transformation, history: three basic models of change

These commonalities are, evidently enough, of a very general nature and rest on analogies between models which work on the basis of quite different assumptions and notions of causality. It is certainly not possible to address these differences in a satisfactory way in an introduction – and a serious treatment would go beyond my own disciplinary competence and intellectual capacities. I would like to confine myself here to taking a closer look at the structure of the processes of change that these various models describe, without discussing the different properties of the latter.

In the taxonomy that follows, I will distinguish between different processual patterns that describe change – as opposed to equilibrium or reproduction. A specific model may rely on one main processual pattern or may combine several of them. The patterns thus might be understood as an elementary grammar that underlies the different languages of change.

All patterns are at least partially probabilistic and are time dependent. They can thus be described with the help of stochastic matrices. The most prominent of these matrices are those based on Markov chains, the properties of which I will now briefly introduce. The starting point is the simple idea of time as a succession of instances. Each instance can be characterized by a certain state (say A, B, and C). Thus, instance 1 may be characterized by A, instance 2 by C, and instance 3 again by A.

Transition probabilities express the likelihood that upon A follows B or C. These probabilities can be arranged in a matrix of all possible transitions, called the transition probability matrix. A matrix can contain deterministic parts (with transition possibilities of 1) and probabilistic ones (with probabilities between 0 and 1). Let me illustrate these characteristics with an often cited weather example that uses discrete time (days). Weather can only be sunny, foggy, or rainy. Contrary to his habits, the Creator has informed us about how he constructed the weather system and has provided us with the transition probabilities for these different states. We can thus draw the following matrix (see Figure 1.1). In this example, a sunny day follows on a sunny day with a probability of 0.3, a foggy day on sun with probability 0.5. There is never rain after fog.

The three patterns of change can now be exemplified with such matrices.[2] Maybe I should clarify that I use them to describe the probabilistic path through different states of *one individual system* – and not, as in many other applications, to describe the distribution of a large number of systems over the space of possible states. In order to emphasize the illustrative character of the matrices, I will not give numerical values to transition probabilities but indicate with an arrow where a transition is possible (i.e., with a probability between 0 and 1).

The first process is driven by contingency. As mentioned before, contingency is a feature of several of the models that will be discussed in this book.

		Tomorrow's weather		
		Sun	Fog	Rain
Today's weather	Sun	0.3	0.5	0.2
	Fog	0.2	0.8	0
	Rain	0.3	0.3	0.4

Figure 1.1 A basic transition probability matrix

The genotypical mutations that are at the center of biological variation follow, as the chapter by Walter Fontana will show, a structured, but principally aleatory pattern. Structure in this context means that not all transitions (mutations) have the same probabilities; the system thus 'drifts', over time, towards certain states. Contingency also appears in other, more drastic forms, such as the famous asteroid hit that changed the course of evolution – a highly improbable event that would show up only in one cell in a vastly expanded matrix with an infinite state space. The matrix may or may not show different transition probabilities, i.e. contingency may be more or less structured. Note that contingent processes may entail both reversible and irreversible transitions (from 2 to 4 but never from 4 to 2 in the left matrix of Figure 1.2).[3] A special case is a cyclical chain with only two possible states, such as the famous bifurcations of chaos theory, where the system 'jumps' back and forth, in a non-probabilistic way, between two possible states, as shown in the matrix on the right hand side (see Figure 1.2).

A second process is that of transformation. It occurs if a particular state opens up to a new subset of possible states, in other words if it leads to a qualitative change of the system (cf. Abbott 2001: 246f.). In the matrix of Figure 1.3, the system can move from the area of states 1 to 4 to the area of states 5 to 8 when it has reached state 4. Note that once the system has moved into this new area, it will not go back, the transition has a one way sign.[4] I call this process 'transformation' since the new areas of states may represent a qualitatively different state of the system or may even be described as a new system altogether.[5] An example for this type of process is the transition from one phenotype to another through what Fontana calls genetic drift in a 'neutral network'. Another example are chemical reactions, where the

Non cyclical

	1	2	3	4
1	→	→	→	→
2	→	→	→	→
3	→	→	→	→
4	→			→

Cyclical

	1	2	3	4
1			→	
2				
3	→			
4				

Figure 1.2 Contingency

combination of certain substances produces new substances with new characteristics and further possibilities of transformation (see Chattoe, in this volume). Many sociological macro-theories of change could be described by a similar matrix: The transitions are between different 'levels of modernity' that would be triggered by crucial constellations of power at the transition points in the matrix. Several such transition points would lead to Eisenstadt's multiple paths of modernization and modernities represented by different subsets of communicating states. The different paths may end in different states that would be immune to further modifications or outside perturbations.

Other variants could be described: It is conceivable to have cyclical patterns, such that state 10 would feed back to state 1, or open ended, fully irreversible processes within an infinite space of possible states, or a process which comes to an end point, such as in the matrix shown in Figure 1.3, where the process will end at what is called the 'absorbing' state 10. Imagine the infamous 'end of history' declared by Francis Fukuyama would come true; or an institutional transformation leading to an economic equilibrium.

The third pattern of change has, again, entirely different properties. Now the states are defined as events. The transition probabilities are highly unequally distributed among states and the transitions are fully non-recurrent: never does something happen twice. This matrix (Figure 1.4) adequately describes event chains as they are analysed by the neo-historical approaches discussed above. Events are seen as almost fully determined by previous events (indicated by an arrow in the matrix of Figure 1.4, with a very high transition probability), but leave room for the existence of less probable, but nevertheless possible events, which may be explored by constructing a counterfactual argument. The degree of 'historical openness' may change over the course of time and even include moments (transition from 4 to 5 in the matrix below) where probabilities are more equally dispersed over several states, thus

	1	2	3	4	5	6	7	8	9	10
1	→	→	→	→						
2	→	→	→	→						
3	→	→	→	→						
4					→					
5					→	→	→	→		
6					→	→	→	→		
7					→	→	→	→		
8								→		
9									→	→
10										→

Figure 1.3 Transformation

opening windows of contingency in the historical process. Please note that in the matrix there are events (x through $x+2$) that may have taken place if earlier events would not have happened, but will never be reached by the most probable course of history because these states are too far removed from the area of likely states. This obviously implies that we assume an infinite state space (as indicated by adding the states $x+$).[6]

Perhaps surprisingly, the patterns described by chaos theory look similar to a fully deterministic history with all transition probabilities set at 1. The somewhat paradoxical beauty of chaos theory is to demonstrate that a pattern of apparently random successive states is *de facto* fully determined by the function that defines the system – an interesting parallel to the intellectual enterprise of historians who show that what appears to be the product of pure coincidence or the free will of Cleopatra and Marc Anthony, can be understood as a chain of events necessarily succeeding each other. While the causal mechanisms leading from one state to the next are certainly conceived in different ways by chaos theory – where a single equation produces the whole sequence – and conventional history, which evokes different causes for each transition, the patterns of change they describe are strikingly similar. The abstract grammar of these matrices thus allows us to describe similarities between apparently unrelated models such as climate change and neo-historical analysis of institutional change.

	2	3	4	5	6	7	...	x	x+1	x+2
1	→			→						
2		→		→						
3			→							
4				→	→				→	→
5					→	→				
6						→				
7							...			
...										
x										
x+1										
x+2										

Figure 1.4 History I: event chains

Another special case of history is path dependency. The sequence starts with a set of probabilistically related states which represent initial conditions. Once the system reaches a certain state (or two such states, as in the example) within that subset, a fully deterministic path is 'triggered' off, which is fully irreversible. The path may or may not end in stable states, such as in the matrix below where 7 and 10 are absorbing states; or it may again 'open up' to a subset of various probable states, i.e. the path is unlocked at a certain state (as discussed in Castaldi and Dosi's chapter).

Contingency, transformation and history are the three basic post-mechanistic patterns of change that I have identified here. Others may be added. More complex matrices would allow for continuous time, for changes unequally dispersed over time periods (such as in Poisson processes), and for 'deeper chains' where not only the current, but also past states influence the future, a very important modification for the social sciences that deal with systems that have memories. I offered these matrices for strictly heuristic and illustrative purposes: To suggest in which direction one could search for

	1	2	3	4	5	6	7	8	9	10
1	→	→	→							
2	→	→	→				→			
3	→	→	→	→						
4					→					
5						→				
6							→			
7							→			
8									→	
9										→
10										→

Figure 1.5 History II: path dependency

an elementary grammar of change which underlies the various post-mechanistic paradigms discussed in this volume and beyond.

Concept migration between disciplinary fields

I should now like to shift perspective, and look at how these paradigms have been applied across disciplines. Each originated in specific fields, from physics to chemistry, biology, economics to history. Their success has often drawn attention from scholars working in other fields who then used them to answer questions specific to their own disciplines. The problems and prospects of such concept migration will be the topic of this section.

It will be a general discussion drawing on the philosophy and history of science and making references to the chapters whenever appropriate. There is a small, not yet well connected literature on how to understand under which conditions and with what consequences model migration occurs. So far, this literature has generated various typologies, which I should like to synthesize in the following. Five different modes of what has variously been termed 'borrowing', 'exchange', 'import' and 'export' (or assuming the perspective of the concepts: 'transfer', 'migration', or simply 'move') will be distinguished.

The typology differentiates between the various types of intellectual goods that trespass the boundaries between disciplines.

Tool transfer, model migration, methodological analogies, and metaphor move

The first type is the transfer of a research tool, such as a statistical technique, or a mathematical model, or a computer program. Renate Mayntz (Mayntz 1990: 58) lists Thom's mathematical catastrophe theory or Haken's synergetic as examples of mathematical models that have been adopted by the social sciences. Other instances would be the spread of Bayesian logics to different fields, including sociology (Ragin 1998), the use of optimal matching methods originally developed for DNA sequences by historical sociologists (Abbott 2001), or the cladistic method for determining the historical relation between species applied to language history (see Cracraft, this volume).

A second, more demanding type is to integrate not only a mathematical/ statistical technique, but to make sure that the theoretical propositions as well as the empirical terms, i.e. an entire model, find their corresponding propositions and terms in the importing field (see the definition by Morgan and Morrison 1999). There are two variants of such model depending on whether or not the model is respecified in the new field. Accordig to Mayntz (1990) re-specification begins with theoretical generalization, during which a model is stripped of any empirical specifications, and is completed successfully when it has been linked to the new empirical field through new operationizable terms. She cites the sociologist's Niklas Luhman's adoption of general systems theory as an example of this type of model transfer.

In a more literal translation of a model without respecification, the importing researcher looks for one-to-one analogues for each of the terms of the model and makes sure that the causal connections between the terms remain intact. This is what an ample literature in the philosophy of science from Duhem to Campbell to Harré and Hesse describes as an analogy (for an overview see Bailer-Jones 2002: 110–14). Both the less and the more strict forms of model migration may lead to a complete 'assimilation' of the imported model, to a degree where its disciplinary origin may no longer even be remembered (see Klein 1996: 63).

The third mode of borrowing is much less demanding: fewer conditions have to be met for a successful transfer. It concerns methodological strategies rather than models that specify causal connections between empirical terms. A prominent example is the role that non-linear physics played in reshaping the notion of causality in the social sciences, which have been the last to depart from the epistemological ideal of Newtonian physics and full determination. The search for corresponding 'laws' governing the social world has now been abandoned, since it is assumed that if the natural world is full of probabilistic processes or non-linear phenomena, there is a high probability that similar processes govern the social world as well (Mayntz 1990; Urry 2004). According to Kellert (2000: S464), even such loose transfer of methodological principles has to rely on a quite precise analogical operation: Only when we can be sure that the principle characteristics of two fields are sufficiently similar can we assume that the methodologies successfully applied in one field will yield the expected return in the other field as well. A good example is Kiel's plead (in this volume) for searching for non-linear phenomena in the social sciences similar to those of chaos theory in physics and biology, given that the social world is structured in a similar way as the natural world. Another example would be the methodology of contrafactual thinking that Ellen Immergut is introducing in this volume and that may be of importance to other disciplines where single events shape the course of change in a non-experimental setting.

The most controversial form of cross-disciplinary borrowing concerns metaphors. Metaphors are often used to illustrate complex causal models. The 'butterfly effect' or 'emergence' in complex systems are frequently cited contemporary examples. The use of the Judeo-Christian and other powerful metaphors of time in geology (Gould 1988) or sociology (Nisbet 1969) represent well studied cases. Darwin borrowed the metaphorical image of the 'survival of the fittest' from the social scientist Herbert Spencer.

The borrowing of metaphors is discussed, in this volume, by Ghadakar, Kiel, Chattoe, Stichweh and others. Authors are divided, as is the literature, about the worth of metaphor migration is fruitful. Ghadakar points to the dangers of misinterpretation when the normative implications of a metaphor (such as genetic 'fitness') is transposed to another field (such as human society). Kiel is more optimistic and assumes that metaphors from other disciplines may help to overcome routinized patterns of thinking and thus stimulate innovation. This is an argument also presented by Kellert (2000), who describes the effects of metaphor transfer as one of 'deformalization' and thus creative confusion.

Cognitive research on metaphor use helps to understand why migration of metaphors may stimulate innovation. Metaphors provide a new perspective on a topic because they bring to the foreground less salient properties of an empirical object by linking them to the primary properties of the metaphorical image (see the 'salience imbalance' theory of Ortony 1993). In other words, a new metaphor allows us to see an empirical field with new eyes and may thus stimulate new research strategies (see the 'interaction view' on metaphors developed by Black 1993: 35–8). Brüning and Lohmann (1999) have shown, building on Peirce and a case study from oceanography, how new metaphors may develop into models of causal relationships which then are specified, loose their metaphorical quality and may be subjected to empirical tests. Metaphor import can represent, in other words, a 'soft' initial stage in the process of scientific discovery.

Risks and obstacles

Tool transfer, model migration, the borrowing of methodological strategies, and metaphor migration are the four modes of cross-disciplinary exchange that I distinguished in the previous section. All share some problems and risks that are rarely mentioned in this scarce literature dominated by enthusiasts of interdisciplinary co-operation and that are advocates of disciplinary unbounding. I should like to discuss the most obvious ones here.

First, most ideas are transferred long after they have become established in the original field. It takes further time for the new methodology, model or metaphor to be mainstreamed into normal science of the importing field. It may well be that a model, methodology or metaphor is most popular in the new field when it has already been abandoned as a consequence of a paradigm shift in the original field. Many have observed, including Fontana and Chattoe in this volume, that much of mainstream social science still tries to imitate a Newtonian model of physics that has long been revised in favor of a probabilistic approach by physics itself. Another example, within the social sciences, is the current popularity of anthropology's traditional concept of culture, which anthropology has abandoned almost a generation ago (Wimmer 1996).

A second, equally obvious danger is that of misunderstanding. One of the most prominent and obvious examples is that of path dependency, which has often been reduced to a vague notion that 'history matters'. Economics, as Castaldi and Dosi make clear in their contribution to this book, has a much more precise idea of how exactly Clio steers the flow of events. Perhaps even more misused is the notion of chaos, which borrowers have understood as representing indeterminacy and pure stochasticity (cf. Kellert 2000). Some of these misunderstandings are simply based on poor scholarship and thus may not provide enough ground for a general critique of concept migration. Ghadakar's warning against the undesirable implications of metaphor migration should certainly be taken seriously, especially by the social scientists in whose hands concepts such as the 'selfish gene' may produce dubious results (cf. Segerstrale 2001). However, his caveat is clearly not directed against concept transfer as such.

A more serious danger is that of misapplication. Several examples have been identified. According to Lorenz, in this volume, chaos models have been applied to economics without a proper re-specification of the underlying causal propositions, about economic behavior many of the implicit assumptions. As a result, on which the models rest do not make sense from an empirical point of view. In addition, chaos often appears in value domains which are beyond those actually observed in empirical reality – the model thus describes a theoretically possible behavior with little chances of actual occurrence.

In addition to such mis-specification, a model transfer may be criticized as not capturing the relevant aspects of change in the new domain. In his contribution to this volume, John Harriss criticizes the use of neo-institutional approaches to explain social and political phenomena. He argues that social and political change are effects of the transformation of power structures and the cultural patterns linked to them. Both are, however, treated as exogenous variables in the institutionalist approaches. According to his view, adopting a neo-institutionalist frame of analysis therefore adds little value to the sociological enterprise.

Finally, model migration can be risky because differences in the properties of the importing and exporting fields may make a successful re-specification improbable. The analogical operation discussed in the previous section may fail. The most prominent example that comes to my mind is the use of evolutionary analogies in the social sciences. It has been argued time and again (cf. Chattoe, this volume) that the 'environment' which selects variations is not independent, in social systems, from these variations themselves, basically because humans may intentionally manipulate environmental conditions and co-operate with each other to do so.

Mis-specification, irrelevance, and misfit thus represent the major risks that models, metaphors, and methodologies encounter in new disciplinary territories. Despite these risks, traffic on the cross-disciplinary roads is dense. It seems that the barriers to such traffic cited in the literature – e.g. different intellectual cultures and epistemologies (Bauer 1990) – are no longer, if they have ever been, substantial enough to prevent such flows.

The reader of the following chapters will discover, however, that not all roads are traveled in both directions: Chaos theory emanated from mathematics and physics and moved to the natural sciences, economics and the social sciences. Evolutionary biology inspired economics and the social disciplines. Path dependency moved from economics to sociology and political science. Game theory (not discussed in this volume) was originally developed by mathematicians and economists. In the meantime, it is widely applied in evolutionary biology and political science as well. We are not aware, however, of any major social science concept having been adopted over the past decades by economics (with the possible exception of the 'trust') or the natural sciences – the days when Spencer inspired Darwin seem to be gone by now (again with one possible exception: small world theory [Watts, 2004]).

In other words, the dialogue that this book documents exhibits a rather asymmetrical character. The reader will notice that the natural scientists and economists commenting on the papers by Shmuel Eisenstadt and Ellen Immergut had to overcome considerable difficulties in finding an adequate point of view from which to discuss possible links between comparative historical sociology and institutionalist political science. The same holds true for the natural scientists discussing the two economics papers. Reinforcing this impression of asymmetry, the editors have to admit that it has been challenging to find natural scientists who were prepared to comment papers far removed from their disciplinary domain and area of expertise – we are all the more grateful for the excellent contributions, written by outstanding scholars, that we eventually received.

Several explanations for this asymmetry have been put forward, of which I will mention only three here. First, Pantin (1968) has observed some time ago that the natural sciences are more 'restricted' disciplines with very strong linkages between research areas within their disciplinary domains and weak and few ties with other disciplines. The social sciences, by contrast, are 'unrestricted' disciplines with more fuzzy cognitive borders and greater openness to exchange with other disciplines. Economics would be situated somewhere in the middle. Secondly and related to this, there seems to be a flow gradient of borrowing from the more mathematical to the less mathematical disciplines, which may be explained by simple intellectual economy: It is easier to re-specify a model that contains an abstract mathematical core than to first generalize the usually context specific, discursive models of the social sciences into a mathematical language and then re-specify it. Finally, we should mention the asymmetry of power and prestige between disciplinary fields (cf. for France Bourdieu 1988). Concepts emanating from the most highly ranked disciplines, such as theoretical physics, enjoy a nimbus of truth and relevance that those for example from administrative studies will never have. Conformingly, the likelihood that a specialist in administrative science will learn, through the media or the feuilleton, of the latest revisions of the theory of black wholes is much higher than that a theoretical physicist will ever come across the advances in the theory of institutional learning – although it is probably safe to say that the latter may be of much greater importance for the daily life of both individuals than the former.

This last point may help to understand why even the more formal models of the social sciences that would offer themselves as an import good remain unnoticed by economics and natural sciences. An apt example are the advances that have been made, over the past decade, in formalizing the traditional historical method and to develop more rigorous models of the unfolding of events (see Mahoney, this volume). These models (e.g. Abbott 1995; Heise 1989; Abell 1993) are suited to explain event chains, some of them in a comparative way, and thus go beyond the descriptive story of 'one damn thing after the other', as a popular saying describes traditional history. These developments have not been, as Mahoney points out, noticed by economists and natural scientists, although we find plenty of evidence for historical processes in their fields – for chains of events which influence the systems in question in a quite fundamental way and yet have to be treated as noise or contingency by most existing models.

Three examples may suffice to illustrate this point: In the ecological analysis of biota – the combination of species in one particular natural environment – geological events such as volcano eruptions greatly influence the possible migration of species across ecological space and thus the composition of a particular biota. In evolutionary biology, Eldredge and Gould's famous essay on the role of events, such as the appearance of a dramatically fitter mutation or a drastic change in the environment, has sparked a lively debate between gradualists and adherents of the 'punctuated equilibrium' theory (Eldredge and Gould 1972). This debate has not taken notice, to my knowledge, of the arsenal of models and methodologies that the social sciences offer for analyzing event chains. In the economics of path dependency, as Castaldi and Dosi note in their chapter, researchers struggle to deal with the fact that not only initial conditions, as in the original path dependency model, but also subsequent external events shape the development trajectory and can even lead to the abandonment of a given path.

Innovation in the trading zone

Despite the various risks and obstacles to import models, metaphors and methodologies from other disciplinary fields, it remains one of the major sources of innovation in all branches of the sciences. While there is no quantitative study, to my knowledge, that would establish this point, there is a small, yet growing qualitative literature in its support. The romantic legacy of viewing the 'context of discovery', in contrast to the 'context of justification', as the domain of a genius' flashes of insight or of pure luck has long obscured the patterns governing innovative processes in the sciences (Meheus and Nickles 1999). Concept transfer from one domain to another represents one important element of this pattern, together with abduction, thought experiments and heuristic rules governing exploratory research in uncharted terrain. Tool, model, and metaphor transfer each have contributed to major innovations.

The best evidence for the importance of tool transfer comes from physics. Rebaglia (1999) shows that major breakthroughs were achieved by importing mathematical tools and applying them to the physical world. The literature is more ample when it comes to model migration (Bailer-Jones 2002: 110–14; Klein 1996: 61–6). A large number of examples of model import in the hard sciences have been discussed: Bohr's atom model developed through analogies with the solar system; electromagnetic waves were modeled after d'Alembert's vibrating strings equation; Coulomb's law was applied to gravitation, electrostratics, and magnetism; nuclear fission was conceived in analogy to the division of a liquid drop. Examples from economics and the social sciences abound as well: the structuralism of Lévi-Strauss borrowed models from Jacobson's linguistics; anthropological structuralism then moved to psychology (Lacan), sociology and philosophy (Althusser) and political economy (Rey). Game theory models traveled, as mentioned before, from mathematics to economics and from there to political science, sociology, and evolutionary biology. The list of examples seems to be endless. We are left wondering, lacking a more systematic study of the subject, if we could find any process of innovation without some sort of analogical reasoning.

The innovative capacities of metaphor borrowing are less well documented in the literature, partly because the definitional boundaries between models and metaphors has become more and more blurred recently. Many philosophers of science now to look more closely at the metaphorical qualities and functions of all models, even highly formalized and mathematical ones (cf. Bailer-Jones 2002). At least we dispose of case studies on the innovative effects of the transfer of metaphors (in the more restricted sense of the term) (Brüning and Lohmann 1999). It seems that metaphor migration plays a far more limited role in the natural sciences – again due to their 'restricted' character – where model import from neighboring fields or disciplines is much more common than borrowing metaphors from completely different areas of research (see Dunbar 1995 on 'local, regional and long-distance analogies').

Concept borrowing thus represents a core element of innovation and discovery within a discipline. At the same time, it changes the relationship between areas of research by providing new intellectual contact points and avenues for cross-disciplinary co-operation. In order to adequately grasp these effects, we may refer back to the metaphor of a 'trading zone', coined by Galison (1997) to describe the intersections of the different professional cultures of experimenters, instrument makers and theorists in experimental microphysics. In a trading zone, people from mutually incomprehensible cultures come together to trade objects of interest. They develop a highly-restricted proto-language or pidgin for these negotiations. This pidgin allows them to reach agreement about objects of trade even though outside of the zone, within their own cultures, their understandings and uses of these objects differ radically.

The objects of such minimal understanding may be techniques, devices, and most importantly in the context of this introduction, shared concepts, models and metaphors. Löwy (1992) has developed the notion of 'loose concepts' which 'help to link professional domains and to create alliances between professional groups' (Löwy 1992: 373), such as immunologists and epidemiologists. Similarly, Leigh Star and Griesemer (1989) have identified 'boundary concepts', 'adaptable to different viewpoints and robust enough to maintain identity across them' (ibid.: 387) as crucial elements that bind together different disciplines and professional groups.

According to Galison and Löwy, the pidgin may further differentiate and evolve into a shared medium of communication, a 'creole language'. Examples of such highly integrated zones are quantum field theory where particle cosmology, mathematics, and condensed matter physics interact (Galison, forthcoming) or molecular genetics, where micro-extraction and micro-dissection, advanced combinatories, statistics, thermionic optics and the chemistry of enzymes coalesce around the model of the double helix (Canguilhem 1984: 148).

These are large research enterprises of experimental physics or the research departments of big museums, where representatives of different disciplines and professions are co-operating on an institutionalized basis. Examples of fully integrated creole languages to understand change are still rare. Perhaps closest to such fully co-operative and institutionalized research communities are the climate change research discussed in the opening chapters of this book. Another example, not represented in this volume, are those programs where economists, biologists, neuroscientists, psychologists and anthropologists co-operate, often using advanced game-theoretic models, to understand the emergence and further development of co-operation in animal and human societies. So far, these endeavors are comparatively loosely organized in research networks (such as the McArthur Preferences Network) or conferences (see Hammerstein 2003). They focus on very specific behavioral phenomena such as reciprocity in small groups and other small scale social patterns.

This book pursues the more modest aim of both documenting and furthering the cross-disciplinary dialogue around shared models, metaphors, and methodologies for understanding change. It contains examples of all the different types of exchange discussed in this introduction: tool transfer, model migration, the borrowing of methodological strategies, and metaphor move. It illustrates and discusses the various risks involved with conceptual borrowing, namely misunderstanding, misapplication and misfit. Most importantly, it sheds some light on the innovative potential of trading metaphors, models and methodologies. Finally, it offers some goods for future exchange: To apply non-linear systems dynamics to large-scale modernization processes (Somdatta Sinha); to use contra-factuals (Ellen Immergut) or event-chain analysis (Mahoney) for the study of historical events in the natural and economic sciences; to research discontinuous social processes with the model of neutral networks (Walter Fontana; Rudolf Stichweh); to use models of chemical reactions to understand institutional transformations of human societies (Edmund Chattoe); to export the cladistic method for studying phylogenetic change to the social sciences (Joel Cracraft). The remainder of this introduction is dedicated to a preview of each individual chapter.

The chapters

Paul Higgins focuses on the relevance of chaos theory for understanding macro-level climate change. Analysis and prediction of climate phenomena depend on particular spatial or temporal scales. In contrast to short term fluctuations in weather, longer-term climate characteristics such as the seasonal cycle are primarily determined by regular periodic forcing (e.g., the earth's orbit) and are generally predictable. However, interactions between sub-units of the climate system (e.g., ocean, atmosphere, cryosphere, and biosphere) do sometimes lead to complex behavior such as abrupt change or multiple equilibria not evident when each sub-unit is viewed in isolation. These characteristics of the climate system (unpredictability or chaotic dynamics occurring at some scales, but not precluding deterministic projections at other scales; complex behavior resulting from interactions between sub-units of the system) are likely critical for studying other processes of change as well. Thus, the analysis of anthropogenic climate change could benefit from and contribute insights to other, empirically unrelated studies of change in complex macro-level systems.

L. Douglas Kiel takes up the discussion where Paul Higgins leaves it and evaluates the prospects of transferring chaos theory to the social sciences. While social scientists have for many decades recognized the nonlinear nature of social phenomena, they have lacked the appropriate theoretical and methodological tools. The chapter looks at three modes of 'paradigm export': (1) The use of advanced mathematics for discovering chaos in time series, which, however, does not help much in explaining why such phenomena occur. (2) Chaos has also been used in a more metaphorical sense to understand change in complex social systems – a potentially powerful way to overcome linear thinking so prominent in the social sciences. (3) Agent-based modeling as a way of approaching emergence and complex change in the social sciences, represents an alternative way.

Hans-Walter Lorenz reviews what experiences economics has made with chaos theory over the past two decades. He cautions that while it is hardly difficult to discover chaotic behavior in economic systems described by standard differential equations, this behavior is often not relevant from an empirical point of view: sometimes chaos emerges on the basis of empirically unrealistic ad hoc assumptions or of parameter values beyond any empirical scope. Even when there are no doubts about the empirical relevance of chaos, technical problems such as the low number of observations in time series and problems of interpretation (such as misreading 'Monday' effects in stock markets as chaos) remain. In the second part of his essay, Lorenz moves beyond model export to more generally discuss the prospects of interdisciplinary research on shared empirical problems. The multi-system approach to climate change does indeed offer an opportunity to establish a 'tracking zone' between environmental economics and to natural and social science research, despite different degrees of formalization and different normative definitions of the aim of trade.

Evolutionary theory remains the core paradigm of change for the biological sciences. It has seen a dramatic development and expansion since the formulation of the modern synthesis combining Darwinian principles with the insights from molecular genetics. Walter Fontana's paper focuses on genetic variation as one particular aspect of the overall evolutionary dynamics. He offers a model of the genotype–phenotype relation that illuminates how genetic change produces phenotypic change. The model uses a simple molecular instance of such a relation based on the shapes of different RNA sequences that can fold into different forms, the equivalent to a phenotype. The genotypical changes are described as movements in a multi-dimensional space of possible mutations that at certain points result in a shift of phenotype. Some genotypical mutations thus are 'neutral' with regard to phenotype, while others are leading to change in the appearance of the species. The result is a discontinuous, punctuated process of evolutionary change. In a final section, Fontana suggests to export this model into economics and the social sciences by relating genotypical change to modifications in behavioral rules and phenotypical change to institutional and organizational change.

Is this concept of 'neutrality' fruitful for thinking about change in social systems? Rudolf Stichweh's paper discusses two possible applications of Fontana's model in the social sciences. Structural changes, e.g. in the class system of a society, may be neutral with regard to the basic principles of social organization, such as functional differentiation. Secondly, semantics and culture can drift through spaces of meaning without any changes in social structures immediately resulting from this. Even if these are not exact analogues, further exploring the similarities and differences is a promising avenue for future research, the chapter concludes.

Edmund Chattoe's chapter begins with a general discussion of the role of analogy in the history of thought – in the way the term was introduced by Hesse, thus broadly synonymous with what I have termed model import without re-specification and described as, the most demanding form of concept transfer. He then considers the potential benefits of evolutionary analogies for social sciences, and of economics in particular: their non-teleological character, their ability to understand endogenous variation (instead of introducing outside 'noise' from the space they provide for the emergence of new forms. The main body of the chapter presents two case studies inspired by Fontana's work. The first applies the concept of neutral networks to the analysis of social change. He concludes that the model misfits the specifities of the social world because, the classic problem to find an analogon to a selecting environment cannot be overcome. The second case study uses 'algorithmic chemistry' to explore the problems of industrial diversification and of the emergence of classes. He again notes important problems but concludes that this might be a more promising example of model export.

Carolina Castaldi and Giovanni Dosi introduce the concept of path dependency as it originally developed in economics. The chapter opens by appraising the potential for path dependencies and their sources at different levels of observation and within different domains. It then gives an overview of the modeling tools available economics. They note that during the last decade, the metaphorical use of the path dependency argument has become very popular. However, challenging questions remain regarding when and why path dependency effects do indeed occur. Usually, only one of the many possible paths that some 'initial conditions' would have allowed is actually realized – opening up the problematic space of contra-factual reasoning. Moreover, is path dependency shaped only by initial conditions or also by the unfolding of events that happen further down the road? How do socio-economic structures inherited from the past shape and constrain the set of possible evolutionary paths? And finally, what are the factors, if any, which might de-lock socio-economic structures from the grip of their past?

James Mahoney first discusses the principle of increasing returns as the core of path-dependency models in economics. He goes on in exploring the particular combination of determinacy (once a path is chosen) and indeterminacy (in the initial choice of a path) that characterizes this model and shows how similar reasoning has prevailed for a long time in social sciences, where path dependency may be much more frequent than in economics. The mechanisms that produce increasing returns, however, are different in non-market contexts and include the self-reinforcing character of political power and the functional interlocking of institutions. The social sciences have developed modes for analysing path dependency that include the study of de-locking and reversible trajectories. He specifically discusses models of 'reactive sequences' and 'event chains' and concludes by hoping that these new developments in the social sciences may inspire economists to explore similar avenues.

Eörs Szathmáry's chapter discusses how the evolutionary mechanism of natural selection can lead to various forms and varying degrees of path dependency. He describes different aspects, situated on different scales from the palaentological to the microbiological and includes different points of view. Special attention is given to how different hereditary mechanisms (genetic, chemical, epigenic, cultural) determine the degree of replication/variation as well as reversibility/irreversibility. He then shows that evolution is not always fully irreversible: some genes and traits can be resurrected if relatively little time has elapsed since their disappearance. However, the so-called major transitions in evolution, such as from cloning to sexual reproduction or from single-celled organisms to animals, illustrate the awesome power of path dependency in biological evolution. He explains how the apparent contradiction between such historical contingency and evolutionary convergence, e.g. towards analogous organs such as the eyes of squids and humans, can be resolved by looking at engineering constraints and the details of the convergent traits.

Jeffrey B. Nugent introduces new institutional economics as an ensemble of several different, though interrelated approaches. All are relatively recent developments that are only now being added to the standard tool box of 'neoclassical' economics. The most important of these are: transaction and information costs analysis, property rights theory, and the theory of collective action. Thus far, all three models have focused largely on static issues, explaining 'why institutions are the way they are'. The main purpose of his chapter, however, is to evaluate their potential for understanding change and development. He identifies the difficulties in applying new institutional economics to this task, but also offers some examples of at least partial success such as a neo-institutionalist account of why property rights developed differently in North and South America. He concludes by pointing to what has been learned about the relationship between institutional and other dimensions of change.

John Harriss, however, questions even these modest claims to explanatory power. Referring to the example of differing developments in various Indian provinces, Harriss argues that new institutional economics may serve to highlight the importance of power and of politics in understanding these differences. However, it does not in itself explain power and politics but treats them as exogenous variables. The new institutional economics thus represents a useful heuristic device that directs our attention to particular facts that then need to be explained by taking recourse to the analytical tools of the 'old' institutional analysis of a political economy type. A similar point is made with regard to cultural habits of thinking and acting which are closely related to power structures and yet find no place within the neo-institutionalist framework.

Raghavendra Gadagkar follows up with more general reflections on the prospects and dangers of cross-disciplinary borrowing. The first part of the chapter explores some parallels in the institutional set-up of human society and social insects. It specifically deals with the honey-bee dance used to indicate location of food sources, with fungus agriculture among ants, and with the division of labor between queens and workers among social insects. He shows that similar questions as those raised by new institutional economics have been asked by natural scientists studying these phenomena – which leads him to plead for more interaction between natural scientists, economists and social scientists.

In the second part of his chapter, he qualifies this plead by distinguishing between exporting methodologies, concepts, and metaphors. Exporting methodologies, especially those based on direct observation and measurement such as behavioural experiments, is usually fertile, especially for the importing social sciences. Exporting concepts, such as those developed by new institutional economics, may prove to be productive, including for the importing natural sciences. However, a transfer of metaphors (such as 'survival of the fittest') from one field to the other entails great risks because metaphors are usually loaded with value judgments that are misleading when transferred across disciplinary boundaries.

Shmuel Eisenstadt explores the importance of the idea of multilinearity and path dependency for the social sciences. His point of departure are the teleological assumptions of most classic theories of change in this field. Modernity, defined by a high degree of cultural openness combined with the politics of protest and contestation, has indeed spread to most of the world. However, it did not give rise to a single civilization, or to one institutional pattern, but rather to several differing cultural and institutional forms. He identifies the main reasons for this multilinearity: differing initial cultural conditions; specific power constellations between established and protesting elites; initial institutional frameworks that influence future institutional arrangements; and differing ways of incorporation into the global system. Finally, Shmuel Eisenstadt also questions the optimistic tone of much modernist writing about change, pointing to certain cultural and political variants of modernity that may lead to unseen mass violence and suffering.

Oded Stark opens his chapter by picking up on Eisenstadt's pessimistic concluding note. According to Stark, a major difference between social sciences and economics is that the former lack a clear basis for a comparative evaluation of different societies, while the latter can rely on measurements of economic efficiency or overall output levels to judge which of the various 'modernities' is preferable. Contrary to what many sociologists and economists like to think, however, those variants of modernity that favor trust among unrelated individuals need not be more efficient as Stark argues with the help of an example from game theory. The chapter also offers an economic explanation of why modern societies are, according to Eisenstadt, characterised by the politics of protest. They integrate greater numbers of individuals into a communicative space and thus enlarge the reference group for comparing one's own economic standing. As a result, dissatisfaction – and hence the propensity to protest – may increase despite increasing incomes.

Does the development of 'multiple modernities' bear any resemblance with evolutionary processes in the natural sciences? Somdatta Sinha shows that though the language and argumentative styles in these two research fields are quite different, there is a convergence of models and metaphors converge. According to nonlinear dynamical systems theory in biology and physics, systems with multiple variables and nonlinear interactions behave similarly to Eisenstadt's modern societies within the world system. She specifically discusses three ways in which multiplicity emerges first, as bifurcations in a system's behavior when an internal variable reaches a certain value; secondly, as diverging reactions of only minimally different systems to identical external stimuli; and finally, as different responses to different stimuli, depending on which variable is most affected. She concludes that most, yet not all of these properties can also be found in Eisenstadt's account of the history of the modern world. Emphasizing the second mechanism for the production of multiplicity, i.e. that small differences in internal structure may make a big difference in reactions to outside stimuli, she warns against oversimplifications such as the contrasting of a 'Muslim' versus a 'European modernity'.

While the preceding part dealt with the long term macro trends of social change, Ellen Immergut's chapter focuses on short term developments – the daily weather, as it were, in contrast to climate change. What is the balance between the continuity of self-reproducing political institutions and path breaking events? Immergut pleads for a pragmatic, case by case approach and for the use of historical methods to address this question. Historical methods are especially suited to elucidate three crucial problems in the study of the political: the question of how actors change their preferences and definition of their interests; the interplay of changing institutional rules and chains of micro-political events that produce 'contextual causality'; and contingency as it interacts with institutional routines. The partial reforms of the Swedish constitution in 1968 and 1969 represents an ideal case study to explore the potential of this approach. Why did members of the Social Democratic Party agree to eliminate constitutional provisions that guaranteed their hegemonic position at a time when they held the parliamentary majorities necessary to veto any and all legislation, including constitutional reform? Ellen Immergut relies on a historical counterfactual and the study of actor's perceptions and motives in order to answer this question. The case study illustrates the hazardous, unpredictable nature of institutional change and therefore the importance of historical methods for its proper understanding.

Roger Congleton takes up the problem of contingency and chance in human history but arrives at different conclusions. While historical research aims at understanding the particular, e.g. how exactly Swedish constitutional change came about, the social sciences explain general trends and patterns, such as the emergence and spread of democracy in Western Europe and beyond. They are therefore unable to make sense of individual events, which are not entirely determined by the general mechanisms. Such contingency is introduced into the historical process because actors do not have complete information about the future and therefore are prone to take sub-optimal decisions with regard to their rational interests. The Swedish constitutional change is a case in point. Such examples do not, however, contradict the rational choice model of decision making which remains, the author implies, the most powerful model of explaining change in the social sciences.

Joel Cracraft opposes the notion of contingency on similar grounds as does Congleton – and in quite striking contrast to the other evolutionary biologist writing in this volume, Eörs Szathmáry. Perhaps the most prominent argument in favor of contingency and of contrafactual thinking in biology is Gould's point that evolution would have taken a different course if a major asteroid would have missed the earth some 65 million years ago. However, Cracraft argues, contingency only matters for micro-level phenomena and not for the large scale systematic changes in the structure of species or the institutional makeup of society. These systematic changes can actually be explained with a covering law model. In the social sciences, these laws would certainly be of a probabilistic nature and would have to be based on a better identification of the units that change than it has been the case so far. Even in explaining micro-changes, however, contingency and contra-factuals are of limited significance. They may help to explain the effect that a defacto event actually did have. But it is futile to construct alternative versions of future developments assuming that one particular event had not occurred, since we never know if future events would have 'undone' the effects of changing this one link in the historical chain; if in other words, the hit of a second asteroid would have reversed the effects of the first – an argument which seems to lead the author back the classical historiographic approach of Leopold von Ranke, who exhorted his colleagues to exclusively focus on history 'wie es eigentlich gewesen (how it really was)'.

Acknowledgments

This book is the product an unusually long journey, the perhaps inevitable consequence of its scope and ambition. Two thirds of the chapters are based on the papers given at a conference that the two editors had organized at the Center for Development Research of the University of Bonn in May 2002. The list of persons who have helped and encouraged us to realize the conference is long.

Almost two years before it took place, Yehuda Elkana had given me the advice to first read seriously across the various research fields before starting to put together a conference program. This conversation in a coffee shop in Budapest proved to be a crucial initial event for the further – path dependent? – development of the project. I owe him a long and rich experience of intellectual discovery and excitement. On the other hand, he is also to blame for having me realize the limitations of my intellectual horizon and, more painfully, of my cognitive capacities. In this case, the usual disclaimers do not apply.

Most of the reading was done during my stay at the Wissenschaftskolleg Berlin, which provided an ideal environment for this exercise in disciplinary unbounding. Many co-fellows amusedly observed my first steps into their fields and put me back on my feet when I had fallen. Leticia Avilez was particularly patient and sympathetic.

Reinhart Kössler soon joined me in the preparatory reading and the organization of the conference. He proved to be an ideal companion and partner to put the project on track. We have received advice and many valuable suggestions regarding conference and book participants from too many people to name them all. Joachim von Braun, Luis Mata, Paul Vlek, and Peter Wehrheim from the Center of Development Research helped to orient ourselves in the fields of ecology and economics. We are especially grateful to Luis for identifying approaches and persons in the domain of climate change. Raghavendra Gadagkar from the Wissenschaftskolleg and Hans-Jörg Rheinberger of the Max-Planck-Institute for the History of Sciences provided critical inputs. Markus Beiner from the Volkswagen-Foundation ensured financial support for the conference. The Center for Development Research has upheld earlier commitment to provide financial assistance to the book project. Valuable research assistance came from Karina Waedt. We thank them all.

Notes

1. I thank Rick Grannis for helping to avoid some initial confusion regarding the Markov chains used in the second section of this chapter. Bettina Heintz's suggestions put me on the right path searching for literature on model transfer. Special thanks go to Somdatta Sinha and Giovanni Dosi who both offered detailed

comments and suggested important improvements to make the matrices 'work'. Giovanni patiently explained the properties of chaos theory until I finally got it. Walter Fontana and Paul Higgins have edited parts of this chapter and saved me from many imprecisions and misunderstandings. I am afraid the remaining ones are my sole responsibility.
2. Such matrices can have a variety of characteristics, some of which are relevant for my purpose. In the standard matrices such as the weather example above, all states can be reached from other states in a finite number of steps. This is an *irreducible chain*. If the path always leads back to a state through a determined number of moves, we speak of a *cyclical* chain or a periodic chain. Chains without such cyclical moves are called *aperiodic*. If there are states that do not lead to any other states, i.e. with transition probabilities to all other states of 0, we call this an *absorbing state* (imagine that the first sunny day would be followed by sunny days forever). If there is a group of states that only lead to the states within that group but nowhere outside, mathematicians speak of an *ergodic set* (or chain, if the states comprise all possible states). The number of possible states can be finite (*a finite state space*) or infinite. A state space may contain a subset of spaces that communicate with each other with much higher probability than with all other states. The chain is then '*nearly completely decomposable*'. For some chains, we know where to start, i.e. the initial probability for a certain state is 1. In others there are several possible initial states.
3. This chain would be described as irreducible; it has a finite state space with no absorbing states or decomposable subsets; it is fully ergodic; and the initial state probabilities are not known: the process can start anywhere.
4. This chain is a periodic and not irreducible. I assume that this could be described, in mathematical terms, as a nearly completely decomposable Markov chain.
5. Note that transition probability matrices are not the adequate tool to describe the nature of these qualitative changes. They only characterize the probabilities, the pathways and the time necessary to achieve such changes. Nonlinear system dynamics may be a more adequate tool to model the actual transformations of the system's behavior by referring to changing internal and external parameters.
6. Note also that history is fully aperiodic, non-recurrent, has no absorbing states and is not irreducible. In other words, something new always has to happen.

Bibliography

Abbott, Andrew. 1995. 'Sequence analysis', in *Annual Review of Sociology* 21:93–113.
Abbott, Andrew. 2001. *Time Matters. On Theory and Method*. Chicago: University of Chicago Press.
Abell, Peter. 1993. 'Some aspects of narrative method', in *Journal of Mathematical Sociology* 18:93–134.
Arthur, Brian. 1994. *Increasing Returns and Path Dependence in the Economy*. Ann Arbor: University of Michigan Press.
Bailer-Jones, Danjela M. 2002. 'Models, metaphors and analogies', in Peter Magnani and Michael Silberstein (eds), *The Blackwell Guide to the Philosophy of Science*. Malden: Blackwell. 108–27.
Barth, Fredrik. 1995. 'Other knowledge and other ways of knowing', in *Journal of Anthropological Research* 51:65–8.
Bates, Robert H., Avner Greif, Margaret Levi, Jean-Laurent Rosenthal and Barry R. Weingast. 1998. *Analytical Narratives*. Princeton: Princeton University Press.

Bauer, Henry H. 1990. 'Barriers against interdisciplinarity: Implications for Studies of Science, Technology, and Society', in *Science, Technology & Human Values* 15 (1):105–19.
Black, Max. 1993. 'More about metaphor', in Andrew Ortony, *Metaphor and Thought*. Cambridge: Cambridge University Press. 19–41.
Bourdieu, Pierre. 1988. *Homo Academicus*. Stanford: Stanford University Press.
Brüning, R. and Lohmann G. 1999. 'Charles S. Peirce on creative metaphors: A case study on the conveyor belt metaphor in Oceanography', in *Foundations of Science* 4 (3):389–403.
Canguilhem, Georges. 1984. *Wissenschaftsgeschichte und Epistemologie*. Frankfurt: Suhrkamp.
Coase, Ronald H. 1990. *The Firm, the Market, and the Law*. Chicago: University of Chicago Press.
Collier, Rugh B. and David Collier. 1991. *Shaping the Political Arena: Critical Junctures, the Labor Movement, and Regime Dynamics in Latin America*. Princeton: Princeton University Press.
Darden, Lindley. 2002. 'Strategies for discovering mechanisms: schema instantiation, modular subassembly, forward/backward chaining', in *Philosophy of Science* 69:S354–S365.
David, Paul A. 1985. 'Clio and the economics of QWERTY', in *American Economic Review* 75:332–7.
David, Paul A. 1994. 'Why are institutions the "carriers of history"?: path dependence and the evolution of conventions, organizations and institutions', in *Structural Change and Economic Dynamics* 5:205–20.
Dunbar, Kevin. 1995. 'How scientists really reason: scientific reasoning in real-world laboratories', in Robert J. Sternberg and Janet E. Davidson (eds), *The Nature of Insight*. Cambridge: MIT-Press. 365–95.
Eldredge, Niles and Stephen J. Gould. 1972. 'Punctuated equilibria: an alternative to phyletic gradualism', in T.J.M. Schopf (ed.), *Models in Paleobiology*. San Francisco: Freeman, Cooper and Co. 82–115.
Fearon, James D. 1991. 'Counterfactuals and hypothesis testing in political science', in *World Politics* 43:169–95.
Fergunson, Niall (ed.). 1997. *Virtual History: Alternatives and Counterfactuals*. Basingstoke: Macmillan.
Fifth, Raymond. 1992. 'A future for social anthropology', in Sandra Williams (ed.), *Contemporary Futures*. London: Routledge. 208–24.
Fowles, Jib. 1974. 'Chronocentrism', in *Futures* 6 (1):65–8.
Galison, Peter. 1997. *Image and Logic: A Material Culture of Microphysics*. Chicago: Chicago University Press.
Galison, Peter. Forthcoming. *Theory Machines*.
Gould, Stephen J. 1988. *Time's Arrow/Time's Cycle: Myth and Metaphor in the Discovery of Geological Time*. Harvard: Harvard University Press.
Gould, Stephen J. 2002. *The Structure of Evolutionary Theory*. Cambridge: Belknap.
Hammerstein, Peter. 2003. *Genetic and Cultural Evolution of Cooperation*. Cambridge: MIT Press.
Hawthorn, Geoffrey. 1991. *Plausible Worlds: Possibility and Understanding in History and the Social Sciences*. New York: Cambridge University Press.
Heise, David R. 1989. 'Modelling event structures', in *Journal of Mathematical Sociology* 14:139–69.
Kellert, Stephen H. 2000. 'Extrascientific use of physics: the case of nonlinear dynamics and legal theory', in *Philosophy of Science* 68 (3):S455–S466.

Klein, Julie Thompson. 1996. *Crossing Boundaries: Knowledge, Disciplinarities, and Interdisciplinarities*. Charlottesville: University Press of Virginia.

Leigh Star, Susan and James R. Griesemer. 1989. 'Institutional ecology, "translations" and boundary objects: amateurs and professionals in Berkeley's Musum of Vertebrate Zoology, 1907–39', in *Social Studies of Science* 19:387–420.

Löwy, Ilana. 1992. 'The strength of loose concepts – boundary concepts, federative experimental strategies and disciplinary growth: The case of immunology', in *History of Science* 30 (4):371–96.

Mahoney, James. 2000. 'Path Dependence in Historical Sociology', in *Theory and Society* 29:507–48.

Mayntz, Renate. 1990. 'Naturwissenschaftliche Modelle, soziologische Theorie und das Mikro-Makro-Problem', in Wolfgang Zapf (ed.), *Die Modernisierung moderner Gesellschaften. Verhandlungen des 25. Deutschen Soziologentages in Frankfurt am Main 1990*. Frankfurt: Campus. 55–68.

McDonald, Terrence J. (ed.). 1996. *The Historic Turn in the Human Sciences*. Ann Arbor: University of Michigan Press.

Meheus, Joke and Thomas Nickles. 1999. 'The methodological study of creativity and discovery: Some background', in *Foundations of Science* 4 (3):231–5.

Morgan, Mary S. and Margaret Morrison. 1999. *Models as Metaphors. Perspectives on Natural and Social Sciences*. Cambridge: Cambridge University Press.

Nisbet, Robert A. 1969. *Social Change and History. Aspects of the Western Theory of Development*. Oxford: Oxford University Press.

North, Douglas. 1994. 'Economic performance through time', in *American Economic Review* 84 (3):359–68.

Ortony, Andrew. 1993. *Metaphor and Thought*. Cambridge: Cambridge University Press.

Pantin, Carl F.A. 1968. *The Relations between the Sciences*. Cambridge: Cambridge University Press.

Prigogine, Ilya. 1997. *The End of Certainty: Time, Chaos and the New Laws of Nature*. New York: The Free Press.

Ragin, Charles. 1998. 'The logic of qualitative comparative analysis', in *International Journal of Social History* 43 (6):105–25.

Rebaglia, Alberta. 1999. 'Scientific discovery: between incommensurability of paradigms and historical continuity', in *Foundations of Science* 4 (3):337–55.

Rostow, Walt W. 1991 (1960). *The Stages of Growth: An Anti-Communist Manifesto*. Cambridge: Cambridge University Press.

Segerstrale, Ullica. 2001. *Defenders of the Truth: The Sociobiology Debate*. Oxford: Oxford University Press.

Sewell, William H. 1996. 'Three temporalities: Toward an eventful sociology', in Terrence J. McDonald (ed.), *The Historic Turn in the Human Sciences*. Ann Arbor: University of Michigan Press. 245–80.

Steward, Julian Haynes. 1955. *Theory of Culture Change: The Methodology of Multilinear Evolution*. Urbana: University of Illinois Press.

Tetlock, Philip E. and Aaron Belkin (eds). 1996. *Counterfactual Thought Experiments in World Politics: Logical, Methodological, and Psychological Perspectives*. Princeton: Princeton University Press.

Thelen, Kathrin. 1999. 'Historical institutionalism in comparative politics', in *The Annual Review of Political Science* 2:369–404.

Urry, John. 2004. 'Small worlds and the new "social physics"', in Global Networks 4 (2):109–30.

Wallerstein, Immanuel. 1996. *Open the Social Sciences: Report of the Gulbenkian Commission on the Restructuring of the Social Sciences*. Stanford: Stanford University Press.

Watts, Duncan. 2004. 'The "new" science of networks,' in *Annual Review of Sociology* 30: 243–70.
Wimmer, Andreas. 1996. 'Kultur Reflexionem über einen ethnologischen Grundbegriff,' *Zeitschrift für Ethnologie* 48 (3): 401–25.
Zapf, Wolfgang. 1996. 'Die Modernisierungstheorie und unterschiedliche Pfade der gesellschaftlichen Entwicklung', in *Leviathan* 42 (1):63–77.

Part I
Chaos and Order in Climate Change

2
Climate Change: Complexity, Chaos and Order
Paul Higgins

The climate system constantly changes due to internal variability and natural external perturbation. Anthropogenic climate change (due to human caused emissions of greenhouse gases) must be distinguished and characterized from this constantly changing background. Controlled experiments of the actual climate system are not possible but controlled simulation experiments using representative models examine how the natural system will respond. Observations and models together detected a novel change in climate over the past century and attribute that change as likely due to human activities. Analysis and prediction of climate system phenomena often depend on particular spatial or temporal scales. The short-term fluctuations in weather are heavily influenced by changes in irregular internal processes and small-scale interactions (e.g. between the ocean and atmosphere). Chaotic responses result, which precludes long-term forecasts in the absence of perfect observations and formulations that perfectly predict atmospheric dynamics. In contrast, longer-term climate characteristics such as the seasonal cycle are primarily determined by regular periodic forcing (e.g. the earth's orbit) and are generally predictable. However, interactions between sub-units of the climate system (e.g. ocean, atmosphere, cryosphere, and biosphere) do sometimes lead to complex behavior not evident when each sub-unit is viewed in isolation. Such 'emergent properties' include the occurrence of abrupt changes and multiple equilibria.

These four characteristics of climate change research, (1) a constantly changing system, (2) need for representative models, (3) unpredictability or chaotic dynamics occurring at some scales (e.g. weather) but not precluding insights and projections at other scales (e.g. climate), and (4) complex behavior resulting from interactions between sub-units of the system, likely are equally critical for other fields focused on change. Thus, the study of anthropogenic climate change could likely benefit from and contribute insights to other unrelated studies of change.

Introduction

The climate system changes constantly. The earth's axial rotation causes daily changes in temperature, precipitation, cloud cover, humidity and other climate variables. Over short time scales (ranging from minutes to several days) movement of clouds, air masses, and fronts contribute to rapid, often unpredictable, changes in weather. The earth's annual rotation about the sun causes large seasonal changes in climate. Multi-annual to decadal mode transitions such as the El Niño/La Niña cycles in the Pacific (Diaz and Markgraf, 2000) and the North Atlantic Oscillation in the Atlantic (Hurrell, 1995) cause additional variability in the climate system. On millennial time scale changes in the earth's orbit cause the ice-age/interglacial transitions (Imbrie et al., 1993).

It is against this backdrop of constant natural climate variability and change that researchers must distinguish anthropogenic climate change resulting from greenhouse gas (GHG) emissions. This is a two-part problem consisting of observing whether a novel change in climate has occurred (detection) and then determining whether an observed climate change is the result of human activities (attribution). Detection is resolved primarily through observations of the climate system. Knowledge of current and past climate variables (temperature, precipitation, humidity, sea and land ice extent, etc.) reveals unequivocally that climate has changed over the past century (IPCC, 2001). Attributing these observed changes in climate to increases in atmospheric greenhouse gas concentrations – approximately 30% for carbon dioxide since the start of the industrial revolution (Keeling, 1986) – or to natural sources of climate variability requires the use of controlled experiments.

Typically, scientists conduct empirical experiments that control all confounding factors (e.g. natural climate variability resulting from volcanic eruptions or changes in solar output) to isolate and characterize the effects of a single perturbation (in this case the anthropogenic emission of GHGs). Such controlled manipulative experiments of the actual climate system are not possible since we have only one earth and therefore cannot simultaneously test cases that differ only by changes in anthropogenic emissions of the GHGs. One solution is to conduct tests on models of the climate system. Provided these models accurately represent the important processes and interactions of the study system they can be used to determine how the actual system will respond using controlled simulations (e.g. comparison of model simulations which differ in GHGs but in which all other components of the modeled climate system are held constant).

Most models used to study anthropogenic climate change are mathematical models based on the physical laws governing the actual climate system. The most sophisticated of these include equations for the conservation of mass, moment, and energy, an equation of state (the ideal gas law in the case of the atmosphere), and the hydrostatic equation (Washington and Parkinson, 1986). Some models also include detailed interactions between sub-units of the climate system (e.g. ocean, atmosphere, cryosphere, and biosphere). The models are then validated using the seasonal cycle, natural variations caused by changes in the sun's energy (solar radiation) or volcanic eruptions, and previous climate states (e.g. ice ages). Model success in simulating climate under these different conditions increases confidence that they can accurately project responses to anthropogenic perturbations of the climate system.

Recent improvements in models and observations increase our ability to distinguish a human caused signal in the climate system. The Intergovernmental Panel on Climate Change (IPCC) summarizes the current scientific consensus as follows. 'In the light of new evidence and taking into account the remaining uncertainties, most of the observed warming over the last 50 years is likely to have been due to the increase in greenhouse gas concentrations' (IPCC, 2001).

Having first detected a climate change and then attributed that change as likely due to human activities, the next task for climate models is to determine how climate will change in the future, particularly over the next 100 years. Uncertainty in future climate change results from several social, economic and political factors along with natural characteristics of the climate system. In particular, rates of economic and technological development, population growth, and social and political choices will all influence the amount of GHGs emitted to the atmosphere. Climate sensitivity to a given increase in GHGs is also uncertain with current best estimates ranging from 1.5–4.5°C warming for a doubling of carbon dioxide (IPCC, 2001). Taking each of these sources of uncertainty into account, the IPCC currently estimates an increase of 1.4–5.8°C by the year 2100 (IPCC, 2001).

Complexity, chaos and order

Some readers likely remain skeptical that models can project climate over century long time scales as we fail to predict weather accurately even a few days in advance. This apparent paradox results from the simultaneous existence of both predictable and unpredictable sources of variability in the climate system. Over some spatial and temporal scales, internal irregular interactions between components of the climate system (e.g. between the atmosphere and ocean, or between vegetation and wind) heavily influence climate variables such as temperature, precipitation, wind, and humidity. These internal irregular interactions can produce chaotic dynamics (i.e. minute variations in specific atmospheric variables such as temperature, humidity, wind speed, cloudiness, and pressure, even under otherwise identical conditions, lead to large differences in the state of the atmosphere over short periods of time). This sensitive dependence on initial conditions makes prediction beyond a few days highly inaccurate. In sharp contrast, the state of the climate system is generally determined by regular external periodic forcing such as the earth's rotation about its axis and the earth's rotation around the sun, which produce the diurnal and seasonal cycles, respectively (Lorenz, 1993).

This is most clearly illustrated by the distinction between weather and climate. Weather – the short-term atmospheric state of temperature, wind, precipitation, and clouds – is heavily influenced by irregular internal interactions. As a consequence, weather is chaotic and unpredictable beyond a few days. Long-term weather prediction would require knowing the exact state of all climate variables at a single point in time along with having physical equations that flawlessly predict atmospheric dynamics. In contrast, the average weather (including variability) over longer time periods (i.e. climate) is generally predictable as regular periodic forcing, caused by the earth's orbit around the sun, dominates this variability.

Lorenz's famously explained the chaotic nature of weather with his description of 'the butterfly effect' (Lorenz, 1993). The flap of a butterfly's wings in a distant location could result in a tornado occurring locally (or equally likely, the absence of a tornado) at some future time. However, the sensitive dependence on initial conditions (in this case whether the butterfly flaps its wings) generally only alters that portion of the climate system that is influenced by the internal irregular interactions and would not alter the component of variability caused by the regular periodic forcing. Thus, a tornado may or may not occur in Texas as a consequence of small changes in initial conditions in the tropics but a similar change in initial conditions would not change the underlying nature of the seasonal cycle.

Of course, both periodic and irregular forcings influence the climate system simultaneously. Whether irregular internal processes have a large or small influence depends on the scale of interest. Over the short temporal and spatial scales that characterize weather prediction, irregular internal variability heavily influences the state of the climate system, while the long-term responses of climate are influenced primarily by regular periodic forcing. As a result, the impossibility of long-range weather forecasts does not preclude future climate projections.

All this being said, the general predictability of climate is not absolute as complex responses including abrupt changes and multiple equilibria result from internal feedbacks and interactions. Thus, some elements of climate do appear sensitive dependent to small changes in initial conditions (e.g. orbital characteristics, atmospheric composition, sea surface temperatures, distribution of land masses, land-sea fraction, vegetation, and the amount of surface covered by ice), which can produce large changes over time as a result of feedbacks in the system.

The climate system consists of multiple interacting sub-systems (e.g. atmosphere, oceans, ice-covered areas, biosphere, and even social and economic systems). Viewed in isolation (often done along distinct disciplinary lines), each sub-unit may produce internally stable and predictable behavior. However, real-world coupling between these sub-systems often causes the set of interacting systems to exhibit new collective behaviors – often called 'emergent properties' – not demonstrable by models that do not also include such coupling (for greater technical detail see Higgins et al., 2002, from which the remainder of this section is adapted).

One emergent property, increasingly evident in climate and biological systems, is that of hysteresis – multiple equilibria are possible under identical forcing depending on past history. This behavior can be a consequence of feedbacks between sub-units that amplify a given change. In such cases, a change in external forcing moves the climate system to a new equilibrium, which may be self-sustaining even after the original external forcing is restored. The remainder of this paper considers several cases in which apparently complex climate behavior results from the interactions between sub-units of the climate system.

Thermohaline circulation

Interactions between the ocean and the atmosphere produced abrupt temperature changes of 5–15°C throughout the previous ice age (Jouzel, 1999; Dansgaard et al., 1993). These temperature swings appear to result from transitions between multiple equilibria that exist in a density driven form of ocean circulation called thermohaline circulation (THC) (Broecker, 1997; Rahmstorf, 1996, 1999), though changes in ocean and atmosphere interactions at other locations are also possible (Clement et al., 2001). Under present conditions, THC warms Greenland and Western Europe by roughly 5–8°C relative to comparable latitudes (Stocker and Marchal, 2000; Broecker, 1997). Warm surface waters (Levitus, 1982) transport heat northward in the Atlantic and release heat to the atmosphere. Surface water density increases as these waters cool, eventually becoming sufficiently dense to sink. The sinking surface water draws additional warm tropical surface water northward while the resulting deep water flows south. The cold deep water eventually returns to the surface, thereby completing the pattern (for greater technical detail see Higgins et al., 2002, from which this section is adapted).

A schematic stability diagram of THC (Figure 2.1), demonstrates three possible THC equilibrium states under different levels of freshwater forcing, and the theoretical mechanisms for switching between them (Rahmstorf, 1996). These include three classes of deep water formation, (1) sinking both in the Labrador sea and North of the Greenland, Iceland, Scotland (GIS)

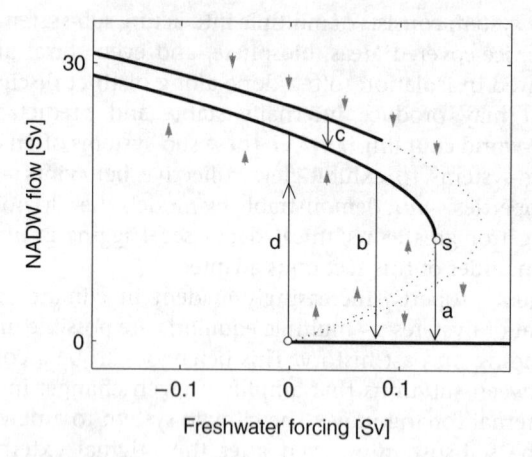

Figure 2.1 Schematic bifurcation diagram for thermohaline circulation as measured by North Atlantic Deep Water (NADW) production in Sverdrups (1 Sv = 1 million m³/s)*

*Solid lines show stable equilibrium states; unstable states are dotted. The upper two solid lines correspond to equilibria with deep water formation at two different sites (the Labrador Sea and north of the GIS ridge). The solid bottom line corresponds to an equilibrium without NADW formation (i.e. THC collapse). Examples of the four basic transition mechanisms are shown by dashed lines: (a) local convective instability, a rapid shutdown or startup of a convection site, (b) polar halocline catastrophe, a total rapid shutdown of North Atlantic convection, (c) advective spin-down, a slow spindown process triggered when the large-scale freshwater forcing exceeds the critical value at the saddle-node bifurcation S while convection initially continues, and (d) startup of convection in the North Atlantic from a THC collapse state. The convective transitions a, b, d can be triggered either when a gradual forcing pushes the system to the end of a stable branch or by a brief but sufficiently strong anomaly in the forcing. cf. Rahmstorf, 1996.

ridge, (2) sinking North of the GIS ridge alone, and (3) a complete overturning shutdown. Switching between the equilibria can occur as a result of temperature or freshwater forcing, each of which can decrease surface density in the North Atlantic beyond a critical value necessary for THC to be stable (Rahmstorf, 1996). In the past, rapid and repeated switching between equilibria occurred on the order of years to decades (Bond et al., 1997; Alley et al., 1993). At the same time, modeling results suggest that future anthropogenic climate change could cause similar changes in THC overturning (Wood et al., 1999; Manabe and Stouffer, 1993) as a consequence of the precipitation increases, glacial melting, and disproportionate high latitude warming that is often projected (IPCC, 2001). Additional research suggests that both the atmospheric greenhouse gas (GHG) concentration and also the rate of GHG increase determine how the coupled ocean-atmosphere system switches between these equilibria (Schneider and Thompson, 2000; Stocker and Marchal, 2000) (Figure 2.2).

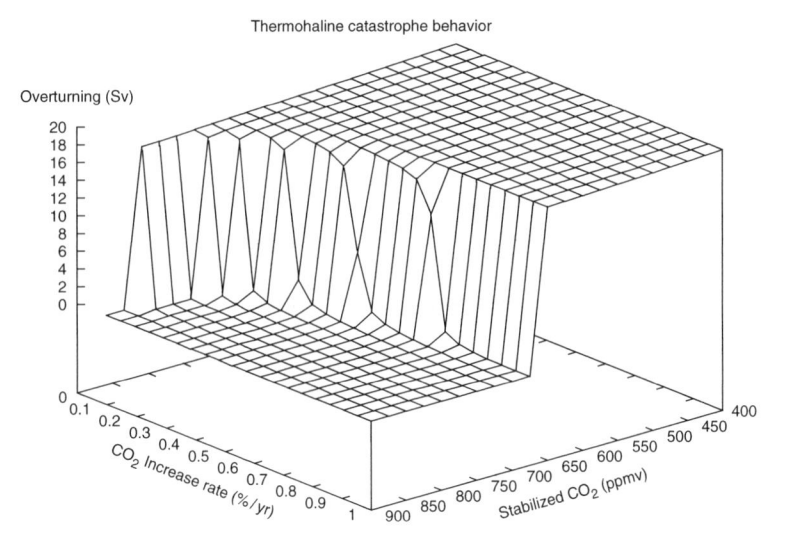

Figure 2.2 Equilibrium results of a simple climate model under different forcing scenarios*

* THC overturning in Sverdrups (1 Sv = 1 million m³/s) is shown on the vertical axis as a function of the rate of carbon dioxide increase in the atmosphere and the stabilization concentration. Higher stabilization levels and more rapid rates of carbon dioxide increase make a THC collapse more likely. From Schneider and Thompson, 2000.

Vegetation cover and climate dynamics

Multiple equilibria in the climate system also occur in response to interactions between the atmosphere and the biosphere (for greater technical detail see Higgins *et al.*, 2002, from which this section is adapted). Changes in either vegetation or the atmosphere constitute feedbacks that can either amplify or damp an initial perturbation to the other sub-unit of the coupled system. Vegetation influences climate by changing (1) albedo – the amount of incoming solar radiation that is reflected by the surface of the earth (snow, ice, and bare ground reflect more radiation back to space than does vegetation), (2) atmospheric moisture (leaf area, rooting depth and photosynthesis all influence rates of transpiration, the plant mediated conversion of surface water into atmospheric water vapor), (3) the friction between the atmosphere and the land surface (the height of vegetation influences the roughness of the surface of the earth at the atmosphere boundary), and (4) the carbon cycle (e.g. trees tend to store more carbon than grasses which influences the atmospheric carbon dioxide concentration). At the same time, climate determines vegetation characteristics such as leaf area, rooting depth, fraction of trees or grasses in the landscape, and seasonal display of leaves.

Historical evidence suggests that two equilibria in the coupled vegetation and climate system may exist for the Sahel region of West Africa (10°N–17.5°N, 15°W–15°E) (Wang and Eltahir, 2000c), where an extended period of drought has persisted since the 1960s (Wang and Eltahir, 2000b). Experiments suggest this drought constitutes a shift from a self-sustaining wet climate equilibrium to a self-sustaining dry equilibrium (Wang and Eltahir, 2000b). The shift was triggered by a SST anomaly that initially altered precipitation in the Sahel. The grassland vegetation then shifted to that of a drier equilibrium state (Figure 2.3), causing the reduction in atmospheric moisture to persist over the longer term (Wang and Eltahir, 2000c).

In the model (Wang and Eltahir, 2000a), moist (i.e. favorable) growing seasons facilitate greater root growth of perennial grasses while dry (i.e. unfavorable) growing seasons lead to less root biomass. During subsequent

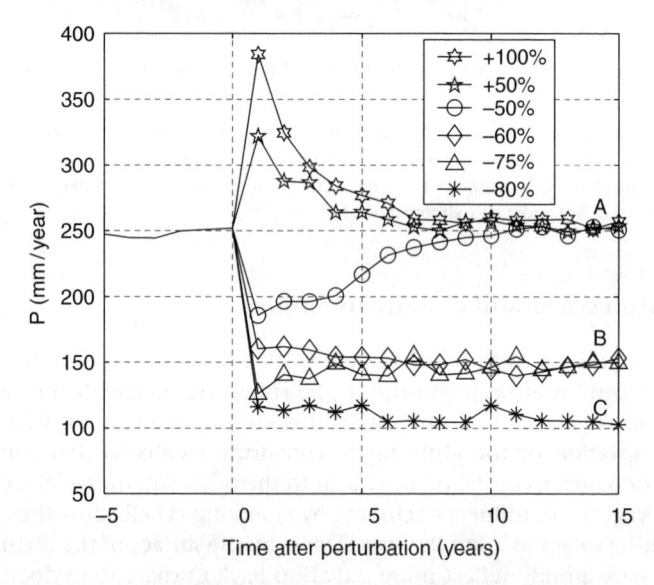

Figure 2.3 Model response of the coupled atmosphere–biosphere system to vegetation perturbations for the Sahel (Wang and Eltahir, 2000c) and starting from a vegetation distribution for West Africa close to today's*

* Leaf area index shows a similar pattern as that shown here for precipitation. Three equilibria are obtained depending on the magnitude of the vegetation perturbation. (A) The coupled climate–vegetation system is stable to perturbations in which vegetation is degraded by 50% or increased by 50, and 100% thus vegetation and precipitation recover to pre-disturbance values. (B) Perturbations in which 60 and 75% of vegetation is degraded result in a second equilibrium in the coupled climate–vegetation system. (C) Perturbation in the form of 80 percent degradation of vegetation results in a third equilibrium. Figure reproduced from Wang and Eltahir, 2000c.

years, the smaller (larger) amount of root biomass leads to less (more) above ground growth resulting in a smaller (larger) leaf area. The smaller (larger) leaf area then transpires less (more) water to the atmosphere, providing less (more) atmospheric moisture for precipitation. This causes the atmosphere–biosphere system to remain in the drier (wetter) equilibrium. In another series of experiments (Claussen, 1998), a coupled global atmosphere–biome model produces two separate equilibrium solutions for North Africa and Central East Asia when initialized with different land-surface conditions (Figure 2.4). The first equilibrium, obtained when the model is initialized with bright sand desert on all ice-free land, is drier, with a distribution of subtropical deserts similar to present day. The other equilibrium, obtained when the model is initialized with forest, grassland, or dark desert on all ice-free land, is moister with a concomitant northward shift of vegetation in to what are now arid regions. The rest of the world is insensitive to changes in initial conditions, producing a single equilibrium in the modeled atmosphere–biosphere system (Claussen, 1998).

A similar study compared simulations with vegetation initialized as either forest or desert (Kleidon et al., 2000). The comparison between these 'green' and 'desert' worlds again illustrates that some regions are sensitive to the initial vegetation while other regions have only a single equilibrium under

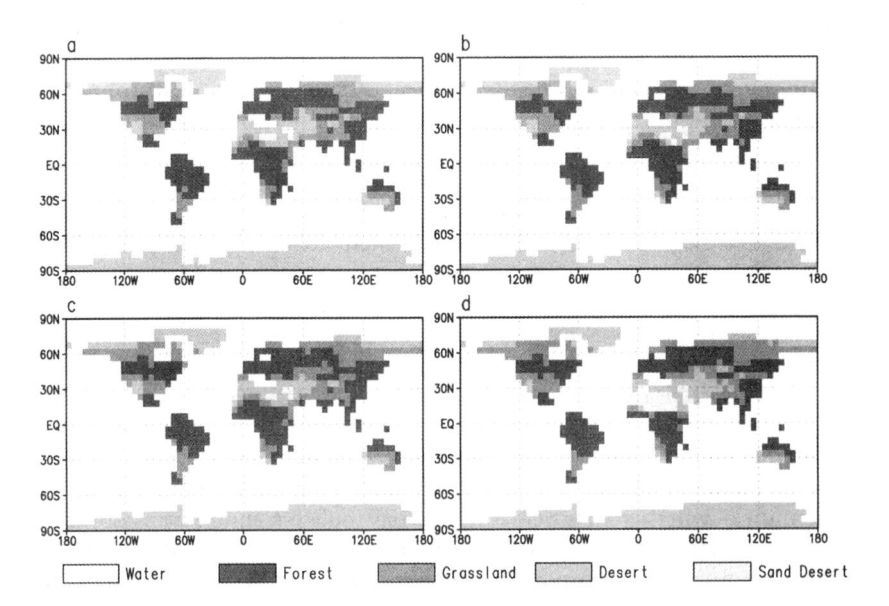

Figure 2.4. Sensitivity of modeled biome distributions to initial vegetation*

* Experiments initialized with (a) forest, (b) grassland, and (c) dark desert produce similar climate–biome equilibria in the Sahara region of the African continent. Initialization with (d) bright sand desert produces a different equilibrium for the region. From Claussen, 1998.

current conditions. In particular, regions of Africa, South Asia, and Australia produced different stable atmosphere–biosphere equilibria, depending on whether the initialized vegetation was forest or desert. In contrast, initialization with either 'green' or 'desert' conditions produces the same equilibrium solution throughout the rest of the world.

But it also must be kept in mind that results from all such models depend on how the model aggregates over processes that can occur at smaller scales than is implicit in the simulation – e.g. local variations in soils, fire regimes, or slope and elevation variability may all be neglected. The extent to which it is necessary to explicitly account for such processes, or to which such processes might influence conclusions about stability, remains a major debate point in all simulations that, for practical necessity, must parameterize the effects of processes occurring on small time and space scales. This suggests that a hierarchy of models of varying complexity (and observations to test them) is the approach most likely to determine the implications of the degree of aggregation in various models.

Other regions, currently possessing a single equilibrium in the climate-biosphere system, demonstrated complexity under past climate conditions. The Sahara was heavily vegetated 6000 ago but over the next 1000–2000 years an abrupt change in vegetation and climate occurred (Claussen *et al.*, 1999). Model simulations revealed that an atmosphere-ocean-vegetation coupling was better able to represent climate of the Sahara, with the addition of vegetation increasing precipitation four fold (Ganopolski *et al.*, 1998); a strong positive feedback between climate and vegetation distribution. Then, as orbital forcing caused a slow and steady decline in summer radiation, the Sahara abruptly underwent desertification as a consequence of interactions between the orbital changes and the atmospheric and biospheric sub-systems (Claussen *et al.*, 1999). Thus, the Sahara of the mid-Holocene crossed a threshold beyond which an abrupt and irreversible change to a new stable equilibrium occurred.

Many factors complicate interpretation of model results such as these, however. Natural variability and ecosystem disturbance – both human and natural – are often not realistically incorporated into vegetation models. Whether different modeled equilibria remain stable under the more complicated conditions of the natural world requires additional exploration for many regions. Furthermore, natural ecosystems are rarely – if ever – at equilibrium at the particular spatial and temporal scale of interest. Therefore, determining whether a particular region has multiple equilibria as opposed to an incomplete recovery from disturbance will require testing across a hierarchy of models incorporating different processes at different scales. To the extent that modeled behaviors remain insensitive to such tests, robust conclusions about complex system behaviors may be drawn with increasing confidence. Of course, the level of disturbance itself is an endogenous component of the climate–vegetation system and any change in the disturbance regime could generate new equilibria in the coupled system.

Discussion and conclusions

Efforts to study anthropogenic climate change could benefit from and contribute insights to other unrelated studies of change due to four similar characteristics: (1) the need to identify and characterize one cause of change from a system undergoing constant change, (2) the impossibility of conducting controlled experiments of the actual study system and the consequent need to use representative models, (3) the importance of scale, specifically that unpredictability or chaotic dynamics occurring at some scales (e.g. weather) does not preclude insights and projections at other scales (e.g. climate), and (4) complex behavior, such as abrupt change and multiple equilibria result from interactions between sub-units of the system that are often not evident when each sub-unit is viewed in isolation. Of these four, the importance of scale and interactions between sub-units may prove particularly insightful.

At virtually any spatial and temporal scale at least some phenomena are likely to exhibit complex responses, abrupt changes, and multiple equilibria for a given forcing while other phenomena will change linearly and predictably. The review of multiple equilibria in the climate system, detailed above, is focused at the broadest scales of ecosystem structure and function as they relate to climate (e.g. albedo, transpiration, carbon storage, and roughness). At other biological scales (e.g. genetic, species, and population) different processes and characteristics may have multiple equilibria and be prone to abrupt changes. Species or population extinction and loss of genetic diversity may occur without transitions in the climate system. Such changes clearly constitute different equilibria that may be profoundly important biologically but these different equilibria are not relevant at the scale of the climate system (this paragraph and the remainder of this section adapted from Higgins *et al.*, 2002).

Along with these scaling issues, assessments of climate change must examine interactions between sub-units of the climate system. In many regions, the climate-biosphere system currently resides in a single stable equilibrium (at least at the scale most relevant to the climate system). Even in these regions, global changes in general and climate change in particular will cause changes in the equilibrium state of the climate-biosphere system. However, if the changes in these regions are largely linear and relatively predictable, then changes that result from increasing greenhouse gas concentrations may be relatively reversible (assuming no unforeseen thresholds are crossed and that such regions do not develop multiple equilibria as a result of those changes).

This is not to suggest that the changes in regions that experience smooth and relatively slow shifts from one equilibrium to another would be benign. Indeed, such changes could constitute large perturbations to biological communities (resulting in range shifts, population or species extinctions, and changes to ecosystem structure and function) and may also dramatically impact social and economic conditions, particularly in the developing world. Moreover, even if the changes appeared smooth and reversible when aggregated over many decades and large regions – the scale of a continent for example – locally there could be abrupt changes and even complex dynamical behaviors. Linear and predictable changes to biological systems could also dramatically influence social and economic conditions through alterations of the flow of goods and services provided by those biological systems. Such goods and services include provision of food, maintenance of the atmospheric composition, mitigation from flood and droughts, provision of pollinators for crops, decomposition of wastes, and the provision of clean water (Daily, 1997). Thus, the atmosphere-biosphere system may change smoothly but social and economic systems that are also coupled to the climate-biosphere system may lead to additional emergent properties.

More complex are the possible changes to come in regions that currently have multiple equilibria in the climate-biosphere-economic system. Changes in these regions may be more rapid or less predictable. Furthermore, changes from one equilibrium to another may not be reversible. Thus, even small future climate changes may require significant environmental and social adaptation if these systems are close to thresholds and therefore forced to other stable equilibria.

Societal adaptation may prove difficult as a consequence of the abrupt, unpredictable, and irreversible changes that may occur as a consequence of climate-ecosystem dynamics. The advanced warning and slowness of transition that would accompany gradual linear changes will not apply for abrupt change. Indeed, efforts to anticipate adaptation to climate change may even prove counter productive in regions that experience such rapid transitions between equilibria. Thermohaline circulation collapse, for example, could occur after the North Atlantic region has made substantial efforts toward adaptation to the warming that accompanies increasing GHG concentrations. Indeed, plausible 'surprises' such as THC collapse pose a quandary to those countries or regions deciding how best to prepare and adapt to climate change. Should the North Atlantic region take steps to fully adapt to smoothly occurring warming, hedge bets by planning for THC collapse or aggressively attempt to mitigate greenhouse gas emissions to levels below calculated thresholds that might trigger THC collapse?

Acknowledgements

Generous financial and intellectual support came from the Global Change Education Program (GCEP). Stephen Schneider and Michael Mastrandrea provided thoughtful insights, suggestions, and comments throughout the manuscript. Paul Armsworth offered valuable suggestions and comments on an earlier draft. This manuscript benefited greatly from questions, comments and discussion arising at the Paradigms of Change workshop. I am especially grateful to Andreas Wimmer and Reinhart Kößler for organizing the workshop and for inviting me to attend. Thanks also to Luis Mata for chairing the climate change session and for valuable discussion throughout the workshop.

References

Alley, R. B., Meese, D. A., Shuman, C. A., Gow, A. J., Taylor, K. C., Grootes, P. M., White, J. W. C., Ram, M., Waddington, E. D., Mayewski, P. A. and Zielinski, G. A. (1993) 'Abrupt increase in Greenland snow accumulation at the end of the Younger Dryas event'. *Nature* 362, 527–9.

Bond, G., Showers, W., Cheseby, M., Lotti, R., Almasi, P., deMenocal, P., Priore, P., Cullen, H., Hajdas, I. and Bonani, G. (1997) 'A pervasive millennial-scale cycle in North Atlantic Holocene and glacial climates'. *Science* 278, 1257–66.

Broecker, W. S. (1997) 'Thermohaline circulation, the Achilles heel of our climate system: will man-made CO_2 upset the current balance?' *Science* 278, 1582–8.

Claussen, M. (1998) 'On multiple solutions of the atmosphere–vegetation system in present-day climate'. *Global Change Biology* 4, 549–59.

Claussen, M., Kubatzki, C., Brovkin, V., Ganopolski, A., Hoelzmann, P. and Pachur, H. J. (1999) 'Simulation of an abrupt change in Saharan vegetation in the mid-Holocene'. *Geophysical Research Letters* 26, 2037–40.

Clement, A. C., Cane, M. A. and Seager, R. (2001) 'An orbitally driven tropical source for abrupt climate change'. *Journal of Climate* 14, 2369–75.

Daily, G. C. (ed.) (1997) *Nature's Services*. Washington, DC: Island Press.

Dansgaard, W., Johnsen, S. J., Clausen, H. B., Dahl-Jensen, D., Gundestrup, N. S., Hammer, C. U., Hvidberg, C. S., Steffensen, J. P., Sveinbjornsdottir, A. E., Jouzel, J. and Bond, G. (1993) 'Evidence for general instability of past climate from a 250-kyr ice-core record'. *Nature* 364, 218–20.

Diaz, H. F. and Markgraf, V. (eds) (2000) *El Niño and the Southern Oscillation*. Cambridge, UK: Cambridge University Press.

Ganopolski, A., Kubatzki, C., Claussen, M., Brovkin, V. and Petoukhov, V. (1998) 'The influence of vegetation–atmosphere–ocean interaction on climate during the mid-Holocene'. *Science* 280, 1916–19.

Higgins, P. A. T., Masterandrea, M. D. and Schneider, S. H. (2002) 'Dynamics of climate and ecosystem coupling: abrupt changes and multiple equilibria'. *Philosophical Transactions of the Royal Society of London series B-Biological Sciences* 357, 647–55.

Hurrell, J. W. (1995) 'Decadal trends in the North Atlantic Oscillation: regional temperatures and precipitation'. *Science* 269, 676–9.

Imbrie, J., Berger, A. and Shackleton, N. J. (1993) 'Role of orbital forcing: a two-million-year perspective'. In: *Global Changes in the Perspective of the Past* (Eddy, J. A. and Oeschger, H., eds.). New York, NY: John Wiley and Sons 263–76.

IPCC (2001) *Climate Change 2001: The Scientific Basis*. Contribution of Working Group I to the Third Assessment Report of the Intergovernmental Panel on Climate Change. Cambridge, UK: Cambridge University Press.

Jouzel, J. (1999) 'Calibrating the isotopic paleothermometer'. *Science* 286, 910–11.

Keeling, C. D. (1986) 'Atmospheric CO_2 concentrations. Mauna Loa Observatory, Hawaii 1958–1986 (NDP-001/R1)'. Oak Ridge, TN: Carbon Dioxide Information Analysis Center.

Kleidon, A., Fraedrich, K. and Heimann, M. (2000) 'A green planet versus a desert world: estimating the maximum effect of vegetation on the land surface climate'. *Climatic Change* 44, 471–93.

Levitus, S. (1982) *Climatological Atlas of the World Ocean*. Washington, DC: US Department of Commerce, NOAA.

Lorenz, E. N. (1993) *The Essence of Chaos*. Seattle, WA: University of Washington Press.

Manabe, S. and Stouffer, R. J. (1993) 'Century-scale effects of increased atmospheric CO_2 on the ocean–atmosphere system'. *Nature* 364, 215–18.

Rahmstorf, S. (1996) 'On the freshwater forcing and transport of the Atlantic thermohaline circulation'. *Climate Dynamics* 12, 799–811.
Rahmstorf, S. (1999) 'Decadal variability of the thermohaline circulation'. In: *Beyond El Nino: Decadal and Interdecadal Climate Variability* (Navarra, A., ed.). Springer 309–32.
Schneider, S. H. and Thompson, S. L. (2000) 'A simple climate model used in economic studies of global change'. In: *New Directions in the Economics and Integrated Assessment of Global Climate Change* (DeCanio, S. J., Howarth, R. B., Sanstad, A. H., Schneider, S. H. and Thompson, S. L., eds). Washington, DC: The Pew Center on Global Climate Change 59–80.
Stocker, T. F. and Marchal, O. (2000) 'Abrupt climate change in the computer: Is it real?' *Proceedings of the National Academy of Sciences of the United States of America* 97, 1362–5.
Wang, G. L. and Eltahir, E. A. B. (2000a) 'Biosphere–atmosphere interactions over West Africa. II: multiple climate equilibria'. *Quarterly Journal of the Royal Meteorological Society* 126, 1261–80.
Wang, G. L. and Eltahir, E. A. B. (2000b) 'Ecosystem dynamics and the Sahel drought'. *Geophysical Research Letters* 27, 795–8.
Wang, G. L. and Eltahir, E. A. B. (2000c) 'Role of vegetation dynamics in enhancing the low-frequency variability of the Sahel rainfall'. *Water Resources Research* 36, 1013–21.
Washington, W. M. and Parkinson, C. L. (1986) *An Introduction to Three-Dimensional Climate Modeling*. Mill Valley, CA: University Science Books.
Wood, R. A., Keen, A. B., Mitchell, J. F. B. and Gregory, J. M. (1999) 'Changing spatial structure of the thermohaline circulation in response to atmospheric CO_2 forcing in a climate model'. *Nature* 399, 572–5.

3
Chaos in Social Systems: Assessment and Relevance

L. Douglas Kiel

From the rise of Social Darwinism in the 19th century to the behavioral revolution in the 20th century, social scientists have looked to both theory and method from the natural sciences to understand the dynamics of change in social systems. Paul Higgins' chapter 'Climate Change: Complexity, Chaos and Order' offers insights from chaos theory and the complexity sciences as means for better understanding change in social systems. These insights have been and, in fact, continue to be applied to our understanding of social systems change (see Jantsch, 1980; Anderson and Pines, 1988; Kiel and Elliott, 1996; Eve *et al.*, 1997).

Social systems have always provided a much more difficult backdrop for understanding change and change processes than have the natural and physical world. The dynamic nature of variables in social systems and the intricate web that connects these variables means understanding the intimate dynamics of social systems change is a daunting task. Consider the success of the natural sciences in space travel and health care relative to the multitude of lingering social problems. We simply understand the dynamics of these natural systems to a larger extent than we understand the dynamics of social systems. Thus understanding social systems change and the dynamics of these processes remains in many ways a mystery more daunting than the current state of knowledge permits. It is this reality that led Nobel Laureate Herbert Simon to often note that it is the social sciences, and not the natural sciences, that should be labeled the 'hard sciences' (Epstein, 1999).

We do know however that the social system reflects what Jay Forrester (1987: 34) has called, '... a highly nonlinear world.' It is this nonlinearity that raises the potential for amplifying events to produce surprise, uncertainty, disorder and change in social behaviors and structures. While social scientists have for many decades recognized the nonlinear nature of social phenomena, they have lacked a theoretical and methodological basis for contending with nonlinearity. The complexity sciences and chaos theory may provide the theoretical and methodological basis for social scientists to better understand the nature of social dynamics and change processes. This commentary examines the problems and prospects for the sciences of complexity and chaos theory to enhance our understanding of change processes in social systems.

Efforts to apply chaos and complexity to change in social systems

Since the late 1980s social scientists have produced thousands of scholarly publications applying concepts from chaos theory and the complexity sciences to social systems change. Chaos theory alone though has served as basis for a large body of social science research over the last two decades (Loye and Eisler, 1987; Kiel and Elliott, 1996; Eve *et al.*, 1997). This body of research reveals two distinct strains in both content and mathematical rigor. The first research project attempts to find evidence of chaotic time series in social science data. These examples range from time series analyses of stock market data to analyses of a variety of organizational data. The goal of this research is largely to obtain improved understanding of the dynamics of complex economic or organizational systems using multivariate analysis. This research paradigm has led to increased knowledge concerning the dynamics of parts of social systems but adds little understanding to the nature of systems level change. The lack of understanding provided by this strain of research is due to the high-dimensionality and networked nature of social systems change. Simply examining changes in the behavior and structure of several variables does little to inform us of systems level or system-wide change. Social systems change is invariably a multivariate, high-dimensional and historical process involving multiple complex relationships between variables that are quite difficult to disentangle. The application of such mathematical applications of chaos theory to topics of social systems change is also exacerbated by the increasingly apparent reality that identifying mathematical chaos is an extremely difficult if not an intractable problem. The challenge of extracting noise from social science data and the problem of extracting what can be defined as deterministic and chaotic data causes many interested social scientists to wonder if mathematical chaos even exists in social science data.

The second strain of applications of chaos theory to social systems is more distinctively focused on social systems change (Loye and Eisler, 1987; Kiel, 1994). This body of research is largely metaphorical and uses Prigogine's theory of dissipative structures as a means for exploring social systems change processes. Chaos theory is applied to these models as a reference to the uncertainty or 'chaos' that occurs during the destabilization of a social system, institution or organization. This view of social systems change views these systems as susceptible to a variety of internal and external perturbations. Since these social systems are viewed as nonlinear systems, these perturbations can, at times, generate positive feedback leading to the breaking of the existing symmetry of the social system. Following this symmetry break is a bifurcation or branching in which uncertainty and 'chaos' dominate as the social system struggles to achieve some new more sustainable and 'complex' form of order and organization or diminishes into a less fit and less organized state. These examples of applying chaos theory to social systems change are rarely supported by mathematical analysis.

This second strain of chaos theory research, better labeled 'Prigoginean change theory', is increasingly serving as a theoretical basis for both understanding and promoting large scale change in a variety of social systems (Prigogine and Stengers, 1984; Marion, 1999). This approach serves as a counterpoint to the many incremental and deterministic models of social systems change. Instead, these applications of chaos theory focus on models of revolutionary or transformational change in which large scale social systems change occurs in a rather rapid time frame (Loye and Eisler, 1987; Marion, 1999). Such applications to social systems range from political revolutions to technological change to organizational change. These metaphorical models again focus on the uncertainty or chaos that occurs during periods of dramatic change in many different social systems. Moreover, these models, contrary to traditional views of social systems dynamics, view instability, disorder and chaos as beneficial qualities for social systems (Stacey, 1992; Kiel, 2002). In short, these qualities of nonlinear dynamical systems best prepare social systems for the change that is not only an historical inevitability, but also a constant in the current historical period of rapid technological change and increasing globalization. These views of social systems change are inherently evolutionary views of change. However, rather than focusing on incremental views of evolution, these views emphasize the punctuated nature of social systems change. Furthermore, these evolutionary perspectives emphasize the concept of autopoeisis, an internally generated process of self-renewal, as essential for the maintenance and further evolution of all social systems. Scholars also see this process as consistent with Prigogine's notion of order through fluctuation as disturbances or fluctuations create new forms of organizational or social order.

Agent-based simulation of social systems change

Since the 1960s, social scientists have produced simulations exploring the dynamics of social systems change (see, for example Forrester, 1969; Dendrinos, 1992). During the last two decades, economists have greatly added to our understanding of the dynamics of change through the insertion and investigation of nonlinear and chaotic dynamics into economic models (Scheinkman and LeBaron, 1989; Hanson et al., 1997). Most recently, social scientists are building computer-based simulations to better understand the dynamics of social systems change. These emerging approaches aim to incorporate an increased level of indeterminacy relative to pre-existing modeling and simulation approaches. Just as meteorologists build weather and climate models, social scientists are now attempting to build animated social worlds or microworlds in computers in an attempt to better understand the dynamics of change (Casti, 1997). This body of social systems investigation is labeled agent-based modeling. Agent-based modeling takes from the complexity sciences several of these sciences' basic assumptions about complex systems. First, complex systems are seen as adaptive systems which leads to the increasing reference to social systems as 'complex adaptive systems' (Holland, 1995). Complex adaptive systems (CAS) also evidence three distinct types of behavior. First, complex adaptive systems evidence nonlinear behavior from which amplifying effects may generate uncertainty and the unexpected. A second assumption of complex adaptive systems is the generation of emergent behavior that defines novel forms of behavior or structure that occur through the multiple interactions of individual actors or agents. Finally, complex adaptive systems are capable of self-organization, meaning these systems possess an internal capacity to re-organize after a major perturbation or change event.

Agent-based modeling thus employs these assumptions of CAS and designs 'would be worlds' in an effort to understand how social systems evolve and change. Agents imbued with various decision rules interact on various 'landscapes' in an effort to examine how these interactions produce systems structure and behavior. Agent-based modeling however assumes that it is the interactions of agents that create social structures and behaviors rather than social structure such as norms or culture pre-existing and thus shaping social structure and behavior. Agent-based modeling thus argues for examining the processes of change 'from the bottom-up' (Epstein and Axtell, 1996). Thus, for agent-based modelers change is a process accruing from the interactions of multiple agents on various social systems landscapes. And, rather than seeking some optimal solution, practitioners of agent-based modeling seek to understand the nature of change and the emergent phenomena that result from social systems change. Examples of agent-based modeling now range from studies of the evolution of markets to studies of the outbreak of civil violence (Berry *et al.*, 2002). Social scientists employing agent-based modeling now have a tool to aid in the process of discovering more about the nature and dynamics of social systems change.

Agent-based modeling has also led to research efforts to examine the factors that effect change in social systems. Agents in such simulations are imbued with 'rules' that shape how they behave on various landscapes. Manipulations of these behavioral rules allow succeeding simulations to present the effects of such change. Policy scientists see such models as a means for developing social systems 'flight-simulators' allowing 'pilots' to manipulate system parameters or interject perturbations in an effort to see the effects on the microworld (Holland, 1995). An appreciation of this effort adds insight into how social scientists view the potential of the complexity paradigm to both understand and produce change. Furthermore this understanding emphasizes assumptions of nonlinearity in social systems behavior. The interjecting of a perturbation intended to create change is consistent with the oft-noted 'butterfly effect' from chaos theory. Some analysts have suggested the application of 'triggers' to produce organizational change with the hope that a minor fluctuation can generate wholesale organizational change at minimal cost (Holland, 1995). These efforts however raise new sets of problems for social scientists. First, even if we know which triggers may cause change, the outcomes may vary due to the changing nature of variables in the social system itself. Unlike the chemical laboratory, each iteration of the experiment may, and perhaps should, produce novel results. Second, it is unlikely that we can decipher which triggers cause specific social change. Unraveling cause and effect in complex social systems hinders the potential for developing a stable set of triggers that work across various systems. Thus theory building becomes a distant hope and social thinkers are left with mere heuristics.

Chaos, complexity, change and humility

What can then be said, given what social scientists understand concerning the applications of the complexity sciences and chaos theory to social systems change? It is clear that both applications of the complexity sciences and chaos theory create serious, and perhaps given current tools, insurmountable measurement problems. Not only is it exceedingly difficult to determine the existence of deterministic chaos, but notions of complexity also raise serious issues about how much complexity exists in a social system. And, how do we measure a social perturbation? Is it possible to place a measurement on a street riot just as chemists calculate moles? But, perhaps the measurement problem is not really the most significant issue when considering paradigms of change in the social sciences. Perhaps social scientists have been tilting at windmills in their efforts to emulate the traditional mathematical methodologies of the natural sciences. Or perhaps what Mirowski (1989) has labeled as the 'physics envy' of some social scientists is simply a mistaken assumption about the relative complexity of the natural and social realms. Given the remarkable complexity of the social realm perhaps a more modest approach to applying models of change to the social realm is required. Attempting to calculate the richness and diversity of the multiple components contributing to social systems change is likely a hopeless endeavor. Yet, this recognition itself lends value to applications of the sciences of complexity to paradigms of change in the social sciences. Rather than relying on specific outcomes or predictable results, the complexity sciences provide insight into the processes of change. It is the understanding of these processes that serves as a valuable tool for understanding and appreciating the dynamics of social systems change. It may simply be that the best scholars can hope for when applying models such as Prigogine's model of change to social systems is the application of effective metaphor. And metaphor is an effective means for learning when the preferred tools of the scientific paradigm are unreliable and inappropriate given the complexities of social systems dynamics.

We can thus look to the works of social theorists such as Niklas Luhmann (1995) to understand the importance of instability, disorder and complexity in social change processes and the creation of new social order. Luhmann's (1995) work provides a societal level perspective of change based on the dynamics of human communication systems and the potentials for autopoetic renewal in social systems. At the organizational level, a multitude of authors are using the complexity paradigm to explore the processes of organizational change (see for example, Kiel, 1994; Guastello, 1995; Marion, 1999). Organizations too are seen as susceptible to multiple fluctuations which may amplify nonlinear effects, generate chaos, but lead to new forms of organization and order. The greatest value of the complexity sciences as a paradigm for change in the social sciences is deeper insights into the processes and the patterns of change processes in social systems.

In his chapter on climate change in this volume, Professor Higgins' notes that climate change research may benefit other fields 'focused on change' via four characteristics employed in climate change research (Higgins, this volume). These characteristics are (1) a constantly changing system, (2) a need for representative models, (3) unpredictability at some scales that does not preclude insights at other scales, and (4) complex behavior resulting from the interactions of sub-units. The application of the sciences of complexity to change processes in social systems meets the four characteristics identified by Higgins. Social systems are clearly dynamic systems that are amenable to representative models such as current agent-based models and produce change and complex behavior resulting from the interactions of sub-units that produce surprise and inhibit prediction. In fact, one might argue that it was the appeal of Lorenz's insights into the nonlinear interactions in meteorological systems that generated the expanding interest in nonlinear dynamics and complexity in the social sciences. Yet, even the complexities of the earth's weather and climate appear to provide better means for prediction and certainty than do the results of change processes in social systems.

While the complexity sciences afford new insights into change processes in social systems, the results of these processes remain subject to the vagaries of interactions that human knowledge is yet prepared to understand. Thus knowledge of the consequences of social change continues to inhibit the development of broad and reaching paradigms of change in the study of social systems. This reality may simply be an unchanging constant of the human condition. For as Peter Allen (1994: 596) has noted, 'Knowing that we cannot know is an important step on the road to wisdom.'

In a nonlinear and complex world even our best efforts defy complete knowledge of the downstream consequences of our actions. Thus Professor Higgins' chapter in this volume helps to emphasize the importance and function of uncertainty in change processes in both natural and complex social systems. Thus, the uncertainty generated by either intentional or accidental human action is significant in both its promise and peril. For while uncertainty produces the promise of creative response and adaptation, it also unleashes the inherent peril of unpredictable outcomes.

References

Allen, Peter M. (1994) 'Coherence, chaos and evolution in the social context'. *Futures* 29 (6): 583–97.
Anderson, P.W. and Pines, D. (1988) *The Economy as an Evolving Complex System* (Santa Fe Institute Studies in the Sciences of Complexity, Vol. 5). Redwood City, CA: Addison-Wesley.
Berry, Brian J.L., Kiel, L. Douglas and Elliott, Euel (eds) (2002) 'Adaptive agents, intelligence and emergent human organization: capturing complexity through agent-based modeling'. *Proceedings of the National Academy of Sciences.* Washington, DC, 14 May 2002; 99 (Suppl. 3).
Casti, John (1997) *Would-Be Worlds: How Simulation Is Changing the Frontiers of Science.* New York: John Wiley.
Dendrinos, Demetrios (1992) *The Dynamics of Cities: Ecological Determinism, Dualism and Chaos.* London: Routledge.
Epstein, Joshua and Axtell, Robert (1996) *Growing Artificial Societies: Social Science from the Bottom Up.* Washington, DC: Brookings Institution Press.
Epstein, Joshua (1999). 'Agent-based computational models and generative social science'. *Complexity* 4 (5): 41–60.
Eve, Raymond, Horsfall, Sara and Lee, Mary (eds) (1997) *Chaos, Complexity and Sociology* (Sage: Beverly Hills, CA), pp. 64–78.
Forrester, Jay W. (1969) *Urban Dynamics.* Cambridge, MA: The MIT Press.
Forrester, Jay W. (1987) 'Nonlinearity in high-order models of social systems'. *European Journal of Operational Research* 30, 104–9.
Guastello, Stephen (1995) *Chaos, Catastrophe, and Human Affairs: Applications of Nonlinear Dynamics to Work, Organizations, and Social Evolution.* Mahwah, NJ: Lawrence Erlbaum Publishers.
Hanson, Lars P., Conley, T.G. and Liu, W.F. (1997) 'Bootstrapping the long run'. *Macroeconomic Dynamics* 1, 279–311.
Holland, John (1995) *Hidden Order: How Adaptation Builds Complexity.* New York, NY: Addison-Wesley.
Jantsch, Erich (1980) *The Self-Organizing Universe: Scientific and Human Implications of the Emerging Paradigm of Evolution.* Oxford, UK: Pergamon Press.
Kiel, L. Douglas (1994) *Managing Chaos and Complexity in Government: A New Paradigm for Managing Change, Innovation and Organizational Renewal.* San Francisco: Jossey-Bass Publishers.
Kiel, L. Douglas and Elliott, Euel (eds.) (1996) *Chaos Theory in the Social Sciences: Foundations and Applications.* Ann Arbor, MI: University of Michigan Press.
Kiel, L. Douglas (2002) 'Knowledge management, organizational intelligence and learning, and complexity'. In: *Knowledge for Sustainable Development – an Insight into the*

Encyclopedia of Life Support Systems. vol. III, UNESCO Publishing/EOLSS Publishers 849–74.

Loye, David and Eisler, Riane (1987) 'Chaos and transformation: Implications of non-equlibrium theory for social science and society'. *Behavioral Science* 32, 53–65.

Luhmann, Niklas (1995) *Social Systems.* Stanford, CA: Stanford University Press, 1995.

Marion, Russ (1999) *The Edge of Organization: Chaos and Complexity Theories of Formal Social Systems.* Thousand Oaks, CA: Sage.

Mirowski, Philip (1989) *More Heat Than Light: Economics as Social Physics.* New York: Cambridge University Press.

Prigogine, Ilya and Stengers, Isabelle (1984) *Order out of Chaos: Man's New Dialogue with Nature.* New York: Bantam Books.

Scheinkman, Jose and LeBaron, Blake (1989) 'Nonlinear dynamics and stock returns'. *Journal of Business* 62, 311–37.

Stacey, Ralph (1992) *Managing the Unknowable: Strategic Boundaries Between Order and Chaos in Organizations.* San Francisco: Jossey-Bass Publishers.

4
Economics, Chaos and Environmental Complexity

Hans-Walter Lorenz

Every scientist who is aware of the possible emergence of chaotic motion in nonlinear dynamical systems should be familiar with the idea that longer-term weather predictions might be impossible. E.N. Lorenz' work on a three-dimensional dynamical system arising in the context of a climate model is almost always cited as the starting point of modern research into nonlinear dynamical systems. The so-called butterfly effect even serves as a vehicle for demonstrating unexpected and dramatic consequences of the presence of non-linearities in dynamical systems. However, for a non-specialist in this field, the theoretical reasons for the emergence of chaotic motion are usually unclear (if one is interested in qualitative justifications of the investigated mathematical nonlinear dynamical systems). Paul Higgins provides fascinating insights into theoretical approaches to the weather and climate phenomena in his paper and concentrates on the idea that the interaction of subsystems (like atmosphere, ocean, biosphere etc.) is responsible for the observable weather and climate phenomena. This global system consisting of interacting subsystems (which reminds of a so-called coupled oscillator system in dynamical systems theory which is known to exhibit chaotic motion when the dimension of the system is sufficiently high) implies the existence of multiple equilibria, with an important consequence. While the coupling of subsystems (either in the form of linearly coupling nonlinear subsystems or of nonlinearly coupling linear systems or both nonlinear variants) can imply the emergence of chaotic motion according to the standard definitions of, say, sensitive dependence on initial conditions, aperiodic motion, ergodicity, etc., Higgins also considers scenarios where, due to external forcing, trajectories leave the basin of attraction of one of these multiple equilibria and enter another basin. Major climate changes are associated with these switches between different basins. I feel unable to comment upon the meteorologist aspects of the empirical issues and the usefulness of the theoretical modeling approaches, but the methodological aspects of the modeling strategy deserve a further discussion. I will concentrate on the connections between climate change and economic behavior and modeling similarities in meteorology and economic theory.

Chaos models in economics

Chaotic motion has been a well-known property of a class of dynamical systems in economic theory for approximately 25–30 years. In traditional economics, the motion of state variables like, e.g. gross domestic product, unemployment rates, product prices, etc. is modeled in a formal fashion very similar to classical mechanics, with the consequence that the evolution of an economic variable is usually described by a (system of) difference or differential equation(s). Early papers on chaotic motion in dynamic economic models therefore looked for genuinely present nonlinearities in models of, say, economic growth or business-cycle theory or searched for economically justified modifications of existing models that resulted in one of the dynamical systems which were known as being able to generate chaotic motion for particular parameter constellations. It has turned out that a large number of existing models either already possesses a formal structure necessary to generate chaotic motion or that the introduction of appropriate nonlinearities cannot be excluded by economic reflections. The very problem associated with this 'immediate' approach toward chaos in economics consists of the question how empirically relevant the (mathematically) necessary nonlinearities really are. In the economic literature many examples of chaotic dynamics in economic models can be found where the desired dynamic property is due to *ad hoc* assumptions without much economic reasoning or to assumptions on parameter values beyond any empirical scope. In addition, a few models (including prominent ones like, e.g. overlapping-generations models with complete information about future states of the economy where the evolution of the economy is governed by the decision of an artificial *representative agent*), obviously do not constitute realistic representations of economic life in actual multi-agent scenarios. It can similarly be argued that the chaotic dynamic behavior of macroeconomic entities in highly aggregated models in the Keynesian tradition (i.e. the standard models of the 1950s and 1960) is not really convincing when no information on the behavior of the underlying micro-behavior is available.

Whatsoever, it is more or less simple to construct theoretic models in economic dynamics which (depending on the assumed time concept) finally imply deterministic, nonlinear difference or differential equations with the chaos property emerging for particular parameter values. However, these theoretical investigations actually only make sense when empirical time series indeed are chaotic. To answer the question whether observable economic time series are chaotic or not, it is therefore mandatory to study actual time series with the help of the available toolbox of non-parametric statistics like largest Lyapunov exponents, correlation dimensions, etc. In the light of the fact that most tools developed for the investigation of the evolution of state variables in the natural sciences assume that the size of the available data set is very large (in the mathematical ideal close to infinity), the typically available sample size of, say, 50 data entries for annual GDP figures in most western economies does not constitute a very promising setup for definitely detecting chaos in actual time series. Investigators encountered other problems in the empirical analyses of huge data sets which, for example, are available for stock markets with permanently up-dated data sets (occasionally every few seconds). These data sets often contain structure according to institutional arrangements like Monday and Friday effects, end-of-the-month or holiday effects. Many statistical tools identify this structure with chaotic motion and dissociate it from pure noise. To sum it up, it can be said that as a result of a more than 15-year long time-span of investigating actual, empirical time series in diverse economic fields, the presence of chaotic motion (in its precise mathematical definition) cannot definitely be established in the majority of empirical time series, but a few data sets still represent promising candidates for detecting chaos in these series. It seems, however, that it is necessary to give up the concentration on the established tools of non-parametric statistics because almost certainly economic time series are not generated by pure deterministic dynamical systems. Stochastic external influences might occasionally even dominate the fate of an economy. It is therefore desirable to develop statistical tools which detect, e.g. a chaotic deterministic kernel in a stochastic cloud of external stochastic influences.

Challenges of interdisciplinary co-operation

Paul Higgins mentions the importance of economic activities on climate change. Nowadays, even a layman in the field will probably not deny the influence of emissions due to economic activities on the atmosphere with diverse subsequent effects. It might be useful to discuss the role of economics in this context, however, because the present economic literature will certainly benefit from an interdisciplinary co-operation.

Economic theory has dealt with environmental issues for a long time when harvesting models (usually in the form of fishery models) have been investigated. In case of a basically renewable resource like the stock of fish in a geographically defined region it is assumed that a central planning agency attempts to solve an optimization problem of the following form: If it is known how a species grows over time with and without harvesting and when the agency is able to define a precise targeting function (i.e. defining what actually should be optimized), then it is in principle possible to calculate a target-function maximizing time path for the harvesting. Possible target functions might be overall profit functions for all harvesters or other welfare functions. The standard reference for these approaches is Clarke (1976). A basic feature of these models consists in their ignorance of *subsequent* effects of harvesting renewable resources on economic welfare, e.g. in the form of suffering (also economically) from possibly extincting certain species. The direct effects of exploiting the environment is dominating the discussion. When non-renewable resources like crude oil, coal, fossille water in a desert oasis, etc. are considered with this kind of reasoning, economic variables like the discount rate of future profits define the exploitation time path, but the question of whether a resource should be exploited or not for other than economic reasons is usually not asked.

In recent years, the economic literature dealing with environmental issues has developed into several directions. For example, the obvious negative influences of economic activities on the ecological systems is systematically studied with the help of so-called *environmental Kuznets curves*. These curves describe the systematic influence of economic activities on emission levels and allow to determine emission minimizing projects. A broad survey of applications of this methodological approach can be found in Vogel (1999). Another part of the literature deals with the question which influences on the economic system can be expected from the ecological environment. It is obvious that such influences exist (cf. the traditional standard literature mentioned above) because non-renewable resource extraction implies an eventual switching to alternative (and possibly more expensive) production techniques, raw materials, etc. In addition to this line of thought, it must be taken into account that ecological standards have changed over time. The realization of new standards usually also requires a switching to different production techniques with changing cost structures. Since Paul Higgins has concentrated on the chaotic effects resulting from the interaction of multiple subsystems, it should finally be mentioned that several recent economic-ecological papers and monographs investigate complex ecological-economic dynamics (cf. Costanza, 1997, Janssen, 2002 and Rosser, 2001). It can be expected that the literature dealing with these approaches will be extended dramatically and that new research topics will show up in the near future.

The paper suggests that an interdisciplinary collaboration among meteorologists, physicists, chemists, biologists, economists, and other scientists is necessary for a satisfactory description of the status quo of global (and certainly also of particular local) climate conditions and the search for activities that allow for an easing of the effects of climate change or even a reversal of observable climate phenomena. In an ideal scenario, representatives of these sciences come together and attempt to construct a global, all-embracing and universal model, with the help of which the particular nuances of climate change and the interactions between these singular aspects can be investigated. Such an approach can probably not be realized in the future (even on a long time scale). The aims and scopes of the different scientific disciplines are too different for a close collaboration. For example, in the present context (traditional) economics is not only interested in learning about the effects of climate change and the environment on economic conditions but also (according to specific interests) about (i) strategies that allow for maximizing, e.g. profits in the face of a changing environment, or, (ii) economic strategies that allow for a change in the environment. The role of positive and normative aspects and the degree of formalization in the before-mentioned disciplines is probably too different for a universal, interdisciplinary model of climate change in which a variety of aspects can be considered. Paul Higgins' paper demonstrates, however, that an intensive communication between representatives of several disciplines can be very helpful and fruitful in each discipline's approach to dealing with the sources and consequences of climate change. In all these fields much too little still known about these reasons and consequences even from their own isolated perspective. If climate change is indeed the result of the interaction of multiple subsystems, then it is mandatory for all disciplines to learn about the advances made in other fields even if this knowledge transfer can probably only mean that specialists talk to laymen. After a (probably long) interactive updating process it might (hopefully) be possible to learn much more about the reasons of a climate change and possible ways to reverse the development before it might definitely be too late to intervene in this process with the help of restoring human activities.

Bibliography

Clarke, Colin W. (1976) *Mathematical Bioeconomics: The Optimal Management of Renewable Resources*. New York: John Wiley.

Costanza, Robert (1997) 'Modeling complex ecological economic systems: toward an evolutionary, dynamic understanding of people and nature'. In: Costanza, Robert (ed.) *Frontiers in Ecological Economics: Transdisciplinary Essays*, 330–40.

Janssen, Marco A. (2002) *Complexity and Ecosystem Management: The Theory and Practice of Multi-Agent Systems*. Cheltenham: Edward Elgar.

Rosser, J. Barkley, Jr (2001) 'Complex ecologic–economic dynamics and environmental policy'. *Ecological Economics* 37 (1): 23–37.

Vogel, M.P. (1999) 'Environmental Kuznets curves'. *Lecture Notes in Economics and Mathematical Systems* 469. Berlin: Springer-Verlag.

Part II
Genetic Variation in Evolution

5
The Topology of the Possible
Walter Fontana

The problem of change

The classical framework of evolutionary change is based on two aspects of biological organization: phenotype and genotype. The notion of phenotype refers to the physical, organizational and behavioral expression of an organism during its lifetime. It emphasizes the systemic nature of biological organizations. The notion of genotype refers to a heritable repository of information that instructs the production of molecules whose interactions, in conjunction with the environment, generate and maintain the phenotype.

In its simplest manifestation, evolution is driven by the selection of phenotypes, which causes the amplification of their underlying genotypes, and the production of novel phenotypes through genetic mutations (Figure 5.1, top). These two factors of evolutionary change are the focal points of distinct research agendas. Selection is a dynamical phenomenon that arises spontaneously in constrained populations of autocatalytically reproducing entities. The research agenda focused on selection aims at characterizing the conditions under which a phenotypic innovation can, once generated, invade an existing population. The classical fields of inquiry concerned with selection are population genetics and ecology. The main variables are frequencies of genes or species representatives whose change is typically described by a nonlinear dynamical system. This description is not concerned with a mechanistic understanding of how change occurs in the first place nor with how the internal architecture of biological entities arises and how it influences their capacity to vary. Rather, evolving entities are treated as black-boxes whose variation or 'innovation' is generated by some stochastic process with simple characteristics in an isotropic phenotype space, typically motivated by the need for mathematical tractability.

The question of how a phenotypic innovation arises in the first place has, so far, been the concern of a different research track. The heritable modification of a phenotype usually does not involve a direct intervention at the phenotypic level, but proceeds indirectly through change at the genetic level

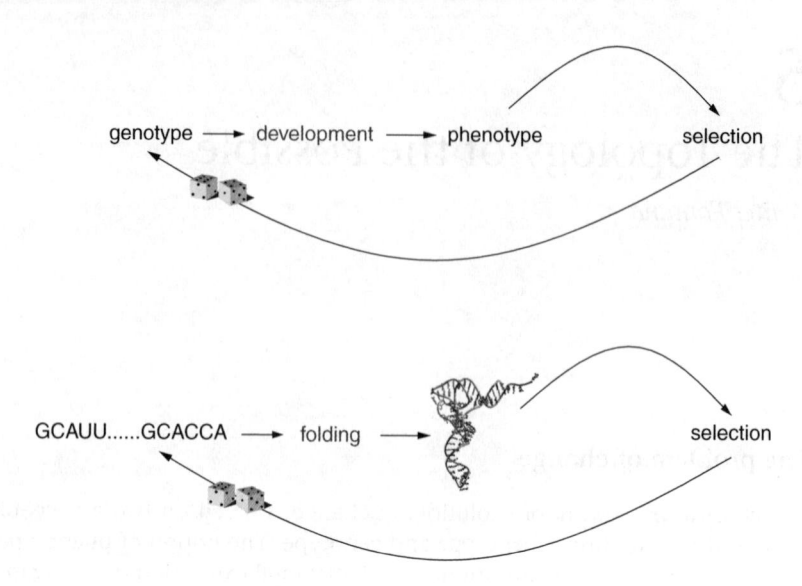

Figure 5.1 The folding of RNA sequences into shapes as a proxy of a genotype–phenotype map. Mutations occur at the genetic level. Their consequences at the phenotypic level are mediated by development, the suite of processes by which phenotype is constructed from genotype. RNA folding is a transparent and tractable model that captures this indirection of innovation within a single molecule. The RNA folding map is characterized by a number of remarkable statistical regularities with profound evolutionary consequences. These regularities may generalize to more complex forms of development.

(Figure 5.1). This forces the processes that link genotype to phenotype into the picture. In biology, these processes are known as development. Evolutionary trajectories, or histories, depend on development, because development mediates the phenotypic effects of genetic mutations and thus the accessibility of one phenotype from another. Aside from constraining and promoting variation, the mechanisms of development are themselves subject to evolution, creating a feedback between evolution and development.

In the early days of the neo-Darwinian school of evolutionary thought, insufficient knowledge of developmental processes justified ignoring the relationship between genotype and phenotype. This pragmatic approach produced a powerful conceptual and formal framework that has been exported – with mixed success – to fields outside biology, particularly 'evolutionary' flavors of economics and the social sciences. The problem remains that important phenomena of phenotypic evolution do not result naturally from a purely adaptationist framework that assumes phenotypes to be completely malleable by natural selection. These phenomena comprise the punctuated mode (the sudden nature) of evolutionary change (Eldredge and Gould, 1972), constraints to variation (Maynard-Smith *et al.*, 1985), the origin of novelties (Müller and Wagner, 1991), directionality in evolution, path-dependency ('frozen accidents') in organismal structure, and phenotypic stability over evolutionary time, such as homology (for a discussion see Gould, 2002).

The origin of the problem is that the arrival of a new phenotype must necessarily precede its survival and spread in a population through selection. While selection is clearly an important driving force of evolution, the dynamics of selection doesn't teach us much about how evolutionary innovations arise in the first place. To say that a mutation is advantageous, means that it generated a phenotype favored by selection, but reveals nothing about why or how that mutation could innovate the phenotype. A model of the genotype–phenotype relation (development) is needed to illuminate how genetic change maps into phenotypic change. In this chapter, I informally discuss a simple molecular instance of such a relation that is based on the shapes of RNA sequences (Fontana and Schuster, 1998a,b; Stadler *et al.*, 2001), see Figure 5.1, bottom. RNA is an extreme case, because genotype (sequence) and phenotype (folded sequence or shape) are different aspects of the same molecule, as illustrated in Figure 5.2. The sequence of an RNA molecule functions as a genotype, since an RNA sequence can be directly copied (or replicated) by suitable enzymes. It is the sequence that is copied, not the shape of the molecule. An RNA sequence acquires a shape through a process in which the sequence folds into a spatial structure guided by many competing interactions between the chemical building blocks (the 'letters of the RNA alphabet') that constitute the sequence. Like a phenotype, the shape conveys biochemical behavior to an RNA molecule and is therefore a target of selection. The mapping from RNA sequences to shapes constitutes the perhaps simplest biophysically grounded example of a mapping from genotypes to phenotypes that is tractable both theoretically and experimentally.

The RNA model allows to study an important question: given a particular 'physics' (that is, a system of causation, which I shall simply call a 'mechanism,' ignoring post-modern worries in other disciplines) that generates phenotype from genotype, what are the statistical characteristics of the relation between genotype and phenotype and how do they affect the dynamics of evolutionary change? In other words, what can we say statistically about how phenotype changes with genotype? It is important to keep in mind the limitations of the model. While RNA is an instance of a genotype–phenotype relation, it is not a model of organismal development. First, the regulatory networks of gene expression and signal transduction that coordinate the production of phenotype (for an overview see Carroll *et al.*, 2001) have no analogue in RNA folding. Second, the molecular processes underlying organismal development are themselves genetically influenced and thus subject to evolution. In fact, major innovations in the history of life are associated with the emergence of new developmental mechanisms (Kirschner and Gerhart, 1998). In contrast, the folding of an RNA sequence into a shape is essentially governed by physico-chemical principles that lie

70 *Genetic Variation in Evolution*

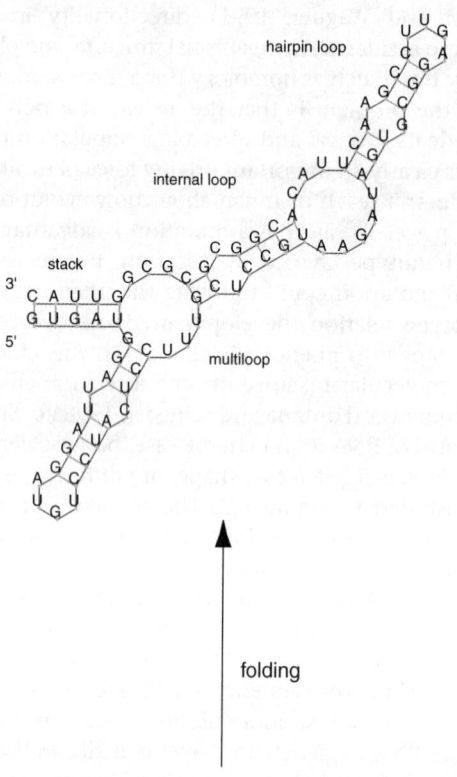

[RNA secondary structure diagram showing hairpin loop, internal loop, stack, and multiloop elements, with the sequence folding from a linear form into a branched structure]

5' GUGAUGGAUUAGGAUGUCCUACUCCUUUGCUCCGUAAGAUAGUGCGGAGUUCCGAACUUACACGGCGCGCGGUUAC 3'

Figure 5.2   RNA shape*

* At the level of resolution considered here, RNA shape is a graph of contacts between building blocks (nucleotides) at positions $i = 1, \ldots, n$ along the sequence. Position 1 is the 5'-end. The graph has two types of edges: the backbone connecting nucleotide i with nucleotide $i + 1$ and hydrogen bonded base pairings between positions.

beyond the control of the sequence. This said, I believe that the RNA model is relevant, because it offers *perspectives* that are potentially generalizable to other, far more complex, situations.

RNA shape as a systemic phenotype

Figure 5.2 sketches all one needs to know about RNA folding for the present discussion. At one level, we are given a sequence of fixed length over an alphabet of four letters: A, U, G, and C. The letters represent certain molecular building blocks. RNA is chemically a very close relative of DNA. As in DNA, specific pairs of building blocks interact with one another preferentially: A pairs with

U, G pairs with C, and G also pairs with U. We call building blocks that pair with one another 'complementary'. The difference to DNA is that RNA occurs single-stranded. Rather than pairing up with a second complementary strand, as in the DNA double-helix, an RNA sequence folds up on itself by matching complementary segments along the sequence, as shown in Figure 5.2. The notion of structure depicted in Figure 5.2 simply consists in the pattern of pairings between positions along the sequence. The structure formation is driven by changes in (free) energy upon pairing. Stretches of pairs (the 'ladders' or stacks in Figure 5.2) stabilize a structure, while loops destabilize it. Notice that the pairing of two segments necessarily creates a loop. In this abstraction, pairings between positions located in different loops are not allowed. Consequently, the formation of a paired stretch (which is energetically 'good') generates not only a loop (which is energetically 'bad'), but it prevents the positions within that loop from pairing with any positions outside of it. A huge number of structures are possible for any given sequence. Yet, one or a few of these structures balance the trade-offs in an energetically optimal fashion. I shall call the energetically optimal fold of a sequence simply its 'shape'. This shape can be computed by clever and fast algorithms (Waterman and Smith, 1978; Nussinov and Jacobson, 1980; Zuker and Stiegler, 1981; Zuker and Sankoff, 1984) based on empirically measured energy parameters (Turner et al., 1988; Jaeger et al., 1989; He et al., 1991; Walter et al., 1994; Mathews et al., 1999).

The actual shape of a sequence is a three-dimensional structure (see, for example, the bottom of Figure 5.1). Today, this three-dimensional structure cannot be computed directly from the sequence. Although the shape shown in Figure 5.2 is a very crude model, it is not a fiction either. The pairings in an actual structure can be established empirically and it turns out that the three-dimensional structure contains many of the pairings predicted by this more abstract notion of shape.

The pairing rules and their energetics establish a map (or a process) that assigns a shape to each sequence. The relevant point for the present discussion is that a shape cannot be changed directly, but requires modification of the sequence which is then interpreted by the folding process to generate the new shape. This situation mirrors the indirection in transforming phenotypes through genetic mutations (Figure 5.1).

The number of possible sequences of small to moderate length is unimaginably huge. For example, there are 4^{100} (that is, 10^{60}) sequences of length 100 – which is not very long for biological molecules. Our goal can only be to characterize the overall *statistical* regularities of the mapping from sequences to shapes.

Shape in RNA (as for bio-molecules in general) is a systemic property. Changing the 'letter' at a given sequence position alters many pairing possibilities throughout the sequence, possibly tipping the optimal pairing pattern. The effect at the shape level of changing a sequence position is strongly dependent on the sequence context. In a four-letter alphabet, there are three possible substitutions, or mutations, at each sequence position. I call a position *neutral*, if it allows for at least one mutation that does not alter the shape of the sequence. Accordingly, such a mutation is a neutral mutation. Although a neutral mutation leaves the shape of a sequence unchanged, it has two important contextual effects illustrated in Figure 5.3. The shape shown at the top left of Figure 5.3 remains the same if C is substituted by G at the position labeled x. Yet, whether x is C or G determines the shape obtained when mutating the position y from G to A (Figure 5.3, lower half). This illustrates that the effect of a mutation at position y depends on the building block at position x. In biology, this dependency is called *epistasis*. I'd like to draw attention to another, more subtle, effect: although the replacement of C by G at x does not alter the shape, it alters the number and location of neutral positions (Figure 5.3, top right). Thus, while a neutral mutation does not change the phenotype, it affects the potential for change.

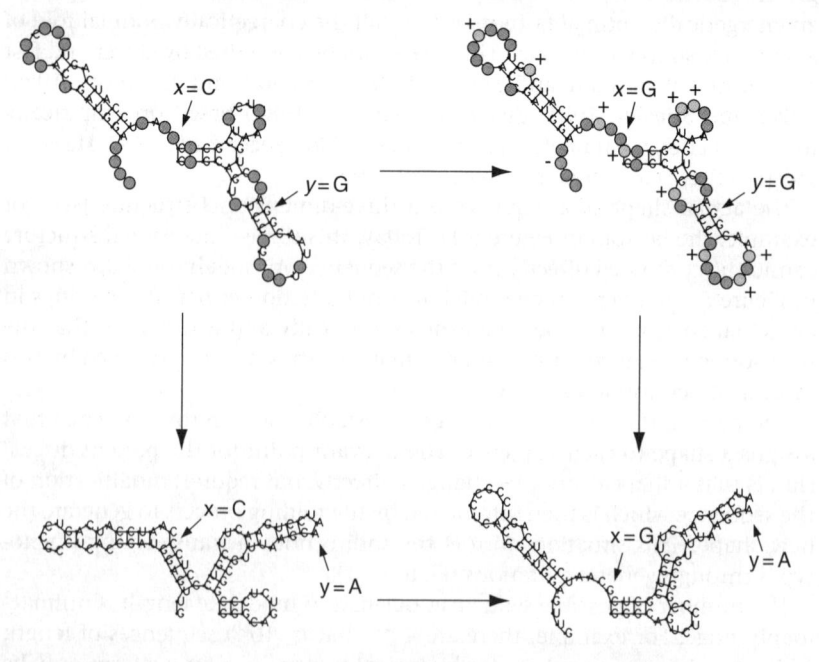

Figure 5.3 Epistasis*

* Bullets indicate a neutral position. In the top left sequence, position x is neutral because the substitution of G for C preserves the shape, as shown in the top right sequence. Yet, neutral *positions* do change as a consequence of a neutral mutation. The dark and open bullets in the top right sequence indicate positions that have gained or lost neutrality, respectively. The lower part illustrates the context sensitivity of mutational effects. The neutral mutation from C to G at x affects the consequences of swapping A for G at position y.

The tendency of a sequence to adopt a different shape upon mutation is called variability and is a prerequisite for its capacity to evolve in response to selective pressures (evolvability). Variability underlies evolvability. Figure 5.3 illustrates that variability (quantified as the number of non-neutral positions) is sequence dependent. Variability can therefore evolve (Wagner, 1996; Wagner and Altenberg, 1996). Note that the change in variability occurs here without the mechanisms of folding (development) themselves changing. This sequence-dependent change of variability is related to Waddington's concept of canalization (Waddington, 1942).

Neutral networks

I focus next on one statistical property of the RNA folding map that I believe to be a 'paradigm of change' of metaphorical import beyond RNA.

In the following it is useful to think of all possible sequences of a given length as related to one another by a notion of distance. The distance between two sequences is the smallest number of mutations required to convert one sequence into the other. The direct neighbors of a sequence are all sequences one mutation away. For example, a sequence of length 100 has 300 neighbors. This results in a very high-dimensional metric space. Despite its abstract nature, this so-called sequence space (Eigen, 1971), see Figure 5.4, is real in the sense that mutations are chemical events interconverting sequences with a certain probability. The probability of going from one

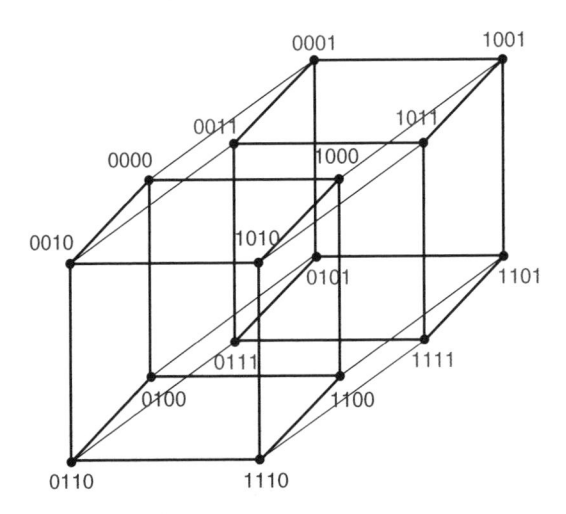

Figure 5.4 Sequence space for sequences of length 4 over the binary alphabet {0,1}*

* The connections correspond to single mutations that change only one position. The picture hints at how the four-dimensional sequence space (a hypercube) is constructed from the three-dimensional space (a cube).

sequence to another in a single event depends on their distance (the number of edges on the smallest path connecting them). The farther apart the sequences, the lower the probability for a direct jump.

Equipped with the notion of a sequence space, consider the sequence and its shape depicted at the top left of Figure 5.3. The positions marked by grey bullets are neutral. At least one mutation at each neutral position leads to a neighboring sequence with the same shape – a neutral neighbor. A given sequence has typically a significant fraction of neutral neighbors at a distance of one or two mutations. The same observation holds for each of these neutral neighbors. (For example, the sequence shown at the top right of Figure 5.3 also has several neutral positions, as indicated by grey bullets.) In this way, we can jump in steps of one or two mutations from sequence to sequence, mapping out a vast network in sequence space on which the phenotype is preserved. For this extensive, mutationally connected, network of sequences, we coined the term *neutral network* (Schuster *et al.*, 1994), Figure 5.5A.

Robustness enables change

The possibility of changing the genotype while preserving the phenotype is a manifestation of a certain degree of phenotypic robustness toward genetic mutations. Yet, at the same time, it is a key factor underlying evolvability, the capacity of a system to evolve. It seems paradoxical, but robustness enables change. To understand this, imagine (with the help of Figure 5.5A) a population with phenotype 'star' in an evolutionary situation where phenotype 'triangle' would be advantageous or desirable. Phenotype 'triangle', however, may not be accessible in the vicinity of the population's current location in genotype space, say, somewhere in the northern portion of the 'star'-network of Figure 5.5A. In the popular image of a rugged fitness landscape, the population would be stuck at a local peak, forever waiting for an exceedingly unlikely event to deliver the right combination of several mutations to yield phenotype 'triangle'. Yet, if phenotype 'star' has an associated neutral network in genotype space, the population is not stuck, but can drift on that network into far away regions, vastly improving its chances of encountering the neutral network associated with phenotype 'triangle' (Huynen, 1996; Huynen *et al.*, 1996; Fontana and Schuster, 1998ab; Nimwegen *et al.*, 1999), see Figure 5.5A. Neutral networks therefore enable phenotypic innovation by permitting the accumulation of neutral (that is, phenotypically inconsequential) mutations. These mutations, however, alter the genetic context, enabling a subsequent mutation to become consequential.

Of course, there is no guidance on a neutral network, only drift. Yet, drift confined to the 'star'-network dramatically increases the chance of eventually encountering the 'triangle'-network. In the absence of neutral networks, an exceedingly unlikely direct jump from an isolated 'star'-location to some far-away 'triangle'-location would be required.

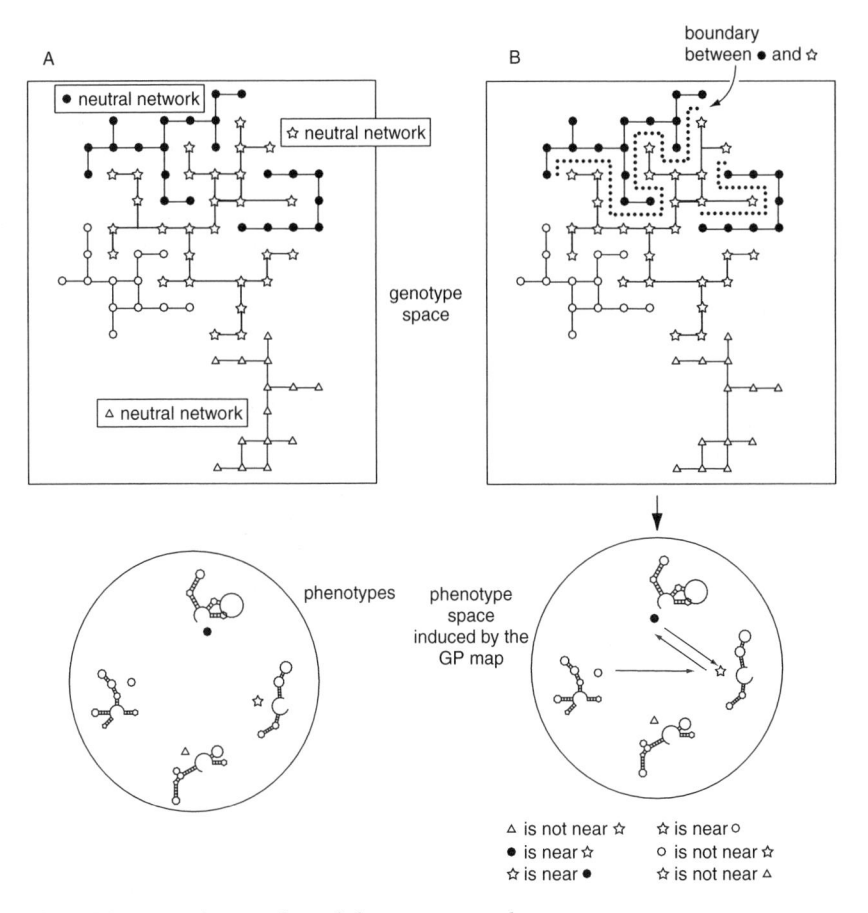

Figure 5.5 Neutral networks and shape space topology.
A: Schematic depiction of neutral networks in sequence space (upper part)*
B: Phenotypic neighborhood as fraction of shared boundary**

* The symbol of a sequence indicates its phenotype (lower part of the figure). A population located in the northern portion of the 'star' phenotype cannot access the 'triangle' phenotype, but it can diffuse on the 'star'-network until it encounters the 'triangle'-network.
** The 'star'-network is near the 'open circle'-network, because a random step out of 'open circle' has a high probability of yielding 'star'. The 'open circle'-network, however, is not near the 'star'-network, because a random step out of 'open circle' has a low probability of yielding 'star'. This effect can result from differently sized networks. In another case, the 'star' and 'triangle'-networks have similar sizes, but they border one another only rarely; 'star' and 'open circle' are not in each other's neighborhood. This is the case for shift transformations like the one shown in Figure 5.6A.

What appears to be a sudden and abrupt change at the phenotypic level becomes the result of several random neutral genetic changes. The existence of neutral paths in RNA sequence space was impressively demonstrated experimentally by Schultes and Bartel (2000). The importance of neutrality for evolution was long recognized by the Japanese population geneticist Motoo Kimura (Kimura, 1968; Kimura, 1983). Population genetics, however, is concerned with the dynamics of gene frequencies. Neutrality is interesting in that context because it leads to a diffusion dynamics in the *frequency* of a neutral gene within the population. In the present context, I'm concerned with something quite different, namely the diffusion of a whole population over a neutral network in gene space. The next consequence of neutral networks that I wish to emphasize is the *organization of phenotype space*.

The topology of phenotype space

A space is formally a set of elements with a structure imposed by the relationships among the elements. A relation of distance gives rise to a metric space. For example, the distance between two RNA sequences is the number of positions in which they differ. For a metric space to be relevant, the distance must be defined in terms of naturally occurring (or technologically controllable) operations that interconvert elements. This is indeed the case with sequences, where the operations are mutations (or technologically targeted changes in the sequence). But what is a space of phenotypes, if phenotypes cannot be modified directly? Although any number of distance (or similarity) measures between phenotypes can be defined, a phenotype space so-constructed is of no help in understanding *evolutionary* histories, because it relates phenotypes in a manner that fails to take into account the indirection required to change them. Rather, what is needed is a criterion of *accessibility* of one phenotype from another by means of mutations in their underlying genetic representation. Such a notion of accessibility can then be used to define a concept of *neighborhood* that underlies the structure of phenotype space. Such a neighborhood does not rest on a notion distance (Fontana and Schuster, 1998b; Stadler *et al*., 2001). This runs against common sense, which conceives 'neighborhood' in terms of 'small distance'. I shall not pursue this here, but a branch of mathematics known as point set topology clarifies the sense in which neighborhood is more basic than distance (see any text book on topology).

Recall that a neutral network is the set of all genotypes adopting a particular phenotype. In Figure 5.5B, I define a relation of accessibility between to (RNA) shapes, symbolized by 'star' and 'open circle', in terms of the adjacency of their corresponding neutral networks in genotype space (Fontana and Schuster, 1998a,b). The *boundary* of a neutral network consists of all sequences that are one mutation off the network. The intersection of the neutral network of 'open circle' with the boundary of the neutral network of 'star', relative to the total boundary of 'star's' network, is a measure of the probability that one step off a random point on the 'star'- network there is a sequence folding into 'open circle'. A rough schematic of such intersections is depicted in Figure 5.5B by a dashed line that traces the apposition of the 'star'-network and the 'filled circle'-network.

To fix ideas with a cartoon, think of Europe as sequence space and of a neutral network as a European state (a 'phenotype'), say, France. Now scan the boundary of France and measure the fraction of boundary it has in common with some other state, say Germany. That fraction measures the likelihood of ending up in Germany when making a random step out of France. Doing this for all states bordering France yields a distribution of relative border lengths. This distribution has an abstract average. Now define the neighbors of France as those states whose relative boundary with France is longer than that average. This construction is needed in actual genotype space because a neutral network is a very high-dimensional object that borders to a huge number of other neutral networks. (The example of geographic states fails miserably in this regard.) Most of these boundaries, however, are tiny and their states are not classified as 'neighbors' in this construction. (The actual distribution of relative boundary sizes is a generalized power-law (Fontana and Schuster, 1998b).) For important technical details the reader should consult Stadler *et al.*, 2001. Notice something funny, though. France is near Monaco, because a large fraction of Monaco's boundary is with France. Yet, Monaco is not near France, since France's boundary to Monaco is tiny. Returning to our original picture, a phenotype may be easily accessible from another phenotype by genetic mutations, but the reverse may not be true. A common sense manifestation of this is the observation that it can be very difficult to build structure (few paths), but very easy to destroy it (many paths). I will not provide RNA examples here, since their discussion requires too much terminology that is of little import to the target audience of this chapter. The important message is that accessibility is *not* a distance. A distance relation is always symmetric, while accessibility is not, as illustrated in Figure 5.5B. The RNA phenotype space is not a metric space, but a much weaker and less intuitive space (Stadler *et al.*, 2001).

Continuous and discontinuous change

One more step is needed to put together a full picture. Obviously, Austria is not near France, because Austria has a zero border to France. We can ask, however, whether there is a route from France to Austria such that at each step we always make a transition to a state in the current neighborhood – for these are the easy transitions. For example, we could step from France into Germany (easy) and from Germany into Austria (easy). Roughly speaking, mathematicians call a path continuous if at any given small step it proceeds along the neighborhood relations of the space the path is traversing; in other words, if you don't have to lift the pencil when drawing the path …

We can now ask an important question about the structure of phenotype space: Given any two phenotypes, is there always a continuous ('easy') path connecting them? That is, can we travel from one neutral network to another that is not a neighbor (in the boundary sense defined above), such that at each step (mutation) along the way we always transit to a neighboring network, until we reach our target network? In the RNA case, the answer is no (Fontana and Schuster, 1998a,b). This means that certain shape transformations are irreducibly difficult to achieve, no matter which route we try. The formal construction I've sketched here, allows us to call such transformations discontinuous in the mathematical sense of the concept. When looked at in detail, these transformations are local changes in shape – nothing fancy – but they require the *simultaneous* rearrangement of several interactions (pairings) among building blocks, because any intermediate resulting from a serial rearrangement would be unstable or physically impossible (Figure 5.6A shows an example).

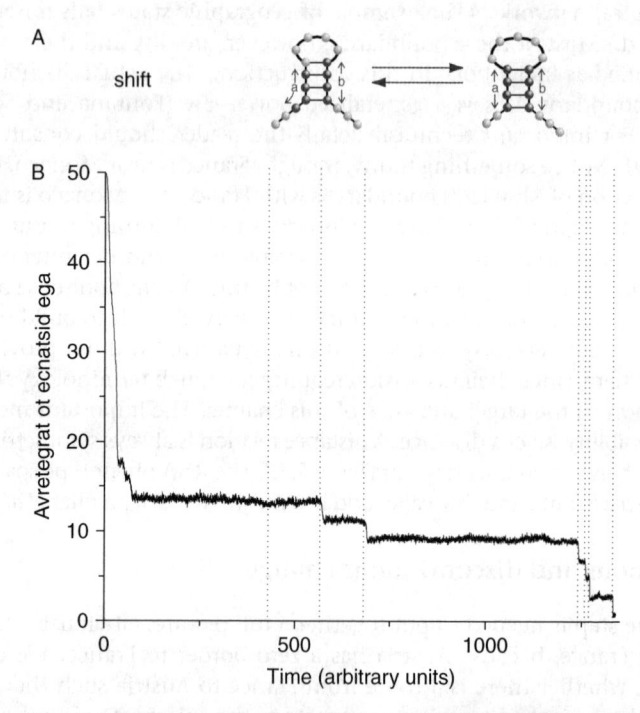

Figure 5.6 A: An example of a discontinuous shape transformation in RNA*
B: Punctuation in evolving RNA populations**

*In this case (a shift) one strand slides against the other in a paired segment. All pairings must slide in one single event, since any partial sliding would create bubbles destabilizing the intermediate. Triggering such a shift by a single mutation requires a special sequence context (Fontana and Schuster, 1998a,b).

**A population of RNA sequences evolves under selection for a specific target shape. The homing in on the target shape shows periods of stasis punctuated by sudden improvements. (Fitness is maximal when the distance to the target shape has become zero.) Yet, the phenotypic discontinuities (marker lines), as revealed by an *ex post* reconstruction of the evolutionary trajectory, are not always congruent with the fitness picture. In this example, the first two jumps in fitness turn out to be continuous in the phenotype topology discussed here and a crucial discontinuous transition is fitness neutral (first marker line).

A few observations deserve emphasis. Analyzing the mapping from genotypes to phenotypes in terms of neutral networks enables a mathematically rigorous definition of continuity (or discontinuity) in evolutionary trajectories. First, this notion of (dis)continuity cross-cuts (dis)similarity. Some transitions between similar shapes are discontinuous (e.g. Figure 5.6A) and some transitions between dissimilar shapes are continuous. Second, the notion of discontinuity defined here is *not* related to sudden jumps in fitness or the discrete nature of the notion of phenotype considered here. The categories of discontinuity are caused by the genotype–phenotype map. They remain the same regardless of the further evaluation of phenotypes in terms of fitness. (Of course, the particular shapes observed at discontinuous transitions in actual evolutionary trajectories will depend on the fitness map. Yet, the nature (type) of these transformations does not depend on the fitness map.) Third, the dynamical signature of this phenotype topology is punctuation (see Figure 5.6B). A population of replicating and mutating sequences under selection drifts on the neutral network of the currently best shape until it encounters a gateway to a network that conveys some advantage or is fitness-neutral. That encounter, however, is not under the control of selection, for selection cannot distinguish between neutral sequences. While similar to the phenomenon of punctuated equilibrium recognized by (Eldredge and Gould, 1972) in the fossil record of species evolution, punctuation as described here occurs in the *absence* of externalities (such as meteorite impact or abrupt climate change in the case of species). Here, punctuation reflects the variational properties of the developmental architecture.

Summary

Development matters

The processes that link genotype to phenotype typically generate a many-to-one mapping, that is, there are considerably fewer phenotypes than genotypes. Yet, this degeneracy also depends on the level of resolution at which we define phenotype. If, in the RNA case, we were to define shape in terms of atomic coordinates, there could be no redundancy; every sequence would have a unique shape and the topic of this paper is mute. I believe, however, that it is precisely the onset of robustness – that is, extended neutral networks – that identifies a meaningful level of resolution for phenotype. The notion of neutrality depends on an appropriate notion of equivalence, which is a notoriously difficult notion to pin down in biological, economic or social contexts.

Robustness enables change

The mathematical and statistical analysis of the mapping from RNA sequences to shapes has produced the concept of a neutral network – a mutationally connected set of genotypes that map to the same phenotype. I believe that this concept holds for many genotype–phenotype mappings beyond the simple RNA case. Neutral networks are key to change. It is hard to understand how evolution could be successful at all, if developmental processes don't give rise to neutral networks. The popular image of a rugged evolutionary landscape, where populations get stuck at local optima, is certainly incomplete and perhaps even wrong. Populations may be pinned at the phenotypic level, but they constantly change at the genetic level, drifting on neutral networks, thereby dramatically increasing their chances for phenotypic innovation.

Variation is not innovation – the limits of selection

A neutral network has many neighbors (in the boundary sense). A transition from a given network to any of its neighbors corresponds to the acquisition of an 'easy' new phenotype. We have seen, however, that certain phenotypic changes *must* go through border-bottlenecks, involving a transition from one network to another that touches the former only rarely. I have distinguished (with some mathematical motivation) such phenotypic changes as being continuous or discontinuous. I'd like to belabor that point by suggesting that the continuous/discontinuous distinction reflects an intuitive distinction between variation that occurs readily (a better mouse trap) and innovation, that is, a new phenotype that is not readily achievable and that requires long periods of drift. The former variation is typically present in any reproducing population that drifts over a neutral network and explores its boundary by mutation. Such omni-accessible variation is therefore easily available to selection for adaptive responses. In the case of an innovation, however, selection can do nothing to guide a population over a neutral network in the direction of the rare boundary segment that marks an innovation, since selection cannot distinguish between genetic variants of the same phenotype. The occurrence of innovations is not under the control of selection, but, of course, their fate is. The economic literature makes a distinction between invention and innovation. Invention denotes the first occurrence of an idea, while innovation refers to its adoption in practice. From this perspective, what I call 'innovation' would be the economist's notion of invention, since my notion of innovation only reflects the constraints imposed by the genotype–phenotype map, independently of the fixation or rejection of an innovation through selection.

Beyond genotype and phenotype

I have described a paradigm of change in a rather specific molecular and biological setting. Underneath the surface, the notions of genotype and phenotype appear as convenient idealizations even in biology (Griesemer, 2000). They appear hardly meaningful in technological, economic and social realms. I'd like to conclude, therefore, with a little translation for readers from other disciplines.

In social, economic and technological domains, we rely on systems consisting of many interacting heterogeneous components whose collective action gives rise to ordered system behavior. When we wish to change that behavior, we hit on a simple fact whose consequences I have described in this contribution: Behavior is not a thing, behavior is the property of a thing. It follows that the only way to change behavior is to change the thing that generates it. The change of a property is necessarily indirect.

As an example, consider a computer program. A computer program implements a function. A function is not something that can be touched and altered directly. To alter the function one must alter the program text. The mapping from program to function is what computer scientists call the semantics of the programming language in which the program is written. Substitute for computer program your favorite social, technological, or economic organization, for function the relevant qualitative behavior of that organization and for semantics the dynamical principles that govern the micro-interactions among the parts of the organization.

At a high level of abstraction, the structure of the situation may be represented by a mapping from systems to behaviors of those systems. An input to this mapping is the specification or description of a particular system configuration. The outcome of the mapping is the behavior or function of that system configuration. The mapping represents the dynamical principles that are characteristic for a particular class of systems – computer programs, molecules, electronic circuit diagrams, urban transportation systems, firms in a given industry. In other words, the mapping represents the rules or processes that are *constitutive* for a given class of systems. In social contexts, institutions would be a component of this mapping. The unfolding of this constitutive dynamics in the context of a particular system configuration yields the behavior associated with that configuration.

When we wish to change behaviors of systems, we often have a spatial metaphor in mind, such as going from 'here to there', where 'here' and 'there' are positions in the space of behaviors. But what exactly is the nature of this space? Who brought it to the party? It is a widespread fallacy to assume that the *space* of behaviors is there to begin with. This is a fallacy even when all possible behaviors are known in advance. How does this fallacy arise? When we are given a set of entities of any kind, we almost always can cook up a way of comparing two such entities, thereby producing a definition of similarity (or distance). A measure of similarity makes those entities hang together naturally in a familiar metric space. The fallacy is to believe that a so-constructed space has any significance. It isn't, because that measure of similarity is not based on available real-world operations, since we cannot act on behaviors directly. We only have configuration-editors, we don't have property-editors. Seen from this operational angle, that which structures the space of behaviors is not the degree of similarity among behaviors but a rather different relation: operational *accessibility* of one behavior from another in terms of system-reconfigurations. This brings the mapping from system configurations to behaviors into the picture. The structure of behavior-space is induced by this mapping. It cannot exist independently of it.

Limits

In the previous sections I have described how this space-structure arises in the context of RNA molecules. With the above paragraphs in mind, substitute 'system' for 'RNA sequence' and 'behavior' for 'RNA shape'. It deserves emphasis, however, that the RNA model is a limiting (and highly idealized) case in which the mapping from sequences (systems) to shapes (behaviors) is exogenous to the system itself. In the game of Go, the rules are not determined by the board configuration; they are given independently. While this limiting case is useful to illustrate the induced character of behavior-space, the most intriguing cases are those in which the mapping itself is endogenously generated by the system. This is the case in all of organismal biology, where genes also code for products that implement development. The endogeneity of the mapping from systems to behaviors is clearly a feature of economic and social systems as well. Nothing prevents the rules of the game (and thus the game itself) from changing. It may be harder to change a rule than to change a game configuration, but this is a question of time scales, not of principle. In the case of an endogenous mapping, the structure of phenotype (behavior) space remains an induced one. However, the concept of space becomes significantly more subtle. Not only can a change in the mapping rearrange the accessibility structure between phenotypes, it can also bring into existence phenotypes that were not possible before. We still lack a formal grip on this radical form of change. In the mid sixties, C. H. Waddington vented his frustration with the mathematical apparatus of theoretical biology: '*The whole real guts of evolution – which is, how do you come to have horses and tigers, and things – is outside the mathematical theory*' (Quoted by (Gould, 2002, p. 584).) The challenge remains of bringing the guts of evolution within the scope of a mathematical theory – that is, of formalizing the emergence of new classes of biological objects.

Acknowledgements

My research at the Santa Fe Institute was supported by Mr. Jim Rutt's Proteus Foundation and by core grants from the John D. and Catherine T. MacArthur Foundation, the National Science Foundation, and the US Department of Energy, and by gifts and grants from individuals, corporations, other foundations, and members of the Institute's Business Network for Complex Systems Research. I would like to thank Andreas Wimmer and Reinhart Kößler for a stimulating meeting on 'Paradigms of Change'.

References

Carroll, S. B., J. K. Grenier and S. D. Weatherfee (2001). *From DNA to diversity*. Malden, Massachusetts, Blackwell Science.
Eigen, M. (1971). 'Selforganization of matter and the evolution of biological macromolecules.' *Naturwissenschaften* 58: 465–523.
Eldredge, N. and S. J. Gould (1972). 'Punctuated equilibria: an alternative to phyletic gradualism.' *Models in Paleobiology*. T. J. M. Schopf. San Francisco, CA, Freeman, Cooper & Co.: 82–115.
Fontana, W. and P. Schuster (1998a). 'Continuity in evolution: on the nature of transitions.' *Science* 280: 1451–5.
Fontana, W. and P. Schuster (1998b). 'Shaping space: the possible and the attainable in {RNA} genotype–phenotype mapping.' *J. Theor. Biol.* 194: 491–515.
Gould, S. J. (2002). *The Structure of Evolutionary Theory*. Cambridge, MA, Belknap/Harvard University Press.
Griesemer, J. R. (2000). 'Reproduction and the reduction of genetics.' *The Concept of the Gene in Development and Evolution: Historical and Epistemological Perspectives*. P. Beurton, R. Falk and H.-J. Rheinberger (eds). Cambridge, MA, Cambridge University Press: 240–85.
He, L., R. Kierzek, Santa Lucia Jr, A. E. Walter and D. H. Turner (1991). 'Nearest-neighbour parameters for {G-U} Mismatches.' *Biochemistry* 30: 11124.
Huynen, M. A. (1996). 'Exploring phenotype space through neutral evolution.' *J. Mol. Evol.* 43: 165–9.
Huynen, M. A., P. F. Stadler and W. Fontana (1996). 'Smoothness within ruggedness: the role of neutrality in adaptation.' *Proc. Natl. Acad. Sci. USA* 93: 397–401.
Jaeger, J. A., D. H. Turner and M. Zuker (1989). 'Improved predictions of secondary structures for {RNA}.' *Proc. Natl. Acad. Sci. USA* 86: 7706–10.
Kimura, M. (1968). 'Evolutionary rate at the molecular level.' *Nature* 217: 624–6.
Kimura, M. (1983). *The Neutral Theory of Molecular Evolution*. Cambridge, UK, Cambridge University Press.
Kirschner, M. and J. Gerhart (1998). 'Evolvability.' *Proc Natl Acad Sci USA* 95(15): 8420–7.
Mathews, D. H., J. Sabina et al. (1999). 'Expanded sequence dependence of thermodynamic parameters improves prediction of RNA secondary structure.' *J. Mol. Biol.* 288: 911–40.
Maynard-Smith, J., R. Burian, S. Kauffman, P. Alberch, J. Campbell, B. Goodwin, R. Lande, D. Raup and L. Wolpert (1985). 'Developmental constraints and evolution.' *Quart. Rev. Biol.* 60: 265–87.
Müller, G. B. and G. P. Wagner (1991). 'Novelty in Evolution: Restructuring the Concept.' *Annu. Rev. Ecol. Syst.* 22: 229–56.
Nimwegen, E. van, J. P. Crutchfield and M. Huynen (1999). 'Neutral evolution of mutational robustness.' *Proc. Natl. Acad. Sci. USA* 96: 9716–20.
Nussinov, R. and A. B. Jacobson (1980). 'Fast algorithm for predicting the secondary structure of single-stranded RNA.' *Proc. Natl. Acad. Sci. USA* 77(11): 6309–13.
Schultes, E. A. and D. P. Bartel (2000). 'One sequence, two ribozymes: implications for the emergence of new ribozyme folds.' *Science* 289: 448–52.
Schuster, P., W. Fontana, P. F. Stadler and I. L. Hofacker (1994). 'From sequences to shapes and back: a case study in RNA secondary structures.' *Proc. Roy. Soc. (London) B* 255: 279–84.
Stadler, B. M. R., P. F. Stadler, G. Wagner and W. Fontana (2001). 'The topology of the possible: Formal spaces underlying patterns of evolutionary change.' *J. Theor. Biol.* 213: 241–74.

Turner, D. H., N. Sugimoto and S. M. Freier (1988). 'RNA structure prediction.' *Annual Review of Biophysics and Biophysical Chemistry* 17: 167–92.

Waddington, C. H. (1942). 'Canalization of development and the inheritance of acquired characters.' *Nature* 3811: 563–5.

Wagner, A. (1996). 'Does evolutionary plasticity evolve?' *Evolution* 50(3): 1008–23.

Wagner, G. P. and L. Altenberg (1996). 'Complex adaptations and the evolution of evolvability.' *Evolution* 50: 967–76.

Walter, A. E., D. H. Turner, J. Kim, M. Lyttle, P. Muller, D. Mathews and M. Zuker (1994). 'Coaxial stacking of helices enhances binding of oligoribonucleotides and improves prediction of RNA folding.' *Proc. Natl. Acad. Sci.* 91: 9218–22.

Waterman, M. S. and T. F. Smith (1978). 'RNA secondary structure: A complete mathematical analysis.' *Mathematical Biosciences* 42: 257–66.

Zuker, M. and D. Sankoff (1984). 'RNA secondary structures and their prediction.' *Bull. Math. Biol.* 46(4): 591–621.

Zuker, M. and P. Stiegler (1981). 'Optimal computer folding of larger RNA sequences using thermodynamics and auxiliary information.' *Nucleic Acids Research* 9: 133–48.

6
Neutrality as a Paradigm of Change
Rudolf Stichweh

Walter Fontana presents in his paper a convincing case for the relevance of neutrality as a paradigm of change for biological systems. Neutrality means a drift of a biological system through a succession of states which do not change the phenotype of the system and which are therefore neutral to natural selection. By such a succession of neutral states the system may accidentally come near to a position in which one further small incremental change in genotype implies the transfer of the system to another phenotype, a transfer which may be perceived to be an improbable one and which perhaps would not have happened in the absence of neutrality. Another twist one can give to the same argument will not look to a temporal succession of states but to the simultaneous occurrence of different genotypes in a population of units. All these different genotypes will produce the same phenotype and can therefore coexist in a neutral space. If a *need* for a different phenotype should arise there will always exist in such a population of genotypes some exemplars which by some few alterations can effect a transfer into the advantageous phenotype. In this way *neutrality* as a theoretical paradigm presents a good case for a continuous evolution going on in a population of genotypes being equivalent towards one another in a phenotypical sense. All of these genotypes stand for possible alterations. At the same time, a concept of *discontinuity* can be formulated on this basis. Discontinuity then means all those variations and the phenotypes generated by them which are not represented in the population of genotypes by genotypes which would only need a few steps for changing into related genotypes which could then bring about the relevant phenotype. Continuous evolution looks *as if* natural selection as the evolutionary force could directly instruct the changes it *needs*. Discontinuities on the other hand mean that natural selection has to wait for a long time until very improbable variations accidentally arise which it then can favour.

Is this concept of *neutrality* a fruitful one in thinking in a more general sense about change in systems, especially change in social systems? First of all, two important difficulties have to be mentioned. In social systems, a precise analogue to the distinction of genotype and phenotype is not easily to be perceived. And there is in social systems no such thing as *development*, that is a class of mechanisms mediating between genotype and phenotype.

In the following I will look at three relevant distinctions of the social sciences to examine if their understanding can be improved upon in making use of neutrality as a theoretical paradigm. A first candidate is the distinction of *system* and *operation* (or *system* and *behaviour*) which in variant formulations will be found in most present-day social theories. This distinction may be seen as a potential analogue to the distinction of genotype and phenotype but it is difficult to relate it to the argument of Walter Fontana. Operations are much more fluid and variable than the systems or structures they realize or implement, and therefore there is no such thing as an evolutionary drift of systems/structures through spaces being neutral in an operational or behavioural sense.

A second candidate which is probably more instructive is the distinction of structure and organization proposed by Maturana and Varela (1980). This distinction is again present in numerous conceptual variants. Structural changes in a social system do not change anything about organization (e.g. autopoiesis as an organizational feature). Therefore they are neutral to organization. But in a succession of structural changes it can happen at any time that a state is achieved in which a new structural pattern is no longer compatible with the organization of the system and this organization changes abruptly and discontinuously. A related kind of argument has been experimented with in the case of sociological differentiation theory which in authors such as Niklas Luhmann is based on the distinction of *structural differentiation* as ongoing process and *forms of differentiation* (such as: functional differentiation, stratification, segmentation) as discontinuous principles of ordering a multiplicity of differentiated systems (Luhmann, 1997, ch. 4). Looking at a specific *form of differentiation* – for example at functional differentiation – one can easily observe ongoing processes of structural differentiation which are *neutral towards the principle of functional differentiation*. And then there exist limits to such a neutrality of ongoing structural differentiation towards functional differentiation. At these limits further structural differentiation will result in a new form of societal differentiation finally being established.

The present author has, in looking for the genesis of functional differentiation, collected evidence for a related argument. He tried to demonstrate that for the society of estates of late medieval and early modern Europe the addition of new corporations such as religious orders, trading companies, universities and cities to the world of hierarchical estates was in some respects a neutral addition, as the corporations were invested with the outer signs of a stratified social order (Stichweh, 1991, ch. II). They were characterized by status, dignity and honour as is pertinent in hierarchical society. But such a stratified social order enriched by ever new corporations which had to be distinguished in relation towards one another along functional lines, too, became unstable at some point, and then a new principle of social differentiation (a lateral, non-hierarchical, functional order) became *accessible* from within the society of estates. Of course, differences between the picture drawn in this argument and Fontana's considerations regarding neutral changes in genotype are easily to be identified. The changes mentioned in the argument about ongoing structural differentiation are not really neutral towards the principle of hierarchical differentiation which they later dissolve. It would be more adequate to say that they are characterized by a certain structural *bipolarity*. They have a neutral side (the status aspect of corporations) and on the other hand they exhibit the features of differentiation along functional lines which are incompatible with a hierarchical order. That is, the parallel between this argument and Fontana's paradigm only consists in social systems succeeding for some time in *neutralizing* structural changes which from a later and retrospective point of view will be seen as signposts of a newly arising principle of social differentiation.

A third interesting and perhaps most apposite case we will briefly examine here is the social scientific distinction of *semantics* and *social structure*[1] which is closely related to the distinction of *culture* and *social system* (Parsons, 1973). Regarding both distinctions, one can observe obvious parallels to Fontana's distinction of genotype and phenotype. Semantics as well as culture can drift through spaces of potential meaning without any changes in social structures and social systems immediately resulting from this drift. This means that in this case we can construct a more exact analogue to Fontana's argument. We may conceive semantic and cultural elements as a population of units which define a space of which it can be said that all the elements in this space are compatible with present social structures and social systems. But for each of these semantic and cultural elements it may be said that they occupy different positions in this contemporaneous space, positions which are characterized by differing contiguities to other potential elements which clearly are located outside of the space of compatible possibilities. For each of these semantic and cultural elements there exists at least one nearby possibility which is outside the space of those possibilities compatible with present-day social structures. From this consideration, one can derive the picture of continuous evolution Fontana sketched in his analysis of a mechanism of biological evolution. If a *need* for structural change should arrive, there is always a semantic/cultural element which is near to a variant from which the respective change in social structures can be established and legitimated. In a paper from 1990, David Sloan Wilson gave a populationist interpretation of historical semantics and he pointed to the polymorphisms in historical semantics which he illustrated by the historical semantics of *self* which freely intermingles positive and negative evaluations as in *self-interest* vs. *selfish* (Wilson, 1990). This illustrates the kind of ambivalence in historical semantics which for some time may neutralize social change and at the same time opens spaces of possibilities from which one may in short time change over into a world of different structures. And in building on this argument, one may establish a concept of *discontinuity* which as it is the case in Fontana's paradigm means those possibilities which are separated by considerable distances from all the individual elements even in a very diverse semantics. What I want to conclude from these brief remarks is that Walter Fontana has established neutrality as a suggestive paradigm of (evolutionary) change. This has to be further examined in interdisciplinary discourse. There are difficulties to be seen such as the obvious disanalogies to social systems regarding the distinction of genotype and phenotype and regarding the biological concept of development. But nonetheless *neutrality* may become a valuable entry in the vocabulary of terms of a theory of evolutionary change (of social systems). In a more complete discussion this concept will have to be related to well-established terms such as *latency* and to concepts such as *pre-adaptive advance* (or *exaptation* according to Gould and Vrba, 1982). This vocabulary is not well formulated yet. This makes it clear how considerable the distance is which separates us from a satisfying and complex theory of evolutionary social change.

Note

1. Cf. for the present usage of this distinction Luhmann (1980, ch. 1); Stäheli (1998, 315–39); Stichweh (2000).

Bibliography

Gould, Stephen J. and Elisabeth S. Vrba. 1982. 'Exaptation – a missing term in the science of form', *Paleobiology* 8: 4–15.
Luhmann, Niklas. 1980. *Gesellschaftsstruktur und Semantik: Studien zur Wissenssoziologie der modernen Gesellschaft*, Vol. 1, Frankfurt a.M.: Suhrkamp.
Luhmann, Niklas. 1997. *Die Gesellschaft der Gesellschaft*, Frankfurt a.M.: Suhrkamp.
Maturana, Humberto and Francisco J. Varela. 1980. *Autopoiesis and Cognition: The Realization of the Living*, Dordrecht and Boston: Reidel.
Parsons, Talcott. 1973. 'Parsons, culture and social system revisited', pp. 33–46 in: Louis Schneider and Charles M. Bonjean (eds), *The Idea of Culture in the Social Sciences*, Cambridge: Cambridge University Press.
Stäheli, Urs. 1998. 'Die Nachträglichkeit der Semantik. Zum Verhältnis von Sozialstruktur und Semantik', in: *Soziale Systeme* 4, 315–39.
Stichweh, Rudolf. 1991. *Der frühmoderne Staat und die europäische Universität. Zur Interaktion von Politik und Erziehungssystem im Prozeß ihrer Ausdifferenzierung (16.–18. Jahrhundert)*, Frankfurt a.M.: Suhrkamp, esp. ch. II.
Stichweh, Rudolf. 2000. 'Semantik und Sozialstruktur: Zur Logik einer systemtheoretischen Unterscheidung', in: *Soziale Systeme* 6, 237–50.
Wilson, David Sloan. 1990. 'Species of thought: a comment on evolutionary epistemology', *Biology and Philosophy* 5: 37–62.

7
Using Evolutionary Analogies in Social Science: Two Case Studies

Edmund Chattoe

This chapter considers the use of evolutionary analogies (EA) in social science. It begins with a general discussion of the role analogy plays and then considers the specific benefits of EA for unsolved problems in social science. Most of the chapter presents two case studies of Fontana's work. The first applies neutral networks to social structure. The second uses 'algorithmic chemistry' to explore industrial diversification and the emergence of classes. The purpose of the case studies is not to build working simulations but to illustrate the process of proposing, developing and criticising analogies systematically.

The role of analogy

Following Hesse (1963) and Brodbeck (1968) we can identify three requirements for effective analogy. First, that one domain is much better understood than the other is. Secondly, that the concepts in one domain can be put into one-to-one correspondence with similarly precise equivalents in the other. Thirdly, that the causal connections between concepts in one domain are preserved between their equivalents in the other. This definition provides a framework for examining arguments about the relevance of EA in social science. In particular, arguments tend to focus on how well the second requirement is satisfied. The implications of the other requirements are often neglected. Regarding the first requirement, it is not only necessary that one domain be better understood, but that critics share that understanding! One standard critique of EA in social science is that social action lacks clearly defined 'genes' as objects of selection. However, this criticism often reveals a belief that genes are simply mapped onto traits and that 'social genes' should share this property. In fact, genes displaying this regularity are untypical and often code for phenotypic traits having little impact on fitness. As Fontana observes, development – the mechanism by which *sets* of genes create phenotypes – is far more crucial. This raises the significance of the third requirement. Empirically tested knowledge in biology does not uniquely identify *a* 'theory of evolution'. Competing theories (like gradualism and punctuated equilibrium) are compatible with 'survival of the fittest' and must be distinguished empirically. Thus, it is necessary to understand causal connections in both domains to judge the 'necessity' of particular aspects of the theory for effective analogy. The fact that Lamarckism is not observed in biology does not necessarily render analogy between non-Lamarckian biology and Lamarckian society invalid. The reason is that there is nothing logically incoherent in adding Lamarckism to natural selection, it just happens to be absent in biological systems for well understood physiological reasons. Its presence or absence therefore does nothing crucial to change the process by which phenotypes adapt to the environment. This raises an epistemological point neglected in the original definition: that *research methods* may also have to transfer between domains. Leaving aside known misunderstandings of biology, a fundamental problem with functionalism (the best known EA in sociology) was that it imported only the *theory* of natural selection and none of the methods used to establish the truth of that theory in biology. Clearly, some methods (like controlled experimentation) may be unethical in social systems but substitutes (like natural comparative experiments) exist. Note that, historically, development of the genetic theory of natural selection (and its empirical testing) preceded an ability to see genes. We may never 'see' postulated equivalents of genes (mental models and organisational routines) in social systems. Nonetheless, other methods for observing them and measuring their properties may prove sufficient for advancing the theory as occurred in biology.

Developing useful analogies may thus be treated as a 'scientific' process. Beginning with well-developed models from other fields, social scientists use expertise to identify problems to which they might apply. They must then ensure that *prima facie* applicability is not compromised by significant differences between mechanisms in the two domains. However, defining 'significant' difference should not be left to ideological predilection. Instead, it should involve two substantive processes. First, building models in the new domain, so it can be *shown* that differences do not materially affect system behaviour.[1] Secondly, importing (with suitable modifications) techniques by which the truth of the theory was established in the original domain and applying them in the new domain.

The benefits of EA in social science

Having provided a framework for more rigorous analogy development, I will now progressively focus the argument. In this section, I discuss unresolved issues, in social science that might benefit from EA. In the following two sections, I will illustrate my approach using case studies. These are 'given' by the purpose of this chapter as a response to Fontana. It is important to stress that the aim of this chapter is not to 'criticise' his models but only to consider their applicability in social science. It seems to me that one model (algorithmic chemistry) is more widely applicable than the other (neutral networks). Fortuitously, having one more and one less successful analogy as case studies helps us evaluate the proposed framework better.

We can distinguish the evolutionary approach from that prevailing in social science along four dimensions.

1. *Absence of system level teleology:* Although evolutionary approaches are compatible with individual decision and *intentions* at the system level, the complexity and autonomy of the environment undermines notions of 'successive stages', 'determinism', 'progress' and 'perfectibility'. These notions still underpin many sociological theories of change (Eisenstadt, this volume) and naïve applications of institutionalism in economics (Nugent, this volume). Since introducing global teleology into social systems is invariably question begging or incoherent, this absence is a positive feature.

2. *Compatibility with deliberation:* Another fundamental dichotomy is that between structuralism and rational choice. Sociology is accused of arguing that attributes like class and gender strictly determine individual action. By contrast, economics substantially neglects structural factors and social groupings in explaining choice. Humanity is an existence proof for the compatibility of natural selection with the ability to deliberate. It seems certain that social behaviour displays a mixture of intentional action and environmental selection (through unintended consequences) in different proportions depending on the domain. This synthetic approach is thus more empirically plausible than either pure structuralism or individualistic rationality. In particular, it explains the importance of micro-foundations (the dynamics of the system are driven by individual actions rather than structures) and imperfect understanding (there can be no 'underlying social fundamentals' because the environment is co-evolutionary, dynamic and reflexive) rightly stressed by Dosi and Winter (2002: 4–5).

3. *Endemic diversity:* Empirically, diversity is the rule and not the exception in social systems. Although one can 'repair' theories based on representative agents or structuralism by appealing to 'noise', 'error' and 'out-of-equilibrium' behaviour, it seems more scientific to develop theories in which convergence is explained endogenously.

4. *Real novelty and increasing complexity:* As Fontana points out, much science based on physics (including some evolutionary biology) has tended to ignore the emergence of new entities as a distinctive phenomenon. The same applies to social science and particularly mainstream economics. The rearrangement of goods through trade is a different phenomenon from the combination of inputs to create wholly new products. Sociology displays the same weakness in its enthusiasm for categories like 'class', usually treated as 'primitives' rather than labels for historically evolved groups. One consequence of genuine novelty is increasing societal complexity. Once a new

product exists, it generates a new possibility horizon for use. Once an identifiable group forms, individuals and groups can take its existence into account in their reasoning. Although there is debate in biology about whether complexity is 'selected for', it is clear that empirically (using measures like occupational diversity) industrial societies *have* increased in complexity and a theory handling this phenomenon in an integrated manner is clearly advantageous.

In the following case studies, I show how these general advantages are instantiated in specific models.

Neutral networks and social structure

Fontana presents a model of neutral networks in RNA. RNA consists of strongly bonded chains made up of four possible 'building blocks'. Only certain pairs of building blocks are disposed to form additional weak bonds *between* chains. To minimise energy and maximise stability, RNA chains 'fold up' in space, maximising the number of these weak bonds they form. Thus, *ceteris paribus* physical conditions, a given genotype (RNA chain) produces a unique phenotype (shape). However, mutation causes blocks to be replaced. Mutation may or may not affect the set of weak bonds between RNA chains. Mutations not affecting the set of weak bonds are called neutral. However, while such mutations do not affect the shape of the chain directly, they *do* affect the set of *other* neutral positions – sites where mutations may occur without changing the folding. As a result, in contrast to the simple gene and trait model, the same phenotype can result from many different genotypes. This set of genotypes is called a neutral network. Perhaps counterintuitively, this arrangement may be helpful in adaptation. The reason is that the 'distance' between two arbitrary RNA chains (in terms of the number of point mutations required to transform one into the other) may be large, even if the change is fitness improving. A neutral network means that the system may be able to 'drift' substantially nearer the target chain while experiencing no external selection pressure since by definition genotypes in neutral networks give the same phenotype.

This model should be tempting to social science. The idea that complex dynamic systems (like organisations) might remain stable despite changes in internal structure (and replacement of individuals) recalls the Weberian analysis of bureaucracy as a stable system of norms and roles (Weber 1968: 965–1005). Similarly, aggregate properties of social networks (the notorious 'six degrees of separation') appear robust to the making and breaking of individual links (Huberman 2001). Unfortunately, differences between biological and social domains rapidly become apparent. Considering organisational stability, there are clear mechanisms of selection within the tradition of evolutionary economics. Environmentally adapted organisations persist and others do not. However, we would need to establish empirically that stable responses *were* an effective adaptation. Furthermore, while firms go bankrupt, bureaucracies are notorious for persistence despite environmental change. Finally, although we might imagine the internal states of organisations represented as neutral networks this seems to be a drastic simplification of the role of power, hierarchy, deliberation, symbolism and negotiation. I think the burden of proof would surely rest on advocates of this analogy to prove that *none* of these things changed the adaptive properties of the system significantly.

By contrast, a neutrality analysis of social networks appears plausible at the individual level. Suppose that only a subset of 'types' forms persistent network links with each other and that these links also affect the neutrality of other positions in the network: consider the effect of an unpopular spouse on an existing friendship network. However, the distinction between strong and weak RNA bonds seems to be important and lacks correspondence in the social system.[2] However, the fundamental problem is that neutrality requires selection pressure on the macrostructure of networks and, by contrast to the case with firms, it is not clear what this might be.[3]

In conclusion, despite appealing features and resonance with sociological theory, neutral networks don't appear to form a basis for analogy. Clearly I could have missed appealing examples but given the best I could envisage, this analogy doesn't seem to reach 'first base': effective mapping of concepts between domains.

Industrial diversification and emergence of class structure

In earlier work, Fontana and Buss (1994, 1996) explore the role of novel entities in dynamic systems. Their 'algorithmic chemistry' describes processes of transformation in which components of the system 'react' to take new forms. (This distinguishes them from physical systems in which components move around but conserve form.) Such systems generate interesting macroscopic properties endogenously. For example, suppose we begin with two components (A and B) reacting to produce C. Only once some C exists can additional reactions occur with A or B to yield further components which may in turn react further. The system thus displays open-ended novelty and increasing complexity. Furthermore, assuming outflow or decay, components that maintain a presence in the system must come to be surrounded by a network of self-maintaining reaction pathways. Such systems can reasonably be defined as 'organisations', taking inputs from the 'environment' and producing 'outputs' while maintaining their internal structure. It can also be shown that these systems are somewhat homeostatic as regards relative concentrations and sets of components.

Again, *prima facie* this is a tempting model. Taking an economic example, new technologies allow the transformation of existing inputs into new outputs. These outputs can, in turn, serve as novel inputs to the system of production. Over time, technologies and products proliferate and societal complexity increases. If products are to survive, they must 'organise' inputs, production and demand. Stable sets of transformations constitute organisations like firms and households. This approach also has sociological implications. Since transformations 'use' labour, labour market positions depend on the set of transformation processes in use. Creating new goods and transformations thus gives rise to the possibility of new classes and new relationships between classes. (For example, the factory system requires alienating, repetitive work that may unite workers making cars and shoes. At the same time, specialisation is likely to engender 'pivotal' groups able to win concessions by their ability to threaten production.)

Clearly, there are important differences between domains. Firstly, in all but the first stages of production, there is deliberation about transformations and outputs produced. Nonetheless, as discussed earlier, system complexity means that outcomes are far from guaranteed by deliberation. Many products find unexpected uses and some worthwhile innovations are not used. The original model does not specify the 'source' of entity and transformation sets.[4] There is no obvious reason why adding new transformations endogenously should be logically incoherent though the behaviour of such systems has not been adequately studied as yet.

A more important difference is the role of 'structure'. Using a 'chemical' analogy, reactions should depend on concentrations of entities in the system. By contrast, treating product sales as a function of ubiquity seems excessively unrealistic. In a nutshell, the issue here is that while Fontana and Buss have filled an important gap in our understanding, society is no more 'pure chemistry' than it is 'pure physics'. Chemical reactions rely as much on physical processes like diffusion as they do on transformation. Thus, the process by which goods are transported should *not* be treated carelessly as a transformation, lest the term become so general that it loses all meaning.[5] Despite this caveat, the role of transformation has clear implications in underpinning a better understanding of open-ended social systems. It gives a clearly defined role to 'waste' – materials that no organisation can use – and displays the 'ecology' of social systems: what is output from one organisation is input to another. Finally, it raises interesting theoretical questions: what is the cost of various forms of organisation in transforming inputs to outputs and maintaining itself? In the next section, I turn briefly to the more general lessons that can be learnt from discussing analogies in this way.

Discussion and conclusions

Given the space available, it was not possible to follow my guidelines through to the final stage of comparing domains using detailed models and empirical validation through transfer of methods. Nonetheless, I hope I have shown how the guidelines can be used to move debate on from the 'four legs good, two legs bad' argumentation often besetting the use of analogy. I am happy to admit that my proposed applications may not show neutral networks in the best light and that my intuitions about which features 'really matter' to system behaviour may be profoundly flawed. (Indeed, this is one of the reasons why I urge the construction of simulation models.) A single author trying to simulate a dialogue about the pros and cons of anything can always be accused of special pleading but I have done my best to present the strengths and weaknesses of each application. However well I have succeeded, the intention is precisely to move debate to the point where other authors can challenge the specific mechanisms and applications I propose as part of a scientific dialogue rather than simply express a personal preference for (or against) reasoning by analogy.

Notes

1. One reason why debates about analogy are unsatisfactory is that the absence of precisely specified models makes emergent behaviours in complex systems impossible to compare.
2. Kinship relations might structure friendship relations but then again they might not.
3. We can speculate that the social capital of different networks helps to sustain them differentially with respect to each other but this is very much in the realm of conjecture.
4. Presumably the initial set of entities is the starting condition and the set of transformations ('laws of nature') is fixed at the outset.
5. It *would* be interesting to extend the Fontana and Buss approach to explicit (if abstract) spatial processes, perhaps combining algorithmic chemistry with cellular automata.

References

Brodbeck, M. (1968) 'Models, meaning, and theories'. In: Brodbeck, M. ed. *Readings in the Philosophy of the Social Sciences.* New York, NY: Macmillan, 579–600.
Dosi, G. & Winter, S. (2002) 'Interpreting economic change'. In: Augier, M. & March, J. eds *The Economics of Choice, Change and Organisation.* Cheltenham: Edward Elgar, 337–53.
Fontana, W. and Buss, L. W. (1994) ' "The Arrival of the Fittest": Toward a Theory of Biological Organisation'. *Bulletin of Mathematical Biology* 56(1) January, 1–64.
Fontana, W. and Buss, L. W. (1996) 'The Barrier of Objects: From Dynamical Systems to Bounded Organisations'. In: Casti, J. & Karlqvist, A. eds. *Boundaries and Barriers: On the Limits to Scientific Knowledge.* Reading, MA: Addison-Wesley, 56–116.
Hesse, M. (1963) *Models and Analogies in Science.* London: Sheed and Ward.
Huberman, B. (2001) *The Laws of the Web.* Cambridge, MA: The MIT Press.
Weber, M. (1968) *Economy and Society: An Outline of Interpretive Sociology,* vol. 3, G. Roth & C. Wittich eds. New York, NY: Bedminster Press.

Part III
Economics of Continuity: Path Dependency

Part II

8
The Grip of History and the Scope for Novelty: Some Results and Open Questions on Path Dependence in Economic Processes*

Carolina Castaldi[†‡] *and Giovanni Dosi*[‡]

Introduction

The very notion of multiple paths of socio-economic change ultimately rests on the idea that history is an essential part of the interpretation of most socio-economic phenomena one observes at any time and place. The property that *history matters* is also intimately related to that of *time irreversibility*. In the socio-economic domain and in many areas of natural sciences as well, one cannot reverse the arrow of time – even in principle, let alone in practice – and still recover invariant properties of the system under investigation. That is, in a caricature, you may get a lot of steaks out of a cow but you cannot get a cow out of a lot of steaks.

Such ideas of irreversibility and history-dependence are indeed quite intuitive and, as Paul David puts it, 'would not excite such attention nor require much explication, were it not for the extended prior investment of intellectual resources in developing economics as an ahistorical system of thought' (David 2001).[1] However, even after acknowledging that 'history matters' – and thus also that many socio-economic phenomena are *path dependent* – challenging questions still remain regarding when and in which fashions it does. In tackling path-dependent phenomena, an intrinsic difficulty rests also in the fact that in social sciences (as well as in biology) one generally observes only one of the many possible histories that some 'initial conditions' would have allowed. Moreover, is history-dependence shaped only by initial conditions, however defined? Or does it relate also to irreversible effects of particular unfolding of events? How do socio-economic structures inherited from the past shape and constrain the set of possible evolutionary paths? And finally, what are the factors, if any, which might de-lock socio-economic set-ups from the grip of their past?

In this chapter, partly drawing on other works by one of the authors (Dosi and Metcalfe 1991 and Bassanini and Dosi 2001) we discuss some of these questions.[2] In the next section we appraise the potential for path dependences and their sources at different levels of observation and within different domains. The following section presents a highly introductory overview of the different modeling tools one is utilizing in order to interpret the history-dependence of an increasing number of socio-economic phenomena. Next, in the penultimate section, we highlight some results and interpretative challenges concerning some path-dependent properties of socio-economic evolution. Finally, in the concluding section we discuss the factors underlying the tension, so to speak, between *freedom* and *necessity* in such evolutionary processes.

Sources of path dependence and irreversibilities

One indeed observes many potential causes for path dependence from the micro level all the way to system dynamics. Let us review a few of them.

For our purposes here, we refer to a broad definition of path dependence as dependence of the current realization of a socio-economic process on previous states, up to the very initial conditions.

Irreversibilities related to the decision-making of individual agents

Start by considering quite orthodox decision settings wherein agents hold invariant choice sets and preferences and are endowed with the appropriate decision algorithms. Suppose however that one of the following holds: (i) decisions are taken sequentially over time; (ii) they reflect uncertainty or imperfect information. Either of these conditions is sufficient for path dependence, in the sense that past decisions or past beliefs determine present and future decision processes.[3]

Individual learning

More generally, a powerful driver of self-reinforcing dynamics for individual agents or collections of them is any process of *learning*. If agents learn, their behaviours depend, other things being equal, also on their memory of the past, i.e. on initial conditions and on the history of their experience. This is a quite general property which holds irrespectively of the purported degrees of 'rationality' attributed to the agents themselves. So it is easily shown to hold under Bayesian learning whereby agents update expectations on some characteristics of the environment or on each other's features.[4] More so, path dependence applies under a wider class of learning processes whereby agents endogenously change also their 'models of the world', i.e. the very interpretative structures through which they process information from the environment (cf. the discussion in Dosi *et al.*, 1996). In all that, path dependence goes hand in hand with irreversibility: all agents with what they know now would not go back to yesterday's beliefs and actions even under yesterday's circumstances.

Local interactions

In many interactive circumstances one is likely to find that individual decisions are influenced by the decision of other agents, in ways that are not entirely reducible to price mechanisms.[5] One famous example concerns segregation phenomena. Suppose that an individual moves to a certain neighborhood only if at least a good proportion of his neighbors is of his same 'kind' (wealth, race, or other). If individuals are influenced by each other's decision in this fashion, homogenous neighborhoods tend to form. Hence, very rapidly, the housing configuration will lock into segregation of the different kinds of agents.[6] Similarly, another example of interdependence of preferences is provided by the way fashions, customs or conventions emerge. It suffices that individuals have some tendency to conform to the behaviour of people around them for a common behaviour to spread within the population of agents.[7]

Increasing returns

A quite general source of path dependence in allocation processes is associated with the presence of some form of *increasing returns* in production or in the adoption of technologies and products. The basic intuition is that production technologies (or collective preferences) in these circumstances entail *positive feedbacks* of some kind.

Recall for comparison the properties of *decreasing returns*, say, in production: in such a case, less input of something – for an unchanged output – means more input of something else, and, if returns *to scale* are decreasing, inputs have to rise more than proportionally with the scale of output. Conversely, under increasing returns, loosely speaking, 'one can get more with (proportionally) less' as a function of the scale of production or of the cumulated volume of production over time. In the economists' jargon increasing returns imply 'non-convex technologies'.

Non-convex production possibility sets may have different origins. They may stem from sheer physical properties of production plants. For example, in process plants output grows with the volumes of pipes, reaction equipments, etc., while capital costs tend to grow with the surfaces of the latter. Since volumes grow more than proportionally to surfaces, we have here a source of *static* increasing returns. Another example, still of 'static' kind, involves indivisibilities for some inputs (for example, minimum scale plants). The point for our purposes here is that under non-convex technologies history is generally not forgotten. The production system may take different paths (or select different equilibria) according to its very history.

The property is indeed magnified if one explictly accounts for the role of information/knowledge 'impactedness', of untraded interdependences amongst agents and of *dynamic* increasing returns.

Properties of information

The way information is distributed across different agents in a system, say a market or any other environment that provides the ground for economic interactions, together with the very properties of information, contribute to shape the consequences of economic interactions themselves.

As the seminal works of Arrow have highlighted (cf. the overview in Arrow, 1974) information is not an ordinary good which can be treated, say, like a machine tool or a pair of shoes. Shoes wear out as one uses them, while information has typically got a high up-front cost in its generation but can be used repeatedly without decay thereafter. Moreover information typically entails a non-rival use, in that it can be used indifferently by one or one million people. These properties entail decoupling of the costs of generation and the benefits of use of information. One could say that the cost of production of Pythagoras' theorem was entirely born by Pythagoras himself, while all subsequent generations benefited from it for free (except for their efforts to build their own *knowledge* enabling them to understand it).

At the same time, information (and more so knowledge[8]) might be appropriable in the sense that other agents might have significant obstacles to access it, ranging from legal protections, such as patents, all the way to sheer difficulty of fully appreciating what a particular piece of information means. This property exerts an influence opposite to the former ones in terms of incentives to profit-motivated investment in knowledge generation. Increasing returns in use and non-rivalry may produce 'under-investment' from the point of view of social usefulness, while conditions for appropriability may provide effective incentives for investment.[9] Together, path-dependent learning is influenced by the trade-off between 'exploitation' and 'exploration' (as Marcn, 1991 put it), that is between allocation of efforts to refining and exploiting what one already knows and investment in search for new potentially valuable information and knowledge (and equally important, by the belief agents hold about them).

Agglomeration economies

A number of case studies have posed the question of why specific production activities have concentrated in certain areas and not in others. The Silicon Valley and many other local industrial districts whose history is associated also with specific economic and technological activities, provide striking examples of 'agglomeration economies'. The common story starts from initial settlements and, possibly, some favorable conditions for specific activities. Then decisions to locate similar activities in the same region are re-inforced via (partly) *untraded interdependences* supported by spatial proximity. These may include stronger technological spillovers among producers (even when competitors), access to specialized labour force that tends to concentrate in the area and easier interactions with suppliers.[10]

Dynamic increasing returns

Technological innovation and diffusion are domains frequently displaying *dynamic increasing returns*, that is nonlinear and self-reinforcing processes that occur over time.[11]

The process of accumulation of technological knowledge

The processes of accumulation of technological knowledge typically display dynamic increasing returns: new knowledge cumulatively builds upon past one, and it does so in ways whereby in many circumstances yesterday's advances make today's improvements relatively easier.[12] The cumulativeness of technological learning is enhanced by the property of knowledge – as distinct from sheer information – of being partly tacit, embodied in the skills, cognitive frames and search heuristics of practitioners as well as in the collective practices of organizations.[13]

Moreover, as one of us argues elsewhere (Dosi 1982), technological innovations are often shaped and constrained by particular *technological paradigms* and proceed along equally specific *technological trajectories*. In all that, initial conditions – including the economic and institutional factors influencing the selection amongst alternative would-be paradigms – as well as possibly small seemingly 'random' events, affect which trajectories are actually explored. It is a story that one reconstructs at length in Dosi (1984) in the particular instantiation of silicon-based microelectronics, but it appears in different variants across diverse technologies.

Finally, throughout the whole process of establishment of new paradigms and the more incremental patterns of innovation thereafter, the emergence of networks of producers, suppliers, etc. together with other organizations (universities, technical societies, etc.) institutionalizes and so to speak 'solidifies' specific paths of technological learning.

The adoption of technology

Somewhat symmetrically, on the demand side of technological change, i.e. on the side of consumers and technology-users, a wide theoretical and empirical literature has emphasized the relevance of positive feedbacks: for the seminal explorations of the choice problem among alternative products that embody competing technologies, cf. Arthur (1994) and David (1985). Dynamic increasing returns and externalities appear to be at the core of the explanation of why the pool of users/consumers may select technologically inferior standards simply because that technology was the first to be chosen. Indeed, interpretations such as the foregoing ones place a good deal of path dependence weight, together with initial conditions, on 'historical accidents', i.e. more formally on small initial stochastic fluctuations that happen to determine the final outcome for the system. The story of the QWERTY typewriter keyboard is a famous one (David 1985). The keyboard was introduced in 1868. Alternative, more efficient, keyboards were brought to the market later, but did not succeed in replacing the initial one: QWERTY remained the dominant standard due to the 'lock-in' induced by the complementarity between installed base and specificities in the skills of the users.[14] Many other examples can be found when it comes to so-called 'network technologies' for which the issue of compatibility among the different components of the system is a crucial one (a discussion of some examples is in Bassanini and Dosi, 2001).

Properties of selection

Selection processes among heterogenous entities, both at the biological and economic levels, are another important source of path dependence. In the economic arena, selection occurs in multiple domains, concerning, e.g. products (and indirectly firms) on product markets; firms, directly, on financial markets; technologies, indirectly through the foregoing processes and directly via the social dynamics of inter-technology competition.[15] In fact selection processes may entail multiplicity of outcomes for the system[16] if different *traits* (or maybe, *genes*, in biology) – i.e. idiosyncratic characteristics of the composing entities of any agent – contribute in interrelated ways to the *fitness* (in biological terms) or to the *competitiveness* (in economic terms) of agents. This happens for example when there are complementarities between specific characteristics.

One way to represent the relationship between traits and fitness is in terms of *fitness landscapes*.

When the fitness contribution of every trait or gene is independent, the fitness landscape looks like a so-called *Fujiyama single-peaked landscape* (see Figure 8.1). Under the assumptions that higher fitness corresponds to evolutionary advantage in the selection process and that the biological or economic agents adapt in a fixed environment, then the system converges to the single maximum peak, whatever the rule of adaptation and whatever the initial condition.[17]

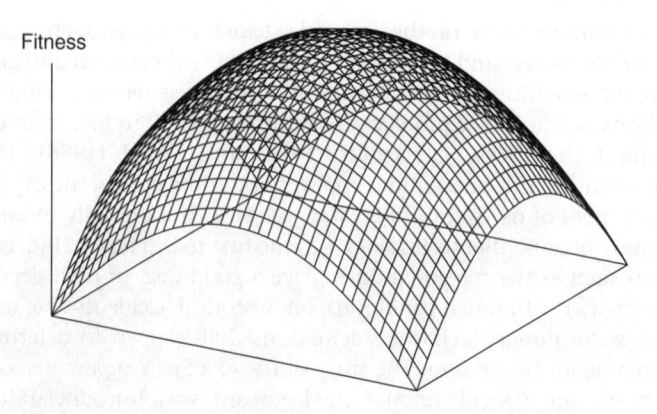

Figure 8.1 The Fujiyama single-peaked fitness landscape

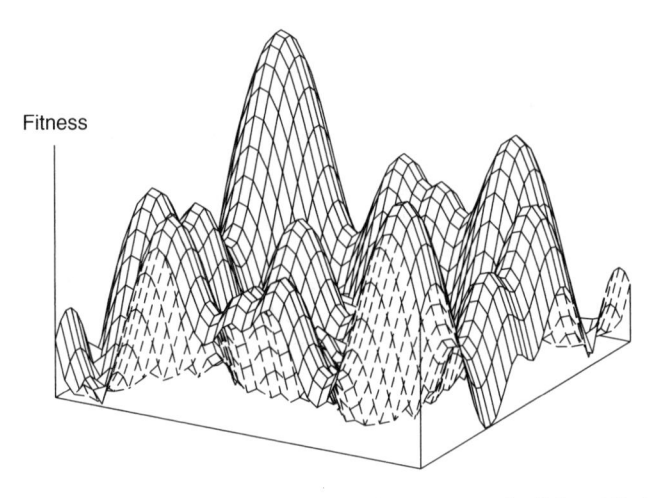

Figure 8.2 A fitness landscape with several local maxima peaks (Schwefel's function)

However, as soon as the fitness contributions of some traits depend on the contributions of other traits, i.e. *epistatic correlations* appear, then the fitness landscape becomes *rugged* and *multi-peaked*, as shown in Figure 8.2. In this case the initial positions in the landscape and the adaptation rules that underly the movement of individual entities in the landscape together determine which (local) peaks are going to be attained by the system.[18]

The nature of corporate organizations

Organizations typically compete on a rugged landscape because of complementarities in the organizational components that contribute to their 'fitness' (or 'competitiveness'). Adaptation over rugged competitive landscapes may often yield lock-ins into different fitness peaks. And, indeed, interrelated technological and behavioural traits are likely to be a primary cause of the path-dependent reproduction of organizational arrangements (Marengo, 1996; Levinthal, 2000).

More generally, an interpretatively challenging view of economic organizations (*in primis* business firms) depicts them as *history-shaped behavioural entities*, carriers of both specific problem-solving knowledge and of specific coordination arrangements amongst multiple organizational members holding (potentially) conflicting interests.

Individual organizations carry specific ways of solving problems which are often hard to replicate also because they have a strong tacit and partly collective component. Organizational knowledge is stored to a significant extent in the organization's *routines*,[19] that is in the operating procedures and rules that firms enact while handling their problem-solving tasks.

Relatedly, the accumulation of technological and organizational knowledge is, to a good degree, idiosyncratic and cumulative.

Business organizations may be viewed as entities which *imperfectly* evolve mutually consistent norms of incentive-compatible behaviours and learning patterns.

Together, (i) the complexity (and non stationarity) of the environments in which firms operate; (ii) multiple 'epistatic correlations' amongst behavioural and technological traits; and (iii) significant lags between organizational actions and environmental performance-revealing feedbacks, all contribute to render utterly opaque the link between what firms do and the ways they are selectively rewarded in the markets where they operate. After all, 'epistatic correlations' on the problem-solving side blur straightforward attributions of blames and credits ('... was it the R&D department that delivered the wrong template in the first place, or did the production department mess it up along the way? ...'). And so do far less than perfect spectacles interpreting environmental signals ('... are we selling a lot, notwithstanding some temporary fall in profitability, precisely because we are on the winning track, or just because we badly forgot the relation between prices and costs ... ?'). In these circumstances path dependence is likely to be fueled by both behavioural ('procedural') and 'cognitive' forms of inertia.

This is also another aspect of the fundamental 'exploitation/exploration' dilemma mentioned above. Within uncertain, ill-understood, changing environments, reasonably favorable environmental feedbacks are likely to reinforce the reproduction of incumbent organizational arrangements and behaviours, irrespectively of whether they are notionally 'optimal' or not.

Institutions

In fact these latter properties are part of a more general point which applies to many other formal organizations, in addition to business firms – e.g. public agencies, trade unions, etc. – and to many institutional arrangements including ethical codes, 'habits of thoughts', etc.[20] As argued by David (1994), *institutions* are one of the fundamental *carriers of history*. They carry history in several ways. First, they carry and inertially reproduce the architectural birthmarks of their origin and tend to persist even beyond the point when the conditions which originally justified their existence, if any, cease to be there. Second, they generally contribute to structure the context wherein the processes of socialization and learning of the agents and their interactions take place. In this sense, one could say that institutions contribute to shape the very fitness landscapes for individual economic actors and their change over time. Third, at least as important, they tend to reproduce the *collective perceptions and expectations*, even when their mappings into the 'true' landscapes are fuzzy at best. At the same time, fourth, institutions also represent social technologies of coordination: as argued by Nelson and Sampat (2001), they are a source of (path dependent) opportunities for social learning.

In brief, institutions bring to bear the whole constraining weight of past history upon the possible scope of discretionary behaviours of individual agents, and relatedly, contribute to determine the set of possible worlds which collective dynamics attain, given the current structure of any socio-economic system.

Such path-dependent properties are indeed magnified by the widespread *complementarities* amongst different institutions which make up the socio-economic fabric of particular countries: cf. the evidence and interpretations put forward from different angles, including 'institutionalist' political economy (Hollingsworth and Boyer, 1997; Hall and Soskice, 2001; Streeck and Yamamura, 2001), game-theoretic inspired institutional comparisons (Aoki, 2001) and historical institutionalism (North, 1990). A thorough discussion of political institutions is in Pierson (2000) who has recently argued that politics is characterized by a prevalence of specific ('political') versions of increasing returns. The major roots are traced, among others, in the collective nature of politics (making for the political equivalent of network externalities), in the complexity of political institutions and in the possibility of using political authority to enhance asymmetries of power. All this, together with the usually short time horizon of political actors and the inertia of political institutions, makes the cost of reversing a specific course of events particularly high, and thus tends to induce widespread lock-in phenomena.

To repeat, complementarities generally induce 'rugged' selection landscapes. So, at this level of analysis, there is no unequivocal measure of any particular mode of organization of e.g. labour markets or financial markets or State/business firm relations. Revealed performances depend on the degrees of complementarity between them. But the other side of the same coin is the frequent presence of 'local maxima' in the admittedly rather metaphorical space of institutional arrangements where countries path-dependently converge.

Take an example among many and consider the institutional arrangements governing national systems of innovation and production. A recent literature has rather convincingly argued that they are major ingredients in shaping growth patterns of different countries and their specialization in international trade (Lundvall, 1992; Nelson, 1993; Archibugi *et al.*, 1999). Moreover, an enormous literature, involving sociology, political science and the political economy of growth, has powerfully emphasized the inertial and self-sustained reproduction of institutions and organizational forms as determinants of specific growth patterns of different nations, showing variegated patterns of 'catching up, falling behind and forging ahead'.[21] Still, political and institutional lock-ins are almost never complete, and what appeared to be 'stable equilibria' for a long period, may be quickly disrupted by a sequence of strongly self-reinforcing, possibly surprising, events. So, even when looking at growth performances across countries, recent history has shown the rise of new (sometimes unlikely) actors in the international economic scene as well as the decline of seemingly unlikely others.

Indeed, secular comparisons between the fates of e.g. the UK, Germany, and the USA; Russia and Japan; Argentina and Korea; etc. entail major challenges to the analysts irrespectively of their theoretical inclinations. So, e.g. while there is hardly any evidence on long-term convergence patterns in e.g. technological capabilities, labour productivities, per capita incomes, etc. equally, there is no easy story on the 'drivers of convergence/divergence' that may be mindlessly applied across different countries. 'History' – both economic and institutional – in our view, most likely matters a lot. But it does so in ways that certainly go well beyond any naive 'initial condition' hypothesis. For example, Korea in the late 40s had educational levels, (population-normalized) capital stock, etc. comparable to the poorest countries in the world and certainly of orders of magnitude worse than Argentina, but also of India. Given that, what are the differences in the socio-economic processes and in their forms of institutional embeddedness, if any, which account for such striking differences in revealed performances?

From micro behaviours to system dynamics, and back

In this section we have tried to flag out a few of the very many likely sources of history- (or, equivalently, in our jargon here, path-) dependence. Some of them straightforwardly pertain to the dynamics of individual agents and, more metaphorically, individual organizations. Conversely, other properties have to do with *system dynamics*, i.e. they concern some properties of the dynamics of *collections of interacting agents*.

The relationship between the two levels of observation however turns out to be a tricky one. Admittedly, economists still do not know a great deal about all that. Two relatively robust properties appear however to stand out.

First, system dynamics is generally shaped by the characteristics, beliefs, expectations of micro actors, even when such beliefs are evidently at odds with any reasonable account of the environment wherein agents operate. Hence, there is often ample room for 'self-fulfilling' expectations and behaviours, obviously entailing multiple *expectation-driven* equilibria or dynamic paths. A good case to the point is the wide literature on 'sunspot equilibria'.[22] The punchline is the following. Suppose some agents in the system hold the view that some 'weird' variables (e.g. sunspots or, for that matter, patterns in beauty contest winners or football scores, etc.) bear lasting influences on economic dynamics. What will happen to the dynamics of the system itself? The answer (obviously overlooking here a lot of nuances) is that the dynamics, or analogously, the equilibrium selection will most often depend upon the distribution of beliefs themselves, no matter how 'crazy' they are.

A *second* robust result concerns aggregation and the general lack of isomorphism between micro- and system-level behaviours. So, e.g. distributions of stationary ('routinized') micro rules may well engender an apparently history-dependent dynamics as a sheer result of statistical aggregation over a multiplicity of agents (Forni and Lippi, 1997). In a similar vein, seemingly 'well behaved' relations amongst aggregate variables – e.g. between prices and quantities – are shown to be the outcome of sheer aggregation over heterogenous, budget-constrained, agents (Hildenbrand 1994).

For our purposes here, these latter properties imply also that one may conceive different combinations between 'flexible', reversible, micro behaviours and powerful system-level path dependences, and vice versa (one proposes a taxonomy in Dosi and Metcalfe, 1991).

Last but not least, note that one ought to account for the importance of *macro-foundations* of micro behaviours. Collective norms, institutions, shared habits of thoughts, etc. have a paramount importance in shaping micro 'mental models', preferences and behavioural patterns: in that, all history frozen in incumbent institutions exerts its self-reproducing effects.

Theoretical representations of path-dependent processes

In the history of the economic discipline one finds lucid early accounts of path-dependent increasing returns. Adam Smith's story on the 'pin factory' is a famous one. In brief, the efficiency of pin production grows with the division of labour, the degrees of mechanization of production and the development of specialized machinery, which in turn depend on the extent of the market, which in turn grows with production efficiency.

Indeed throughout the last two centuries a few seminal contributions have addressed positive feedback processes in knowledge accumulation and economic growth.[23] (Recall also that, as already noted, increasing returns are not necessary for the occurrence of path dependence.[24]) Nonetheless, it is fair to say that increasing returns *and* path dependence have been stubbornly marginalized by the mainstream of economic theory for reasons that is impossible to discuss here.[25] At the same time, a facilitating condition for such a lamentable state of affairs has been for long time the lack of formal instruments accounting for path-dependent processes. However, things have recently changed in this latter respect.

Formal tools

A set of powerful formal results in mathematical modeling tools has provided new ways of representing both nonlinear deterministic dynamics and stochastic ones. Let us provide here a brief, very simple, overview of some helpful formal tools.

Nonlinear dynamics and chaos

Suppose one can represent system dynamics through a *transition function f* that determines the value of the variable at time $t + 1$ in relation to its value at time t:

$$x_{t+1} = f(x_t) \tag{1}$$

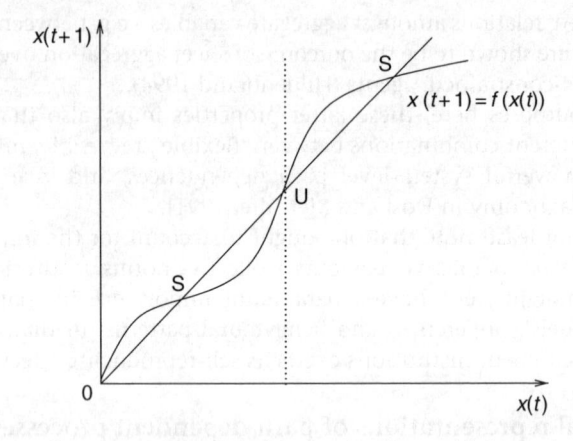

Figure 8.3 A nonlinear transition function that implies multiple steady states*

* Stable and unstable steady states are indicated with S and U, respectively.

Define a *steady state* as a point x^* for which $x^* = f(x^*)$, i.e. a point where the system settles. If the transition function is linear, there exists only one steady state (whether stable or unstable). Multiplicity of steady states occurs as soon as the transition function presents nonlinearities (an example is shown in Figure 8.3). In a deterministic setting, the steady state to which the system will eventually converge is going to be determined solely by initial conditions. An important property of this system is its full predictability. Given the initial condition and the transition function, one in principle knows the final state to which the system will get and also the exact path followed to reach it.

The growing understanding of the properties of nonlinear dynamic systems has brought new insights and tools of analysis.[26] Moreover, as widely shown in fields like physics, chemistry and molecular biology, nonlinear processes can result in 'self-organization' of systems as a far-from-equilibrium property. Highly complex behaviours can arise even with very simple transition functions, the best known example being the logistic function. Such systems however may be highly sensitive to small disturbances in initial conditions and display a multiplicity of patterns in their long-term behaviour. Arbitrarily small initial differences can result in cumulatively increasing differences in the historical trajectories *via* self-reinforcing dynamics. The best known examples are *chaotic* dynamics.

A definition of chaos rests on the sensitive dependence of the underlying dynamical systems on initial conditions, in the sense that arbitrarily small differences lead to increasingly divergent paths in the system dynamics. Hence, chaotic patterns are those whereby the path of the dynamical system is fully unpredictable in the long term, yet with a characteristic structure that differentiates it from a purely casual behaviour and allows short-term predictability.

Stochastic processes

A distinct potential source of path dependence in the dynamics relates to the impact of *ex-ante* unpredictable shocks occurring throughout the process. The property is captured by various types of stochastic models, possibly with time-dependence or state-dependence of probability distributions of shocks themselves.[27]

David (2001) provides two complementary definitions of path-dependent processes, namely:

> *A negative definition:* Processes that are non-ergodic, and thus unable to shake free of their history, are said to yield path dependent outcomes.
>
> ... *A positive definition:* A path dependent stochastic process is one whose asymptotic distribution evolves as a consequence (function of) the process' own history.

The key concept here is that of *ergodicity*. Intuitively, a process is ergodic if in the limit its underlying distribution is not affected by events happened 'along the way' (we provide a more formal definition in the Appendix). This means that in the long run, initial history does not affect the likelihood of the different possible states in which the system may end up. The opposite applies to *non-ergodicity*.

In the theory of stochastic processes, *Markov processes* provide a sort of benchmark for analysis. In a canonic Markov process, the 'transition probabilities' that define the dynamics of the system depend only on the current state of the system, regardless of the whole history of previously visited states. Think of the simple case of a 'random walk', which can be thought to describe the motion of a particle along a line. The particle can either jump up or down, with respective probabilities p and q. These transition probabilities characterize the motion from t to $t + 1$ and only look at one time step, irrespectively to the previous positions occupied by the particle.[28] Conversely, history of previous events is relevant in non-Markovian processes, which have been exploited to model path-dependent economic phenomena. We provide two illustrative examples that also relate to previously discussed sources of path dependence.

The first example concerns *Polya urn processes*.[29] Arthur et al. (1983) have utilized them to model increasing returns in adoption of alternative technologies when the incentive to adopt each technology depends on the number of previous adopters. The setting involves an urn containing balls of different colors. Basically, one can think of different colors as alternative technologies. Each agent draws a ball and then inserts a ball of the same color back in the urn. Then every time that a technology is chosen, the probability that the same technology is chosen at the next time step increases. One can then prove that under rather general conditions the limit state of the system is the dominance of one of the technologies. Being the limit state an absorbing one, this formally defines a process of lock-in in one technological monopoly. The second example is provided by the so-called *voter model*.[30] The model entails random local interactions between agents in a finite population. The basic idea is that agents vote depending on the voting frequencies of their 'neighbours'. In one and two dimensional spaces, it can be proved that the system clusters into an homogenous setting where all agents vote for the same party. Local positive feedbacks represent the key for explaining locking in this irreversible state. At the same time, initial conditions and the particular unfolding of micro-choices determine which of the states is attained within any one 'history' (David, 2001 discusses the socio-economic importance of the model).

Understanding path dependence in economic evolution: some results and challenges

The foregoing examples of formal modeling of path-dependent economic processes ought to be taken as promising even if still rather rudimentary attempts to grasp some fundamental properties of economic dynamics. They certainly fall short of any thorough account of socio-economic evolution (compare for example the 'grand' evolutionary research program as outlined in Dosi and Winter, 2002) which in turn builds upon the seminal Nelson and Winter, 1982). However, they already offer precious insights, and together, interesting interpretative puzzles.

Degrees of history dependence and their detection

David (2001) offers the following categorization of the degree of 'historicity', i.e. of the strength of the influence of the past in economic dynamics:

> *weak history* goes so far as to recognize 'time's arrow' (the rooted sense of difference between past and present) ... ;
>
> *moderate to mild history* acknowledges that instantaneous transitions between discrete states have high and possibly infinite adjustment costs, so that it would take time and a sequence of motions to attain a terminal state (family size, capital stock, reputation, educational or skill level) – whence we have the notion of a dynamic path being an object of choice;
>
> *strong history* recognizes that some dynamical systems satisfy the conditions for path dependence of outcomes, or of transition probabilities and asymptotic distributions of outcomes. (italics added)

Within the broad class of processes displaying *some* forms of history dependence, *how much* does history actually matter? Which one of the foregoing 'degrees of historicity' apply to which phenomena?

In order to address such questions, a major methodological issue immediately springs up. Social scientists (to repeat, as well as biologists) most often observe just one historical path. When very lucky, evolutionary biologists neatly catch – as Darwin was able to do – just some independent (on biological scales, rather short) branches of the same evolutionary process. Social scientists find it even harder to observe and compare alternative sample paths. Hence, how can one be sure that what one seemingly detects in the actual history was not the only feasible path given the system constraints?

For sure, anthropologists have a rich comparative evidence, but it is very hard to bring it to bear on the issues of path dependence discussed here. A comparatively more modest, albeit still daunting, task concerns the analysis of technological and institutional dynamics within the domain of modern, mostly 'capitalist', history. Have they been the only feasible paths given the system constraints? Or, conversely, can one think of other dynamics – notionally feasible on the grounds of initial conditions – whose exploration has been ruled out by any actual sequence of events?

Even more specifically, why have historically observed technologies been chosen? Were they 'intrinsically' better in ways increasingly transparent to the involved actors? Or, conversely, did they become dominant as a result of multiple (mistake ridden) micro decisions, piecemeal adjustments, co-emergence of institutional structures, etc. – irrespectively of the presence of notionally 'superior', relatively unexplored activities? (Of course similar questions apply to the emergence and persistence of particular institutions, forms of corporate organizations, etc.)

Competing answers to this type of questions and competing methodologies of investigation clearly fold together.

At one extreme, a style of interpretation focuses on the final outcome of whatever process, and – when faced with notional, seemingly 'better' alternatives – it tries to evaluate the 'remediability' of the *status quo*. Under high or prohibitive remediation costs and in absence of striking 'irrationalities' along the past decisions history, one next declares the absence of path dependence. This extreme view tries to justify and explain any end-state of the system as being the best possible outcome given the (perceived) constraints by imperfectly informed but fully 'rational' agents along the whole path. The view, emphatically illustrated in Liebowitz and Margolis (1990, 1995)[31] basically aims at rationalizing whatever one observes as an equilibrium and, at the same time, at attributing rational purposefulness to all actions which led to any present state.

On all that, David (2001) and Dosi (1997) coincide in the skepticism about any Panglossian interpretations of history as 'the best which could have happened', mainly 'proved' by the argument that 'rational agents' would not have allowed anything short of the optimal to happen (compared with Voltaire's *Candide* on the virtues of Divine Providence).

Conversely, a distinct perspective rather bravely tries to face the *challenge of counter-factuals* ('... what would have happened if ...'): hence it focuses on the actual thread of events, on the possible amplification mechanisms linking them, and together, on the (varying) potential leverage that individuals and collective actors retain of influencing selection amongst future evolutionary paths.

A good part of such exercises is inevitably 'qualitative', based on case studies, circumstantial comparisons across firms and countries displaying different evolutionary patterns, etc.[32] However, complementary investigations address some path-dependent properties on more quantitative grounds, concerning, e.g. real and financial time series.

A complementary task: detecting nonlinearities

Early examples of statistical tools devised for detecting forms of path dependence are those trying to detect chaos. While *chaos* can be easily obtained out of economic models, not much supporting evidence has been collected so far. Limitedly to high-frequency financial time series, there is some evidence of chaotic behaviour,[33] but there is hardly evidence for other economic series.[34] Brock *et al.* (1991) formally test for chaos in a number of economic and financial datasets. This is done by applying the 'BDS' test, presented in Brock *et al.* (1996) in order to detect low-dimensional chaos. At the same time, as Brock (1993) himself critically discusses '... chaos is a very special species of nonlinearity ...'; so that it is misleading '... to conclude that weak evidence for chaos implies weak evidence for non linearity' (p. 7).

Moreover, note that the apparent linearity/predictability/lack of path dependence on some time scale does not rule out 'deeper', possibly 'slower' path-dependent dynamics. So, for example, under conditions resembling what in biology are called 'punctuated equilibria' (Elredge and Gould 1972) phase transitions between apparent steady states might occur infrequently, rather abruptly, unpredictably triggered by particular chains of events,[35] while possibly still leaving linear structures of the time series in the (quasi) equilibrium phases.

Path dependence in economic evolution

Granted all that, how do such different degrees of path dependence show up in the 'grand' evolutionary interpretation of economic change? The latter, as outlined in much greater detail in Coriat and Dosi (1998a) and Dosi and Winter (2002), entails at the very least as fundamental building blocks:

(i) heterogenous, 'boundedly rational' but innovative agents;
(ii) increasing returns in knowledge accumulation;
(iii) collective selection mechanisms, including of course market interactions;

(iv) multiple forms of social embeddedness of the processes of adaptation, learning and selection.

The late S. Gould has reminded us (originally addressing evolution in the biological domain) that an illuminating angle to interpret evolutionary dynamics is by trying to identify what would remain unchanged if 'the tape of evolution would be run twice' (1977). It is indeed a very challenging question for social scientists too.

Needless to say, 'running the tape' all over again most likely would change the identities of who is 'winning' or 'losing'; who survives and who does not; who is getting at the top and who is getting at the bottom of the social ladder. However, this should not come as such a big surprise. After all, it is much more plausible to think of *system-level* path-*independent* or path-dependent equilibria, irrespectively of individual destinies.

Hence, what about system-level dynamics? One must sadly admit that evolutionary arguments have too often been used as *ex post* rationalizations of whatever observed phenomena: again, the general belief is that 'explaining why something exists' has too often meant showing why in some appropriately defined space 'whatever exists is a maximum of something' and this is the reason why it inevitably exists.

Quite a few applications of 'evolutionary games' to both economics and biology are dangerously near such an interpretative archetype. And so is a good deal of 'socio-biology' (a bit more of a discussion by one of us in Dosi and Winter, 2002). Admittedly, even relatively sophisticated evolutionary interpretations of economic change tend to overlook the possible history dependence of specific evolutionary paths.

Conversely, a few (mostly qualitative) analyses – already mentioned above – of the properties of national systems of production and innovation and of the 'political economy' of growth powerfully hint at underlying path dependencies. And the conjecture is indirectly corroborated by several formal results, from different theoretical fields. They include, as already mentioned, path-dependent selection amongst alternative institutional arrangements (Aoki, 2001), models of selection amongst alternative technologies, and also path dependencies in the statistical properties of growth processes of stylized industries and economies – even under unchanged initial conditions – (cf. Winter *et al.*, 2000) and Dosi and Fagiolo, 1998).

Certainly, most of the work of exploration of the possible properties of history dependence in incumbent models of economic evolution still awaits to be done. At the same time, an equally urgent task regards the development of broader interpretative frameworks explicitly addressing *hierarchically nested evolutionary processes*, allowing for e.g. (on average, slowly changing) institutions which in turn structure (on average, faster) dynamics of social adaptation, technological explorations, etc. (A fascinating template of such an exercise concerning biology is presented in Fontana, this volume.)

Selected histories might be quite 'bad': the painful acknowledgment of the distinction between interpretative and normative analyses

'Evolution' as such, both in the biological and socio-economic domains, does generally involve at least 'weak to mild' history dependence – in the foregoing definitions. However, a much trickier question regards the properties of those very evolutionary processes as judged against any normative yardsticks. Does 'evolution' entail some notion of 'progress' in some appropriately defined space?

As already mentioned, the general notion of history dependence of any socio-economic process is in principle quite separate from any normative evaluation of the 'social quality', however defined, of the outcomes which history happens to choose. As David (2001) argues in detail, one may think of quite a few circumstances easily involving multiple *neutral* equilibria or paths which turn out to be (roughly) equivalent in normative terms. On formal grounds, the original Polya urn example is a good case to the point. The whole set of reals on the interval [0, 1] happens to be fixed points satisfying the condition $f(x) = x$, where x may stand for frequencies of e.g. technologies, behaviours, strategies, organizational forms, etc. and $f(x)$ for the probabilities of their social adoption, without any distinct normative feature attached to them.

Further suggestive examples come from biology hinting at the widespread occurrence of *neutral drifts* in the genotypical space mapping into diverse but fitness-equivalent phenotypical structures.

At the same time, it equally holds that many path-dependent processes do entail the possibility of lock-in into equilibria or paths which are 'dominated' in normative terms (i.e. intuitively are 'socially worse'), as compared to other notionally 'better' ones which *could have been explored* given some initial conditions but ultimately turn out to be unreachable under reasonable switching costs at later times[36] (the argument is forcefully presented in Arthur, 1994). In this respect, a few analyses have focused so far upon rather simple cases of choices amongst technologies and social conventions, often highlighting the path-dependent properties of the underlying selection processes. However, the relevance of path-dependent selection of relatively 'bad' institutional set-ups and technologies remains a highly controversial question. One inclination is to depart from any naive notion of evolutionary dynamics leading – notwithstanding painful detours and setbacks – 'from worse to better'. In many respects such a 'progressive' view is shared by a whole spectrum of scholars, ranging from Karl Marx to contemporary neoclassical economists.[37] Empirically, as Nelson (2002) has recently suggested, it may well be that 'physical technologies' tend to often display more 'hill-climbing' features as compared to 'social' technologies, due to our relatively higher ability in the former domain to test hypotheses and codify solutions to the problems at hand. So, for example, while, electricity or antibiotics or vaccines happened to be rather uncontroversial technological advancements, one seldom finds crisp matching examples in the social domain.

Come as it may, history is full of cases of collective dynamics irreversibly leading *from better to worse*: in our view, the story of Easter Island vividly depicted in Diamond (1995), far from being an odd outlier, is indeed an archetype of common processes of transitions to worse and worse coordination equilibria. The decadence of many civilizations probably belongs to that same class of collective dynamics: institutions and microbehaviours coevolve in ways such as to yield recurrent transitions to 'worse and worse' social arrangements.

Locking and de-locking: some conclusions on the tension between freedom and necessity

It follows from our foregoing discussion that two somewhat complementary mechanisms are always at work. On the one hand, specific histories of competence-building, expectation formation, emergence of particular organizational structures, etc. together yield relatively *unique* and hence *heterogenous* micro histories.

On the other hand, broader mechanisms of alignment of individual and organizational decisions, together with convergence to dominant technologies and institutions, tend to reduce such a diversity among agents and bring about relative consistency of behaviours, practices, expectations.

More precisely, mechanisms at the heart of aggregate 'coherence' include: (i) social adaptation by individual actors; (ii) the path-dependent reproduction of a multiplicity of institutions governing interactions amongst agents; (iii) selection mechanisms (comprising of course market-selection dynamics). These processes contribute to explain locking into specific 'socio-economic paths'. But lock-ins seldom have an absolute nature: the unfolding of history, while closing more or less irremediably opportunities that were available but not seized at some past time, is also a source of new 'windows of opportunities' – using again Paul David's terminology – which allow de-locking and escaping from the past.

Let us outline some of the forces that work as potential factors of 'de-locking'.

First, a straightforward mechanism of 'de-locking' is related to *invasions*. They can be literal ones as it has often happened with past civilizations, and also more metaphorical ones, i.e. the 'contamination' with and diffusion of organizational forms, cultural traits, etc. originally developed elsewhere. Sticking just to organizational examples, think of, for example, the worldwide 'invasion' of Tayloristic principles of work organization – originally developed in the US, or more recently the diffusion of 'Japanese' management practices. *Second*, social adaptation is never complete, at least in modern societies. However, precisely the gap between social norms and prescribed roles, on the one hand, and expectations, 'mental models', identities which agents actually hold, on the other, may be an extremely powerful source of 'unlocking' dynamics: within an enormous literature, see the fascinating comparative analysis of the riots of obedience and revolt by Moore (1978), and also, from the economists' camp, the formal explorations of some implications of 'cognitive dissonance' including Akerlof and Dickens (1982) and Kuran (1987, 1991). *Third*, and relatedly, non-average ('deviant') behaviours may well entail, under certain circumstances, what natural scientists call 'symmetry breaking' and phase transitions to different collective structures (Allen 1988).

More generally, *fourth*, a fundamental role in preventing irreversible socio-economic lock-ins is played by various forms of heterogeneity among agents – in terms of e.g. technological competences, behavioural repertoires, strategies, preferences, etc.[38]

Fifth, 'de-locking' possibilities might be a byproduct of those very mechanisms which tend to induce path dependence in the first place. Indeed many organizational forms, behavioural patterns, etc. tend to be selected over multiple selection domains, possibly characterized by diverse selection criteria. So, for example, as one argues in Coriat and Dosi (1998a), organizational routines entail possibly uneasy compromises between their problem-solving efficacy and their properties in terms of governance of conflicting interests. Complementarity of functions, as discussed above, is likely to induce multiplicity of equilibria and path dependence. However, such equilibria may well be 'meta-stable', entailing the possibility of de-locking induced by increased inadequacies in some of the affected domains (i.e. over some 'selection landscapes'). A good case to the point is the increasing mismatching between formal hierarchies, incentives and actual decision powers driving toward the collapse of centrally-planned economies (Chavance 1995).

Finally, *sixth*, a major de-locking force has historically been the emergence of radical technological innovations, new knowledge bases, new sources of technological opportunities (i.e. what one calls in Dosi, 1982, new technological paradigms): on the powers of 'Unbound Prometheus' of technological change cf. the seminal works of Landes (1969), Freeman (1982) and Rosenberg (1976). Ultimately, as we have tried to argue in this chapter, human affairs always involve a tension between the tyranny of our collective past and the apparent discretionality of our wills. Admittedly, one is still rather far from getting any robust understanding even of the basic mechanisms underlying this tension. However, it is a fundamental exercise if one wants to handle the uncountable problems of collective action we continously face and try to (imperfectly) shake free of the grip of the past and shape our future.

Appendix

Ergodicity in stochastic processes

Take the family of *Markov processes* as baseline reference for the stochastic processes of interest here. Given the set of possible states in which the system may find itself, one is in general interested in the probability distribution over the states, possibly different at different points in time. For (time invariant) Markov processes:

$$Pr(X_t = y | X_{t-1} = x, \cdots, X_0 = x_0) = Pr(X_t = y | X_{t-1} = x) = p_{x,y} \qquad (2)$$

i.e. the probability of being in state x at time t conditional on all states visited in the past reduces to the probability conditional only on the state visited in the previous time $t - 1$. The probability $p_{x,y}$ is called the transition probability from state x to state y[39] and together with the distribution on the initial states, fully determines the joint (unconditional) probability distribution over the set of possible states. Moreover, one can partition the set of all possible states into *transient* and *recurrent* states depending on the probability that the stochastic process returns to the states after a first visit. A stochastic process is *ergodic* if one can obtain a probability distribution over the recurrent states that in the limit does not depend on the initial state of the system.

Non-Markovian processes do not satisfy condition (2), implying that the whole path of previously visited states is relevant in determining the probability of finding the process in a specific state at any given time.

Polya urn processes

Assume an urn of infinite capacity containing balls of two colours, say, white and black. At every draw a number c of balls of the same color as the drawn ball is added to the urn. (In the generalized urn scheme there are k different ball colours.) If c is greater or equal to one, the process entails positive feedbacks: if a color is drawn once, then the likelihood of drawing that same colour at the next time step is higher. If $c = 0$ the process reduces to independent Bernoullian draws, when c is negative the process accounts for negative feedbacks. It can be proved that when $c \geq 1$ such a process converges with probability one to the dominance of one single colour of the balls. This limit state is *absorbing*, meaning a zero probability of leaving it.[40] In the case of an urn with two colors, let X_t be the proportion of white balls in the urn at time t. Consider the case when one ball is added into the urn at time steps $t = 1, 2, \ldots$. The probability that the new ball is white is a function of the share X_t, say $f_t(X_t)$, where $f_t : [0,1] \to [0,1]$. The new ball is then black with probability $1 - f_t(X_t)$.

One can then represent the dynamics of X_t as

$$X_{t+1} = X_t + \frac{\xi_t(X_t) - X_t}{t + n} \quad \text{with } t \geq 1 \quad \text{and} \quad X_1 = \frac{n_w}{n} \qquad (3)$$

where n_w is the proportion of white balls at the initial time and ξ_t are independent random variables with binary outcome defined by

$$\xi_t(X_t) = \begin{cases} 1 \text{ with prob } f_t(X_t) \\ 0 \text{ with prob } 1 - f_t(X_t) \end{cases} \qquad (4)$$

$f_t(X_t)$ represents the average of $\xi_t(X_t)$. Call $\psi_t(X_t) = \xi_t - f_t(X_t)$ the difference between $\xi_t(X_t)$ and its mathematical expectation, so that $E[\psi_t(X_t)] = 0$. Then we can rewrite Eq. 3 as

$$X_{t+1} = X_t + \frac{[f_t(X_t) - X_t] + \psi_t(X_t)}{t+n} \quad (t \geq 1)$$

Under this formulation the realization of the process at time $t + 1$ is given by the realization at time t plus a term with two components. The first component, $f_t(X_t) - X_t$, is a systematic one, the second component is the zero-mean noise $\psi_t(X_t)$. Then the limit points of the sequence X_t have to belong to an appropriately defined set of zeros of the function $f_t(x) - x$ for $x \in [0, 1]$.

The outlined case is indeed the most general one, without conditions on the continuity of the f function. For discussions and formal proofs of the limit results see Arthur et al. (1983), Dosi and Kaniovski (1994), Hill et al. (1980).

Replicator dynamics

Evolutionary theory developed in biology rests on two main elements: (i) perpetual generation of novelty; (ii) selection of 'superior' species, given heterogenous populations.

The original mathematical representation of the selection mechanism via the so-called *replicator dynamics*, is through the Fisher equation (more on economic applications in Metcalfe, 1998). Assume the existence of n types of entities in the population. Call x_i the fraction of the population of type i and call F_i its fitness. Then the Fisher equation in the linear continuous simplification reads:

$$\dot{x}_i = c x_i [F_i - \bar{F}]$$

where $\bar{F}$ is the weighted fitness average in the population:

$$\bar{F} = \sum_i x_i F_i$$

The way such a simple version of the replicator dynamics operates is such that the relative frequency of types with higher-than-average fitness grows, while the proportion of types characterized by below average fitness shrinks. If the fitness measure is constant over time, the system is bound to converge to the dominance of the fittest type. More general formulations allow for non linear interactions amongst traits which contribute to overall fitness and for changes of fitness landscapes themselves (cf. Silverberg, 1988 for a discussion of various selection models).

Notes

* This work is part of an ongoing research project involving from the start Andrea Bassanini. Support by the Sant'Anna School and by the Center for Development Research (ZEF), Bonn is gratefully acknowledged. We also wish to thank comments by Giulio Bottazzi, Uwe Cantner, Ping Chen, Paul David and Willi Semmler.
† ECIS, Eindhoven Center for Innovation Studies, Eindhoven, The Netherlands.
‡ LEM, Sant'Anna School of Advanced Studies, Pisa, Italy.

1. Indeed, it is difficult to find *purely* ahistorical representations even in mainstream economics, except from some breeds of economic theory such as rational expectations or general equilibrium theories.
2. Detailed discussions of some of the issues tackled in this chapter are in Arthur (1994), David (1988, 2001), Freeman and Louçã (2001), Hodgson (2001) and Witt (2005).
3. Of course, path dependence holds, *a fortiori*, if preferences are themselves endogenous (cf. the discussion in Aversi *et al.* (1999) and the references therein). In these circumstances past events irreversibly change the decision criteria agents apply even under an invariant choice set and invariant information from the environment.
4. Within the enormous literature, cf. Kreps and Spence (1985), Hahn (1987), Arthur and Lane (1993).
5. For interaction-based models cf. Brock and Durlauf (2001a), which reports a variety of empirical examples.
6. Cf. Schelling (1971).
7. For formal models in different perspectives cf. Föllmer (1974), Bikhachandani *et al.* (1992) and Young (1998).
8. The distinction between the two is discussed in Dosi *et al.* (1996): see also the references therein.
9. Incidentally note however that the latter investment might turn out of a socially pernicious kind, trading off relatively small private rents against huge collective losses in knowledge accumulation. The current lamentable legal arrangements on so-called Intellectual Property Rights (IPR) are an excellent case to the point (a sophisticated discussion of IPR is in Arora *et al.*, 2001), while a more sanguine but convincing illustration is presented in Coriat and Orsi, 2002).
10. The wide literature includes a variety of models and explanations that also assign different relevance to the initial conditions: see, among others, Krugman (1991a) and (1996), Arthur (1994), Fujita *et al.* (1999).
11. Cf. Dosi (1988) for a detailed discussion of the properties of technological knowledge.
12. This is not to say of course that some forms of 'decreasing returns' never endanger knowledge accumulation. Intuition suggests immediately a few historical cases where technological opportunities appear to progressively shrink. However, at a closer look, what generally happens is that increasing returns may well tend to dry out, but one is still a long way from decreasing returns setting in (that is, more formally, one is basically talking about the properties of second derivatives). This applies even in the case of all those resource-based activities such as agriculture and mining which have been for more than two centuries the menacing reference of the mainstream in the economic discipline.
13. A partly overlapping idea is that learning is typically *local*, in the sense that what agents learn tends to be 'near' what agents already know: cf. the pioneering models by Atkinson and Stiglitz (1969) and David (1975) and within the subsequent literature, Antonelli (1995), among others. All this admittedly involves a highly metaphorical notion of 'nearness', since we still fall short of any robust topology, or anything resembling it, in the space of knowledge.
14. The QWERTY story reports on initial events that may constrain long-term outcomes. In a different example, David (1992) vividly reports about the individual role played by Thomas Edison in the early battle to win dominance in electricity supply market and discusses in general the power of intentionality in determining historical paths. See also the discussion in Section 5.

15. One of the formal representations of such competitive processes is through so-called replicator dynamics: cf. Silverberg (1988), Weibull (1995), Metcalfe (1998), Young (1998), and the pioneering Winter (1971) (we offer a basic intuition in the Appendix).
16. Using the terminology that will be more formally defined in the next section, we can define this property in terms of 'multiplicity of equilibria'.
17. Even in this case, things might not be so simple. For example, the irrelevance-of-initial-conditions property may well turn out to rest on very demanding assumptions, including the presence of the 'best' combination of traits from the very start and its survival throughout the ('disequilibrium') process of adaptation/selection. For an insightful discussion cf. Winter (1975).
18. A general introduction to 'rugged landscape' formalizations is in Kauffman (1989).
19. Cf. Nelson and Winter (1982), Cohen *et al.* (1996), Coriat and Dosi (1998a), Dosi *et al.* (2000), among others.
20. More detailed discussions of the nature of 'institutions' by one of us are in Dosi (1995) and Coriat and Dosi (1998b).
21. For some stylized facts, cf. Abramovitz (1986), Dosi *et al.* (1994) and Meliciani (2001), among others.
22. The original reference is Cass and Shell (1983); for a recent survey cf. Benhabib and Farmer (1999).
23. Outstanding examples include A. Young, N. Kaldor and G. Myrdal. See the discussion in Arthur (1994), ch. 1.
24. Cf. the thorough discussions in David (1988, 1993, 2001). Moreover, in Bassanini and Dosi (2001) one shows that under certain conditions increasing returns are neither sufficient for path dependence (see also below, the final section).
25. On different facets of the epistemology of a paradigm which, for a long time, has stubbornly focused on the properties of history-independent equilibria, cf. Freeman and Louçã (2001), Hodgson (2001) and Nau and Schefold (2002).
26. Cf. Haken (1981), Prigogine (1980), Prigogine and Stengers (1984), Brock and Malliaris (1989), Rosser (1991).
27. A good deal of the formal tools can already be found in classics such as Feller (1971) and Cox and Miller (1965). However their economic application (with some significant refinements) is a more recent phenomenon.
28. An important feature is however worth mentioning: even for the simplest random walk the state at time *t* embodies the full memory of all shocks which drove it from its very beginning.
29. See the appendix for some formal definitions and results.
30. For details, cf. the original model Holley and Liggett (1975) and Liggett (1999).
31. For critical assessment see David (2001) and Dosi (1997).
32. An interesting exercise, involving a few respected historians is Cowley (1999).
33. For a survey of the literature on chaos in macroeconomics and finance see LeBaron (1994), and previously Kelsey (1988).
34. A more optimistic view on the pertenance of chaotic dynamics for economic phenomena is in Chen (1993, 2005).
35. An insightful germane discussion, building on the 'long waves' debate on 'economic growth' is in Freeman and Louçã (2001).
36. Formally, one can show that asymptotic switching costs may well be infinite, *under dynamic increasing returns* even from an 'inferior' to a (notionally) 'superior' technology/organizational form, etc.

37. Indeed, many economists, even among the most sophisticated ones, are inclined to read history as a painstaking process driving – notwithstanding major setbacks – toward market ('capitalist') economies: see for example Hicks (1969).
38. In fact heterogeneity of agents can help also in explaining why 'locking' might not occur: instead one might observe market sharing of different technologies or organizational forms. As shown in Bassanini (1999) and Bassanini and Dosi (2005), convergence to monopoly of a technology, an organizational form, etc. may in fact not occur even under conditions of increasing returns if the degree of heterogeneity of agents is high enough. Similarly Herrendorf *et al.* (2000) prove that heterogeneity of agents is a condition for avoiding multiple or indeterminate equilibria in GE models.
39. Here we take the transition probability to be time-invariant. One could generalize to time-dependent probabilities.
40. For an overview of the state-of-the-art in generalized urn schemes and hints on their economic applications, cf. Dosi and Kaniovski (1994).

References

Abramovitz, M. (1986) 'Catching up, forging ahead and falling behind', *Journal of Economic History*, 86, pp. 385–406.

Akerlof, G.A., Dickens, W.T. (1982) 'The economic consequences of cognitive dissonance', *American Economic Review*, 72, pp. 307–19.

Allen, P.M. (1988) *Evolution, Innovation and Economics*, in Dosi *et al.* (1988), pp. 95–119.

Antonelli, C. (1995) *The Economics of Localized Technological Change and Industrial Dynamics*, Boston: Kluwer Publishers.

Antonelli, C., Foray, D., Hall, B., Steinmuller, E. (eds) (2005) *New Frontiers in the Economics of Innovation: Essays in Honor of Paul David*, Cheltenham, UK and Northampton, MA: Edward Elgar.

Aoki, M. (2001) *Toward a Comparative Institutional Analysis*, Cambridge: MIT Press.

Archibugi, D., Howells, J., Michie, J. (eds) (1999) *Innovation Policy in a Global Economy*, Cambridge: Cambridge University Press.

Arora, A., Fosfuri, A., Gambardella, A. (2001) *Markets for Technology: The Economics of Innovation and Corporate Strategy*, Cambridge, MA: MIT Press.

Arrow, K. (1962a) 'Economic welfare and the allocation of resources for innovation', in R.R. Nelson (ed.), *The Rate and Direction of Inventive Activity*, Princeton: Princeton University Press.

Arrow, K. (1962b) 'The economic implications of learning by doing', *Review of Economic Studies*, 29, pp. 155–73.

Arrow, K. (1974) *The Limits of Organization*, New York: Norton.

Arthur, W.B. (1988) 'Competing technologies: an overview', in G. Dosi *et al.* (eds), *Technical Change and Economic Theory*, London: Pinter Publisher.

Arthur, W.B. (1994) *Increasing Returns and Path Dependence in the Economy*, Ann Arbor: University of Michigan Press.

Arthur, W.B., Ermoliev, Y.M., Kaniovski, Y.M. (1983) 'A generalized urn problem and its applications', *Kibernetika*, 19, 49–56 (republished in Arthur 1994).

Arthur, W.B., Lane, D.A. (1993) 'Information contagion', *Structural Change and Economic Dynamics*, 4, pp. 81–104 (republished in Arthur 1994).

Atkinson, A.B., Stiglitz, J.E (1969) 'A new view of technological change', *Economic Journal*, 79, pp. 573–8.

Aversi, R., Dosi, G., Fagiolo, G., Meacci, M., Olivetti, C. (1999) 'Demand dynamics with socially evolving preferences', *Industrial and Corporate Change*, 8, pp. 353–408.

Banerjee, A.V. (1992) 'A simple model of herd behavior', *Quarterly Journal of Economics*, 107, pp. 797–818.

Bassanini, A.P. (1999) *Can Science and Agents' Diversity Tie the Hands of Clio?: Technological Trajectories, History, and Growth*, mimeo, OECD, Paris.

Bassanini, A.P., Dosi, G. (2000) 'Heterogeneous agents, complementarities, and diffusion: do increasing returns imply convergence to international technological monopolies?', in D. Delli Gatti, M. Gallegati and A. Kirman (eds), *Market Structure, Aggregation and Heterogeneity*, Berlin: Springer.

Bassanini, A.P., Dosi, G. (2001) 'When and how chance and human will can twist the arms of Clio', in Garud and Karnoe (eds) (2001).

Bassanini, A.P., Dosi, G. (2005) 'Competing technologies, technological monopolies, and the rate of convergence to a stable market structure', working paper, forthcoming in Antonelli *et al.* (eds) (2005).

Benhabib, J., Farmer, R.E.A. (1999) 'Indeterminacy and sunspots in macroeconomics', in J. Taylor and M. Woodford (eds), *Handbook of Macroeconomics*, Amsterdam: Elsevier Science.

Bikhchandani, S., Hirshleifer, D., Welch, I. (1992) 'A theory of fads, fashion, custom, and cultural change as informational cascades', *Journal of Political Economy*, 100, pp. 992–1026.

Brock, W.A. (1993) 'Pathways to randomness in the economy: emergent nonlinearity and chaos in economics and finance', *Estudios Economicos*, 8, pp. 3–54.

Brock, W.A., Dechert, W.D., Scheinkman, J.A., LeBaron, B. (1996) 'A test for independence based on the correlation dimension', *Econometric Reviews*, 15, pp. 197–235.

Brock, W.A., Durlauf, S.N. (2001a) 'Interactions-based models', in J.J. Heckman and E. Leamer (eds), *Handbook of Econometrics*, Vol. 5, ch. 54, pp. 3297–380, Elsevier Science B.V.

Brock, W.A., Durlauf, S.N. (2001b) 'Discrete choice with social interaction', *The Review of Economic Studies*, 68, pp. 235–60.

Brock, W.A., Hsieh, D.A., LeBaron, B. (1991) *Nonlinear Dynamics, Chaos and Instability: Statistical Theory and Economic Evidence*, Cambridge: MIT Press.

Brock, W.A., Malliaris, A.G. (1989) *Differential Equations, Stability and Chaos in Dynamic Economics*, North-Holland.

Cass, D., Shell, K. (1983) 'Do sunspots matter?', *Journal of Political Economy*, 92, pp. 193–227.

Chavance, B. (1995) 'Hierarchical forms and coordination problems in socialist systems', *Industrial and Corporate Change*, 1, pp. 271–91.

Chen, P. (1993) 'Searching for economic chaos: a challenge to econometric practice and nonlinear tests', in R. Day and P. Chen (eds), *Nonlinear Dynamics and Evolutionary Economics*, Oxford: Oxford University Press.

Chen, P. (2005) 'Evolutionary economic dynamics: persistent business cycles, chronic excess capacity, and strategic innovation in division of labor', forthcoming in Dopfer (ed.) (2005).

Cohen, M.D., Burkhart, R., Dosi, G., Egidi, M., Marengo, L., Warglien, M., and Winter, S. (1996) 'Routines and other recurring action patterns of organizations: contemporary research issues'. *Industrial and Corporate Change*, 5, pp. 653–98.

Cooper R., John A. (1988) 'Coordinating coordination failures in Keynesian models', *Quarterly Journal of Economics*, 103, pp. 323–42.

Coriat, B., Dosi, G. (1998a) 'The institutional embeddedness of economic change: an appraisal of the "evolutionary" and the "regulationist" research programs', in K. Nielsen and B. Johnson (eds), *Institutions and Economic Change*, Cheltenham, UK: Edward Elgar.

Coriat, B., Dosi, G. (1998b) 'Learning how to govern and learning how to solve problems: on the double nature of routines as problem solving and governance devices', in A. Chandler, P. Hagstrom, O. Solvell (eds), *Dynamic Firm*, Oxford: Oxford University Press.

Coriat, B., Orsi, F. (2002) 'Establishing a new regime of intellectual property rights in the United States: origins, content, problems', *Research Policy*, 31, pp. 1491–507.

Cowley, R. (ed.) (1999) *What If*, New York: Putnam Publisher.

Cox, D.R., Miller, H.D. (1965) *The Theory of Stochastic Processes*, London and New York: Chapman and Hall.

David, P.A. (1975) *Technical Choice, Innovation and Economic Growth: Essays on American and British Experience in the Nineteenth Century*, Cambridge: Cambridge University Press.

David, P.A. (1985) 'Clio and the economics of QWERTY', *American Economic Review*, 75, pp. 332–7.

David, P.A. (1988) *Path Dependence: Putting the Past into the Future of Economics*, Stanford University, Institute for Mathematical Studies in the Social Science, Technical Report 533.

David, P.A. (1992) 'Heroes, herds and hysteresis in technological history: Thomas Edison and "The Battle of the Systems" Reconsidered', *Industrial and Corporate Change*, 1, pp. 129–81.

David, P.A. (1993) 'Path dependence and predictability in dynamic systems with local network externalities: a paradigm for historical economics', in D. Foray and C. Freeman (eds), *Technology and the Wealth of Nations*, London: Pinter Publishers.

David, P.A. (1994) 'Why are institutions the "carriers of history"?: path dependence and the evolution of conventions, organizations and institutions', *Structural Change and Economic Dynamics*, 5, pp. 205–20.

David, P.A. (2001) 'Path dependence, its critics and the quest for "historical economics" ' in P. Garrouste and S. Ioannides (eds), *Evolution and Path Dependence in Economic Ideas: Past and Present*, Cheltenham, UK: Edward Elgar.

Diamond, J. (1995) 'Easter's end', *Discover*, 16, pp. 63–9.

Dopfer, K. (ed.) (2005) *The Evolutionary Foundations of Economics*, Cambridge: Cambridge University Press.

Dosi, G. (1982) 'Technological paradigms and technological trajectories: a suggested interpretation', *Research Policy*, 11, pp. 147–62.

Dosi, G. (1984) *Technical Change and Industrial Transformation*, New York: St. Martin's Press.

Dosi, G. (1988) 'Sources, procedures and microeconomic effects of innovation', *Journal of Economic Literature*, 26, pp. 120–71.

Dosi, G. (1995) 'Hierarchies, market and power: some foundational issues on the nature of contemporary economic organization', *Industrial and Corporate Change*, 4, pp. 1–19.

Dosi, G. (1997) 'Opportunities, incentives and the collective patterns of technological change', *The Economic Journal*, 107, pp. 1530–47.

Dosi, G., Fagiolo, G. (1998) 'Exploring the unknown: on entrepreneurship, coordination and innovation-driven growth', in J. Lesourne and A. Orléan (eds), *Advances in Self-Organization and Evolutionary Economics*, Paris: Economica.

Dosi, G., Freeman, C., Fabiani, S. (1994) 'The process of economic development: introducing some stylized facts and theories on technologies, firms and institutions', *Industrial and Corporate Change*, 1, pp. 1–45.

Dosi, G., Freeman, C., Nelson, R.R., Silverberg, G., Soete, L. (eds) (1988) *Technical Change and Economic Theory*, London: Pinter Publisher.
Dosi, G., Kaniovski, Y. (1994) 'On "badly behaved dynamics:" some applications of generalized urn schemes to technological and economic change', *Journal of Evolutionary Economics*, 4, pp. 93–123.
Dosi, G., Kogut, B. (1993) 'National specificities and the context of change: the co-evolution of organization and technology', in B. Kogut (ed.), *Country competitiveness: Technology and the Organization of Work*, New York: Oxford University Press.
Dosi, G., Marengo, L., Fagiolo, G. (1996) *Learning in Evolutionary Environment*, IIASA Working Paper, WP-96-124, IIASA (International Institute for Applied Systems Analysis), Laxenburg, Austria, forthcoming in Dopfer (ed.) (2005).
Dosi, G., Metcalfe, J.S. (1991) 'On some notions of irreversibility in economics', in Saviotti, P.P., Metcalfe, J.S. (eds), *Evolutionary Theories of Economic and Technological Change*, Harwood Academic Publishers.
Dosi, G., Nelson, R.R., Winter, S.G. (eds) (2000) *The Nature and Dynamics of Organizational Capabilities*, Oxford/New York: Oxford University Press.
Dosi, G., Orsenigo, L., Sylos Labini, M. (2003) 'Technology and the Economy', forthcoming in N.J. Smelser and R. Swedberg (eds), *Handbook of Economic Sociology*, 2nd edn, Princeton University Press.
Dosi, G., Winter, S.G. (2002) 'Interpreting economic change: evolution, structures and games', in M. Augier and J. March (eds), *The Economics of Choice, Change and Organizations: Essays in Memory of Richard M. Cyert*, Cheltenham, UK: Edward Elgar.
Durlauf, S.N. (1993) 'Nonergodic economic growth', *Review of Economic Studies*, 60, pp. 349–66.
Durlauf, S.N. (1994) 'Path dependence in aggregate output', *Industrial and Corporate Change*, 1, pp. 149–72.
Eldredge, N., Gould, S.J. (1972) 'Punctuated equilibria: an alternative to phyletic gradualism' in T.J.M. Schopf (ed.), *Models in Paleobiology*, San Francisco: Freeman, Cooper and Company.
Feller, W. (1971) *An Introduction to Probability Theory and Its Applications*, New York: J. Wiley.
Föllmer, H. (1974) 'Random economies with many interacting agents', *Journal of Mathematical Economics*, 1, pp. 51–62.
Forni, M., Lippi, M. (1997) *Aggregation and the Microfoundations of Dynamic Macroeconomics*, Oxford: Oxford University Press.
Freeman, C. (1982) *The Economics of Industrial Innovation*, London: Pinter Publisher.
Freeman, C., Louçã, F. (2001) *As Time Goes By: From the Industrial Revolution to the Information Revolution*, Oxford: Oxford University Press.
Fujita, M., Krugman, P.R. and Venables, A.J. (1999) *The Spatial Economy: Cities, Regions and International Trade*, Cambridge, Mass.: MIT Press.
Garud, R., Karnoe, P. (eds) (2001) *Path Dependence and Creation*, Mahwah, NJ: Lawrence Erlbaum Associates.
Gould, S.J. (1977) *Ever Since Darwin*, New York: Norton.
Granovetter, M. (1985) 'Economic action and social structure: the problem of embeddedness', *American Journal of Sociology*, 91, pp. 481–510.
Hahn, F.H. (1987) 'Information, dynamics and equilibrium', *Scottish Journal of Political Economy*, 34, pp. 321–34.
Hall, P.A., Soskice, D. (2001) *Varieties of Capitalism: The Institutional Foundations of Comparative Advantage*, Oxford: Oxford University Press.
Haken, H. (1981) *Chaos and Order in Nature*, Berlin: Springer.

Herrendorf, B., Valentinyi, A., Waldmann, R. (2000) 'Ruling out multiplicity and indeterminacy: the role of heterogeneity', *Review of Economic Studies*, 67, pp. 295–307.
Hicks, J.R. (1969) *A Theory of Economic History*, Oxford: Oxford University Press.
Hildenbrand, W. (1994) *Market Demand: Theory and Empirical Evidence*, Princeton: Princeton University Press.
Hill, B.M., Lane, D., Sudderth, W. (1980) 'A strong law for some generalized urn processes', *Annals of Probability*, 8, pp. 214–26.
Hodgson, G.M. (2001) *How Economics Forgot History: The problem of Historical Specificity in Social Science*, London and New York: Routledge.
Hollingsworth, J.R., Boyer, R. (1997) *Contemporary Capitalism: The Embeddedness of Institutions*, Cambridge: Cambridge University Press.
Holley, R.A., Liggett, T.M. (1975) 'Ergodic theorems for weakly interacting systems and the voter model', *Annals of Probability*, 3, pp. 643–63.
Kauffman, S.A. (1989) 'Adaptation on rugged fitness landscapes', in D.L. Stein (ed.), *Lectures in the Sciences of Complexity*, 1, pp. 527–618, New York: Addison Wesley.
Kelsey, D. (1988) 'The economics of chaos or the chaos of economics', *Oxford Economic Papers*.
Kreps, D., Spence, A.M. (1985) 'Modeling the role of history in industrial organization and competition', in G.R. Feiwel (ed.), *Issues in Contemporary Microeconomics and Welfare*, London: Macmillan.
Krugman, P.R. (1991a) 'Increasing returns and economic geography', *Journal of Political Economy*, 99, 484–99.
Krugman, P.R. (1991b) *Geography and Trade*, Cambridge, MA: MIT Press.
Krugman, P.R. (1996) *The Self-Organizing Economy*, Cambridge, MA and Oxford: Blackwell Publishers.
Kuran, T. (1987) 'Preference falsification, policy continuity and collective conservatism', *Economic Journal*, 97, pp. 642–65.
Kuran, T. (1991) 'Cognitive limitations and preference evolution', *Journal of Institutional and Theoretical Economics*, 146, pp. 241–73.
Landes, D.S. (1969) *The Unbound Prometheus; Technological Change and Industrial Development in Western Europe from 1750 to the Present*, Cambridge: Cambridge University Press.
LeBaron, B. (1994) 'Chaos and nonlinear forecastability in economics and finance', *Philosophical Transactions of the Royal Society of London*, A 348, pp. 397–404.
Levinthal, D. (2000) 'Organizational capabilities in complex worlds', in Dosi *et al.* (eds) (2000).
Liebowitz, S.J., Margolis, S.E. (1990) 'The fable of the keys', *Journal of Law and Economics*, 33, pp. 1–25.
Liebowitz, S.J., Margolis, S.E. (1995) 'Path dependence, lock-in, and history', *Journal of Law, Economics, and Organization*, 11, pp. 205–26.
Liggett, T. (1999) *Stochastic Interacting Systems: Contact, Voter and Exclusion Processes*, New York: Springer-Verlag.
Lundvall, B.A. (ed.) (1992) *National Systems of Innovation. Towards a Theory of Innovation and Interactive Learning*, London: Pinter Publisher.
Marengo, L. (1996) 'Structure, competence and learning in an adaptive model of the firm', in G. Dosi and F. Malerba (eds), *Organization and Strategy in the Evolution of the Enterprise*, London: Macmillan.
Marcn, J.G. (1991) 'Exploration and exploitation in organizational learning', *Organization Science*, 2, pp. 71–87.
March, J.G., Simon, H.A. (1992) *Organizations*, 2nd edn, Oxford, UK: Blackwell.

Meliciani, V. (2001) *Technology, Trade and Growth in OECD countries*, London/New York: Routledge.
Metcalfe, J.S. (1998) *Evolutionary Economics and Creative Destruction*, London: Routledge.
Mokyr, J. (1990) *The Lever of Riches*, New York: Oxford University Press.
Moore, B.J. (1978) *Injustice: The Social Bases of Obedience and Revolt*, White Plains, NY: M.E. Sharpe.
Mowery, D., Rosenberg, N. (1998) *Paths of Innovation: Technological Change in 20th Century America*, Cambridge: Cambridge University Press.
Nau, H.H., Schefold, B. (eds) (2002) *The Historicity of Economics*, Berlin: Springer.
Nelson, R.R. (ed.) (1993) *National Innovation Systems: A Comparative Analysis*, New York: Oxford University Press.
Nelson, R.R. (2002) 'Physical and Social Technologies, and Their Evolution', working paper, Columbia University.
Nelson, R.R., Sampat, B. (2001) 'Making sense of institutions as a factor shaping economic performance', *Journal of Economic Behavior and Organization*, 44, pp. 31–54.
Nelson, R.R., Winter, S.G. (1982) *An Evolutionary Theory of Economic Change*, Cambridge, MA: Harvard University Press.
Nicolis, G., Prigogine, I. (1989) *Exploring Complexity*, New York: Freeman.
North, D.C. (1990) *Institutions, Institutional Change and Economic Performance*, Cambridge: Cambridge University Press.
Pierson, P. (2000) 'Increasing returns, path dependence, and the study of politics', *American Political Science Review*, 94, pp. 251–67.
Prigogine, I. (1980) *From Being to Becoming*, New York: Freeman.
Prigogine, I., Stengers, I. (1984) *Order Out of Chaos*, London: Heinemann.
Rosenberg, N. (1976) *Perspectives on Technology*, Cambridge: Cambridge University Press.
Rosser, J.B. (1991) *From Catastrophe to Chaos: A General Theory of Economic Discontinuities*, Boston: Kluwer.
Schelling, T.C. (1971) 'Dynamic models of segregation', *Journal of Mathematical Sociology*, 1, pp. 143–86.
Silverberg, G. (1988) 'Modeling economic dynamics and technical change: mathematical approaches to self-organization and evolution', in Dosi et al. (eds) (1988).
Stadler, B.M.R., Stadler, P.F., Wagner, G.P., Fontana, W. (2001) 'The topology of the possible: formal spaces underlying patterns of evolutionary change', *Journal of Theoretical Biology*, 213, pp. 241–74.
Streeck, W., Yamamura, K. (eds) (2001) *The Origins of Nonliberal Capitalism: Germany and Japan*, Ithaca, London: Cornell University Press.
Young, A.A. (1928) 'Increasing returns and economic progress', *Economic Journal*, 38, pp. 527–42.
Young, H.P. (1998) *Individual Strategy and Social Structure: An Evolutionary Theory of Institutions*, Princeton: Princeton University Press.
Weibull, J.W. (1995) *Evolutionary Game Theory*, Cambridge: MIT Press.
Winter, S.G. (1971) 'Satisficing, selection and the innovating remnant', *Quarterly Journal of Economics*, 85, pp. 237–61.
Winter, S.G. (1975) 'Optimization and evolution in the theory of the firm', in R.H. Day and T. Groves (eds), *Adaptive Economic Models*, New York: Academic Press.
Winter, S.G., Kaniovski, Y., Dosi, G. (2000) 'Modeling industrial dynamics with innovative entrants', *Structural Change and Economic Dynamics*, 11, pp. 255–93.
Witt, U. (2005) 'Path-dependence in institutional change', in Dopfer (ed.) (2005).

9
Analyzing Path Dependence: Lessons from the Social Sciences
James Mahoney

The study of path dependence has become a central research area in the disciplines of sociology and political science. These disciplines build on many insights initially formulated by economists. At the same time, however, sociologists and political scientists offer their own distinctive contributions, some of which challenge the core assumptions of economics. In addition, they apply ideas of path dependence to the study of political and social outcomes that are not usually addressed in field of economics.

Castaldi and Dosi's discussion of path dependence offers an extremely useful synthesis of a large body of literature in economics that has developed around the concept of path dependence, including many important contributions by Dosi himself. Here I consider how this literature has been adopted, modified, and extended in the social sciences.

Path dependence: definitions and sources

For analytic purposes, it is essential to distinguish between *definitions* of path dependence, i.e. the characteristics that constitute path dependence, and *sources* of path dependence, i.e. the factors that are likely to trigger path dependence. Castaldi and Dosi's approach is to define path dependence in a very broad way, which in turn allows them to treat a wide range of factors as potential sources of path dependence. By contrast, most sociologists and political scientists use more specific definitions of path dependence, which in turn restricts the range of sources of path dependence.

Almost all formulations of path dependence begin with the assumption that 'history matters.' For Castaldi and Dosi, in fact, this idea along with the related notion of time irreversibility defines path dependence (pp. 1–2). However, the claim that history matters is neither profound nor well specified. While path dependence arguments do argue that the past shapes the future, so do nearly all other causal arguments in economics and the social sciences.

Sociologists and political scientists prefer to see path dependence as a specific type of sequence in which early contingent events set into motion event chains or sequences that have highly predictable features, culminating in outcomes of interest that could not have been explained in light of an initial set of conditions (e.g. Goldstone, 1998; Mahoney, 2000; Pierson, 2000a). A classic example of this kind of sequence is David's (1985) discussion of QWERTY, which has been highly influential in the social sciences. Drawing on examples like this, social scientists frequently trace outcomes back to critical junctures when unexpected, random, or small events trigger deeply patterned sequences of subsequent events.

Given this more specific definition of path dependence, some of factors that Castaldi and Dosi describe as *sources* of path dependence can be reconceptualized as *types* of path dependence. For example, what they call 'increasing returns' and 'dynamic increasing returns' represent a *type* of path dependence, not a source of path dependence. In other words, an increasing returns sequence *is* path dependence; it is not a source of path dependence. The question of what factors produce increasing returns is a separate issue (in fact, many of Castaldi's and Dosi's sources of path dependence produce increasing returns).

In what follows, I discuss the two major sequences that are analyzed as representing path dependence in sociology and political science: increasing returns sequences and reactive sequences. Castaldi and Dosi's discussion offers valuable insights into the logic of both kinds of sequences. However, I also emphasize ways in which these insights may not be fully appropriate for the analysis of social and political phenomena outside of the marketplace.

Path dependence as increasing returns

In political science, path dependence defined as increasing returns has been discussed most extensively by Pierson (2000a, b), who especially builds on the work of Arthur (1994). The idea of increasing returns implies that preceding steps in a particular direction induce further movement in the same direction, such that it is difficult to reverse course once an early direction has been established. In other words, the probability of further steps along the same path increases with each move down that path. This basic notion underpins Castaldi's and Dosi's description of both increasing returns and 'dynamic' increasing returns (pp. 4–8).

Pierson (2000a: 253) notes that increasing returns sequences merit special attention because they may have the following properties:

(1) Unpredictablity: Early random events may be reinforced over time.
(2) Irreversibility: Once a particular path is selected, it may be difficult or impossible to return to an early point when additional alternatives were available.
(3) Nonergodicity: Accidents do not 'average out' over time.
(4) Potential Inefficiency: Sub-optimal outcomes may become locked-in.

All of these elements are touched on in various ways in the Castaldi and Dosi contribution, with especially insightful comments on the connection between unpredictability (what they call 'stochastic processes') and non-ergodicity (see pp. 16–17).

Castaldi's and Dosi's discussion also points to several sources of path dependence as increasing returns, though they themselves do not explicitly make this link because they treat increasing returns as itself a source of path dependence (rather than a *kind* of path dependence). For example, they note that path dependence is especially likely in settings where large set-up or fixed costs are present, learning is possible, and where individuals adapt their expectations in self-fulfilling ways. Furthermore, they also identify particular contexts where these sources are likely to be present, such as institutional environments or environments that are subject to selection pressures. As Arthur (1994) and other economic historians have suggested, these contexts and sources help explain why increasing returns processes – and thus path dependence – occur when they do.

Increasing return processes are pervasive in the social and political world, likely much more so than in the marketplace. Non-market arenas feature a greater need to coordinate behavior, and collection action in turn often produces path dependence because it involves high start-up costs and triggers adaptive expectations. Likewise, the prevalence of institutions and the absence of market signals makes incremental adjustments that might override patterns of lock-in far less likely in the social and political domains. Hence, although economists initially formalized the concept of path dependence, the phenomenon itself is almost certainly more common outside of the marketplace (Pierson, 2000a).

Many of our best examples of path dependence refer to social and political processes.[1] For example, social scientists have analyzed a range of institutions that have become locked-in through processes of increasing returns, including state bureaucracies, voluntary associations, party systems, and policy orientations (Ertman, 1996; Skocpol, 1999; Lipset and Rokkan, 1967; Hacker, 1998). In addition, they have shown that basic economic institutions such as industrial corporations and even market capitalism itself have become locked-in through increasing returns processes (e.g. Roy, 1997; Wallerstein, 1974). In this sense, social scientists have argued that some of the most fundamental institutions of economics may be the outcome of path-dependent processes.

Non-utilitarian rationales for increasing returns

Economics is a discipline dominated by a single neoclassical theoretical tradition based on utilitarian principles. From this perspective, path dependence by increasing returns occurs because rational actors have incentives to reproduce or expand outcomes, including outcomes that are sub-optimal in light of previously available alternatives. For example, in David's (1985) QWERTY example, rational actors continue with QWERTY because the current benefits of doing so outweigh the costs, even though these actors may have been better off if their predecessors initially started with a different typewriter format.

By contrast, the social sciences are more theoretically eclectic (see Mahoney, 2000). This theoretical eclecticism allows them to conceptualize increasing returns processes from a diverse spectrum of viewpoints. For example, sociologists have explored the ways in which actor power generates increasing returns processes. In these frameworks, an institution may initially empower a certain group at the expense of another group; the advantaged group then uses its power to expand the institution further; the expansion of the institution in turn increases the power of the advantaged group; and the process then repeats itself. In the end, both the institution and the actor's dominant position are locked-in via path dependence. This is precisely the logic that Roy (1997) uses when explaining the dominance of the large industrial corporation and its corporate elite in the United States.

Likewise, sociologists have linked increasing returns processes to functionalist explanations. Here an institution serves a function for some social system, which causes the expansion of the institution, which enhances the ability of the institution to serve the useful function, which leads to further institutional expansion, and so on. These functionalist explanations put explanatory emphasis on the needs of social systems rather than the needs of rational acting individuals. For example, Wallerstein's (1974) explanation of the development and spread of capitalism in Western Europe stresses the functional needs of the world system and the ways in which only Europe could fulfill those needs.

Similarly, works that analyze the effects of culture may adopt legitimation explanations in which codes of appropriateness and legitimacy are reinforced over time. In this case, an initial set of norms may be seen as legitimate, which sets into motion a self-reinforcing process that deeply embeds the norms in culture. For example, Hall and Soskice (2001: 12–14) emphasize the important role of enduring informal rules and norms in producing equilibria in many political–economic interactions.

In the marketplace, a utilitarian theoretical framework will often be appropriate, since the market itself has correcting mechanisms that select against individuals who do not behave in instrumentally rational ways. However, once one moves outside of the marketplace, other theoretical frameworks may be more relevant, and thus non-economic discussions of path dependence may be needed.

Contingency and determinism

Path dependent sequences simultaneously embody claims that could be seen as highly contingent and highly deterministic. On the one hand, these sequences assign a key role to small events, accidents, or other contingent occurrences. On the other hand, the sequences assume that path dependence may lock-in outcomes in quite deterministic ways. The contingency part of path dependence generally corresponds to early events in a sequence; for example, the accidents or small events that initiate a path dependent sequence. The claims about determinism refer to later events in the sequence, once path dependence has already been established.

This combination of contingency at the beginning of a sequence with subsequent determinism leads to a central paradox characterizing path-dependent sequences: these sequences are simultaneously unpredictable in light of a given theoretical framework *and* deterministically reproduced by the variables associated with the same theoretical framework. For example, in economic history, path-dependent analysts are intruiging because they show how certain economic outcomes are 'inefficient,' thereby contradicting the predictions of neoclassical theory. Yet, these same analysts rely fully on mechanisms associated with neoclassical theory to explain the *reproduction* of these inefficient outcomes once they are contingently selected.

To make sense of this paradox, one must recognize that path-dependent arguments contradict prevailing theoretical frameworks only with respect to past options that are no longer viable alternatives. For instance, given the heavy costs of technology reversal, not even Paul David (1985) argues it would be efficient for contemporary economic actors to replace QWERTY with the Dvorak format, even though David believes the more efficient choice would have been to adopt Dvorak from the start. Likewise, when Piore and Sabel (1984) argue that mass production is an inefficient outcome, they are comparing mass production to a possibility that existed in the 19th century: namely, craft production. Piore and Sabel do not believe it would be efficient to abandon mass production for craft production at this point in history. In sum, the contradiction with theory inherent in path-dependent increasing returns sequences applies to options that were available at an earlier critical juncture, not options that are presently available.

Figure 9.1 illustrates the place of contingency and determinism in path-dependent sequences. In this example, three potential options (A, B, and C) are available for selection at Time 1. However, existing theory is unable to explain which of the three options will be chosen (or existing theory predicts the wrong option). Hence, on the basis of initial conditions, the selection of a particular option is a 'contingent' occurrence – i.e. a random event *vis-à-vis* the predictions of a particular theoretical framework. For example, with respect to neoclassical theory, David (1985) argues that the initial selection of QWERTY over a more efficient rival was a contingent event.

Once a particular option is initially favored over the alternatives at Time 2, however, it realizes increasing returns and is stably reproduced during Times 3+. The stable reproduction of the initial choice can be explained in light of the same theoretical framework that could not explain the initial adoption of the option. For example, the stable reproduction of QWERTY is well

```
            A
                        ┌───┐
                        │ B │
            B ──────────┴───┴──────────▶  B, B, B

            C
```
_____

Time 1	Time 2	Time 3+
(Initial Conditions)	(Critical Juncture)	(Self-reinforcement)
Multiple options (A, B, C) are available for selection. Theory is unable to predict or explain which option will be adopted.	Option B is initially favored over competing options. This is a contingent outcome.	Option B capitalizes on initial advantage and is stably reproduced over time. Theory explains well this stable reproduction.

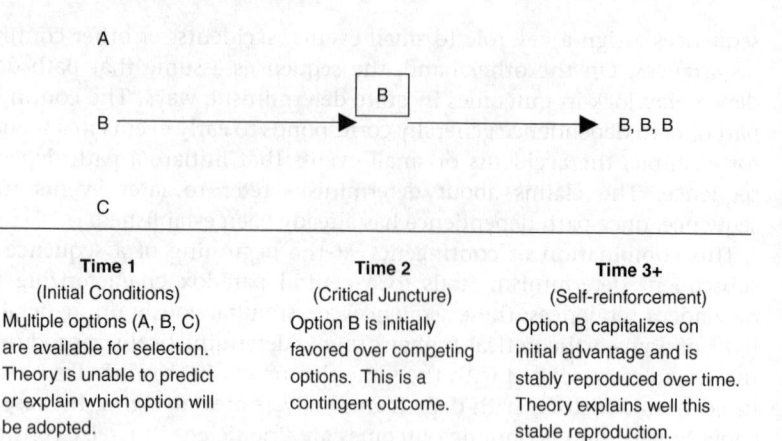

Figure 9.1 Illustration of contingency in self-reinforcing sequence

explained in light of neoclassical theory. Likewise, Roy (1997) explains the stable reproduction of the large corporation in light of the power of the American corporate elite.

The formulation of path dependence presented in Figure 9.1 has been debated within political science and sociology. Most notably, Thelen (1999, 2003) has argued that this formulation leads to too much contingency at the front end of sequences and too much determinism at the back end of sequences. She suggests that social scientists need to recognize that most sequences are characterized by more subtle processes of selection and reproduction, in which only parts of an institution may be selected for adoption, and in which reproduction is tied to processes of transformation. For example, on one level, the US Congress has been stably reproduced over time; on another level, however, the US Congress has undergone enormous change. Hence, it is unclear if this institution has been reinforced by an increasing returns process. Likewise, technological formats like VHS and Microsoft are marked by both profound continuity and profound change. Broad ideas concerning path dependence may not be helpful for capturing these subtleties.

These concerns articulate nicely with Castaldi and Dosi's discussion of degrees of history dependence (p. 18). In David's (2001) formulation, which they quote, path dependence is defined by strong history, in which transitions out of a given pattern are extremely difficult. This is consistent with definitions that see path dependence as defining sequences marked by contingency at the beginning followed by subsequent determinism (e.g. Mahoney, 2000; Goldstone, 1998). The reasoning behind requiring initial contingency is that the most intriguing features of path dependence such as unpredictability, inefficiency, and nonergodicity rely on claims about accidents, small events, and the like at the initiation of path-dependent sequences. In the absence of these claims, path dependence loses much of its analytical bite, potentially devolving into vague assertions such as 'history matters' or 'the past affects the present.'

The deterministic side of path dependence at the back end runs the risk of presenting history as if it were fully locked-in. The corrective involves, as Castaldi and Dosi discuss, theorizing various sources of 'de-locking'. Castaldi and Dosi present a suggestive typology of different ways of de-locking. In somewhat different vein, I have argued that theorizing de-locking requires knowledge of the specific mechanism that sustains lock-in over time (Mahoney 2000). For example, if a sequence is governed by power mechanisms, the challenge of de-locking will center on weakening elite actors who benefit from the status quo and strengthening subordinate actors who are disadvantaged by the status quo. This could involve either what Castaldi and Dosi call an 'invasion' or what they call 'heterogeneity among agents.' However, the key point is that the power resource distribution among actors must be transformed.

Reactive sequences and path dependence

Much of the literature on path dependence in economics has been concerned with understanding the reproduction of a given outcome; that is, with lock-in via self-reinforcement. However, social scientists also consider forms of path dependence that are not self-reproducing. Elsewhere, I have used the expression 'reactive sequences' to characterize these kinds of path dependence (Mahoney, 2000).

Reactive sequences are chains of temporally ordered and causally connected events. In a reactive sequence, each event in the sequence is both a reaction to antecedent events and a cause of subsequent events. Initial events in the sequence are especially important to final outcomes: a small change in initial conditions can accumulate over time and make a great deal of difference by the end of the sequence. The Castaldi and Dosi discussion does not explicitly discuss these sequences, though their important remarks about chaos theory touch on the subject. With chaotic dynamics, sensitive dependence on initial conditions works through a reactive sequence dynamic.

Reactive sequences are significantly different from increasing returns sequences. Whereas increasing returns sequences are characterized by processes of reproduction that reinforce early events, reactive sequences are marked by processes that *transform* and perhaps *reverse* early events. In a reactive sequence, initial events trigger subsequent development not by reproducing a given pattern, but by setting in motion a chain of tightly linked reactions. For example, Isaac et al. (1994) argue that the death of Martin Luther King led to the expansion of race-based poor relief at the expense of a more progressive program of class-based economic reform. To simplify their sophisticated event-structure argument, they show how King's death (Event A) caused the failure of the Poor People's Campaign (B), which in turn led to massive summer riots (C), which heightened welfare militancy (D), which brought about an increase in AFDC applications and court rulings that liberalized AFDC acceptance criteria (E), which fostered an explosion in the AFDC rolls in the late 1960s (F). Event A was critical for Event F, but its impact did not work through an increasing returns process.

Contingency and conjunctures in reactive sequences

With a reactive sequence, it is not self-evident how one should conceptualize the starting point of the sequence, which raises an important problem. Because the decision to select any particular event as the starting point of analysis may seem arbitrary, the investigator is prone to keep reaching back in time in the search for foundational causes that underlie subsequent events in the sequence. In other words, without criteria for identifying a meaningful beginning point, the investigator can easily fall into the trap of infinite regress – i.e. perpetual regression back in time to locate temporally prior causal events.

In a path-dependent reactive sequence, the initial event that sets into motion the overall chain of reactions is contingent. From the perspective of theory, such an event appears as a 'breakpoint' in history – i.e. an event that was not anticipated or predicted. As Sewell (1996) suggests, sequential analysis often begins by focusing on unpredictable events – what he calls 'initial ruptures' – that mark a 'surprising break' with theoretical expectations. By focusing on such breakpoints, analysts of reactive sequences avoid the problem of infinite historical regress. In addition, these breakpoints could be added to Castaldi and Dosi's list of potential sources of de-linking from path dependence. Indeed, a breakpoint can move a system from one pattern of path dependence to another.

The contingent initial event that triggers a subsequent reactive causal chain is often itself the intersection point of two or more prior sequences. Historical sociologists use the expression 'conjuncture' to refer to this coming together – or temporal intersection – of separately determined sequences (Zuckerman, 1997). The point in time at which two independent sequences intersect will often not be predictable in advance. Likewise, the specific event generated by the intersection of the sequences may be outside of the resolving power of prevailing theories. Hence, conjunctures are often treated as contingent occurrences. This is true even though each of the sequences that collide to make a conjuncture may themselves follow a highly predictable causal pattern.

Analyzing chains of events with narrative analysis

Even with knowledge of the contingent breakpoint that launches a reactive sequence, analysts may have difficulty explaining the final outcome of the sequence. As chaos theorists have stressed, final outcomes cannot necessarily be predicted on the basis of early events in a sequence, even if the sequence is governed by rigid mathematical laws. Fortunately, smaller intervals of connected events within the overall sequence often can be predicted or explained. For path-dependent investigators, these smaller sets of intervening steps through which a contingent breakpoint produces a final outcome – not the direct link between the breakpoint and the final outcome – are the central object of analysis.

Historical narrative offers an especially useful method for making sense of the multiple steps in a reactive sequence (e.g. Reisch, 1991). Narrative permits a form of sequential analysis in which the investigator isolates particular links within an overall causal chain. Through a narrative account, the analyst can provide 'a scene by scene description of the particular causal paths' through which an initial breakpoint leads to a final outcome. Furthermore, the step-by-step approach of narrative allows the analyst to use counterfactual methods in exploring specific causal links in the overall chain (see also Castaldi and Dosi's remarks on counterfactual analysis; pp. 113–14). In conjunction with counterfactual analysis, narrative can help the investigator identify what Aminzade (1992: 463) calls 'key choice points' in a reactive sequence – i.e. 'forks in the road ... marked by the presence of alternative possible paths.' These choice points facilitate an exploration of hypothetical 'paths not taken' and alternative futures that could have occurred if particular events in the reactive sequence had been different.

The events that make up a reactive sequence are connected by tight causal linkages, or what Griffin and Ragin (1994) call 'inherent sequentiality.' The basic idea underlying inherent sequentiality is Abbott's (2001) notion that an 'inherent logic of events' characterizes enchained sequences. For example, these chains are often marked by events in which the time order of occurrences is well established, the causal linkages are marked by necessary or sufficient relationships, and the temporal separation of events is minimal. In this context, the narrative analyst can present compelling accounts of how an initially contingent occurrence triggers a sequence of tightly-coupled subsequent occurrences, ultimately culminating in an outcome of interest.

From the perspective of economics, the qualitative orientation of narrative analysis might be seen as a 'soft' method that essentially entails historical story telling. However, the degree to which narrative must be an informal method lacking rigor is actually an open question. Sophisticated new tools exist for formally diagramming narrative structures in terms of explicit causal linkages (e.g. Griffin, 1993). These tools force the narrative analyst to treat each event in the narrative as a variable and specify its causal linkage to all other events. The use of these formal methods entails an aesthetic loss, but it responds to the concern that narrative is nothing more than an 'anything goes' story.

Conclusion

Castaldi and Dosi offer a well-written and highly accessible introduction to issues of path dependence from the field of economics. The literature they discuss has strongly influenced work in the social sciences across multiple substantive domains. Hence, the study of path dependence is a highly synergistic area, where social scientists actively read and use ideas developed by economists. At the same time, I have suggested that social scientists have introduced important modifications to this literature that go beyond what the economics discipline currently offers. An outstanding question is whether these innovations will in turn have an impact within economics.

Note

1. Many or most of the examples of path dependence in economics have been questioned by Liebowitz and Margolis (2001). They in particular dismiss forms of path dependence that are nonremediable – i.e. historical actors acted as best they could with the information available, and a mistake was unavoidable. However, Pierson (2000a: 256–7) argues that outside of marketplace the prevalence of short-sighted actors and the pervasiveness of institutions makes remediable path dependence commonplace. Hence, the Liebowitz/Margolis critique carries less weight when one moves beyond the marketplace. See also Castaldi and Dosi's remarks on p. 113.

References

Abbott, Andrew. 2001. *Time Matters: On Theory and Method*. Chicago: University of Chicago Press.
Aminzade, Ronald. 1992. 'Historical sociology and time.' *Sociological Methods and Research* 20: 456–80.
Arthur, W. Brian. 1994. *Increasing Returns and Path Dependence in the Economy*. Ann Arbor: University of Michigan Press.
David, Paul A. 1985. 'Clio and the economics of QWERTY.' *American Economic Review* 75: 332–7.
David, Paul A. 2001. 'Path dependence, its critics and the quest for "historical economics."' In *Evolution and Path Dependence in Economic Ideas*, edited by P. Garrouste and S. Ioannides. Cheltenham: Edward Elgar.
Ertman, Thomas. 1996. *Birth of the Leviathan: Building States and Regimes in Medieval and Early Modern Europe*. Cambridge: Cambridge University Press.
Gleick, James. 1987. *Chaos: Making a New Science*. New York: Penguin.
Goldstone, Jack A. 1998. 'Initial conditions, general laws, path dependence, and explanation in historical sociology.' *American Journal of Sociology* 104: 829–45.
Griffin, Larry J. 1993. 'Narrative, event-structure, and causal interpretation in historical sociology.' *American Journal of Sociology* 98: 1094–133.
Griffin, Larry and Charles C. Ragin. 1994. 'Some observations on formal methods of qualitative analysis.' *Sociological Methods and Research* 23: 1–12.
Hacker, Jacob. 1998. 'The historical logic of National Health Insurance: structure and sequence in the development of British, Canadian, and U.S. medical policy.' *Studies in American Political Development* 12: 57–130.

Hall, Peter A. and David Soskice. 2001. 'An introduction to varieties of capitalism.' In *Varieties of Capitalism: The Institutional Foundations of Comparative Advantage*, edited by Hall and Soskice. Oxford: Oxford University Press, pp. 1–68.

Isaac, Larry W., Debra A. Street and Stan J. Knapp. 1994. 'Analyzing historical contingency with formal methods: the case of the "relief explostion" and 1968.' *Sociological Methods and Research* 23: 114–41.

Liebowitz, Stan J. and Stephen E. Margolis. 2001. *Winners, Losers, and Microsoft: Competition and Antitrust in High Technology*. Oakland: Independent Institute.

Lipset, Seymour Martin and Stein Rokkan. 1967. *Party Systems and Voter Alignments: Cross-National Perspectives*. New York: Free Press.

Mahoney, James. 2000. 'Path dependence in historical sociology.' *Theory and Society* 29: 507–48.

Pierson, Paul. 2000a. 'Increasing returns, path dependence, and the study of politics.' *American Political Science Review* 94: 251–67.

Pierson, Paul. 2000b. 'Not just what, but *when*: issues of timing and sequence in political processes.' *Studies in American Political Development* 14: 72–92.

Piore, Michael J. and Charles F. Sabel. 1984. *The Second Industrial Divide: Possibilities for Prosperity*. New York: Basic Books.

Reisch, George. 1991. 'Chaos, history, and narrative.' *History and Theory* 30: 1–20.

Roy, William G. 1997. *Socializing Capital: The Rise of the Large Industrial Corporation in America*. Princeton: Princeton University Press.

Skocpol, Theda. 1999. 'How Americans became civic.' In *Civic Engagement in the American Democracy*, edited by Theda Skocpol and Morris P. Fiorina. Washington, DC: Brookings Institute, pp. 27–80.

Sewell, William H., Jr 1996. 'Historical events as transformations of structures: inventing revolution at the Bastille.' *Theory and Society* 25/6: 841–81.

Thelen, Kathleen. 1999. 'Historical institutionalism and comparative politics.' *Annual Review of Political Science* 2: 369–404.

Thelen, Kathleen. 2003. 'How institutions evolve: insights from comparative historical analysis.' In *Comparative Historical Analysis in the Social Sciences*, edited by James Mahoney and Dietrich Rueschemeyer. Cambridge: Cambridge University Press, pp. 208–40.

Wallerstein, Immanuel. 1974. *The Modern World System I: Capitalist Agriculture and the Origins of the European World-Economy in the Sixteenth Century*. New York: Academic Press.

Zuckerman, Alan S. 1997. 'Reformulating explanatory standards and advancing theory in comparative politics.' In *Comparative Politics: Rationality, Culture, and Structure*, edited by Mark Irving Lichbach and Alan S. Zuckerman. Cambridge: Cambridge University Press, pp. 277–310.

10
Path Dependence and Historical Contingency in Biology
Eörs Szathmáry

Introduction: 'between history and physics'

As Smith & Morowitz (1982) once aptly wrote, biology is a discipline 'between history and physics'. On one hand, we do have laws in biology, on top of those coming from physics and chemistry. The model system of population genetics is not very different from laws of theoretical physics, except that it contains more variables and parameters. In mechanics we must know the mass of objects, some key forces and the gravitational constant: in population genetics we must know about population size, allele frequencies, linkage, mutation and migration rates, selection coefficients and so on. But this could still be regarded as 'ordinary' theoretical science, albeit a bit complicated. On the other hand the laws (or rules, if we are more modest) of population genetics are of the 'if A, then B' nature, and often there is nothing *within* the theory that could decide whether A in fact holds or not. That decision comes from physics, chemistry, or – crucially for our present enquiry – history.

Bassanini & Dosi (1999) describe aspects of path dependence that serve as a good general background to the present paper. Following their exposition, it is convenient to distinguish between what I shall call a strong and a weak form of path dependence. The strong form is a result from a combination of irreversibility and the existence of multiple stable attractors for the dynamical system under consideration. In biology, multiple evolutionarily stable strategies (see Maynard Smith, 1982) serve as a good example. Depending on the initial conditions, one, or another, stable strategy may dominate the population, and be resistant against invasion by a mutant strategy. The weak form of path dependence does not invoke multiple stable attractors: if the dimensionality of the system is sufficiently high, and if the dynamics of the system is as described by a Markov process, then path dependence is ensured if the trajectories of two possible realizations do not cross with certainty. Population genetics of small populations is able to generate many such cases.

Furthermore, Bassanini & Dosi (1999) distinguish between two other aspects of path dependence, roughly corresponding to the micro and the macro level. For the biological case, the obvious interpretation is the individual and the population level, respectively. At the individual level we have path dependence whenever changing back the environmental conditions of the population at time $t + \tau$ to those prevailing at time t sees the individuals irreversibly changed. A population of developing, or learning organisms, is an obvious example. At the population level path dependence entails that the probability of going back to some previous state decreases with time, or that switching to a state that could have easily been reached, had the population taken different turn previously, is becoming increasingly improbable as time goes on. I shall argue that macroevolution as such is a prime example such path dependence.

Is it really true that biological evolution is crucially path dependent? This question may seem more surprising since everybody 'knows' the dinosaur story. Had a major asteroid not hit the Earth about 65 million years ago, the dinosaurs may not have passed away and it is quite likely that you would not be reading this text now. Curiously, there is a recent exception to this view, and to the view in general that biology is littered with historical contingency. De Duve (2002) in his book claims that even man had to come and that the Universe was pregnant with us, entirely based on the laws of natural science. Thus he seeks to replace Christian faith and the anthropic principle alike.

As we know it, our present Universe is remarkably 'fit' to host life. Even minor changes of several physical constants would render life, or indeed a structured Universe similar to our own, impossible. How is one to solve that problem? Physicists adhering to the anthropic principle (Barrow and Tipler, 1988) often hold extreme positions, such as the idea that the Universe developed in a manner that had to result in the emergence of minds able to reflect on it (the so-called strong anthropic principle). If this would be so, evolutionary biology would be almost irrelevant. In the end, there would be no crucial path dependency. The difference between the anthropic people and De Duve is that the former place the root of their solution more in physics, the latter more in biology. I hold both positions untenable because they are unsupported by evidence.

In this chapter, I will first discuss the nature of adaptive evolution by natural selection and how it can lead to various forms of path dependence in general. Then I shall look at how different hereditary systems contribute to the determination of biological populations. We shall see that evolution is not always fully irreversible: some genes and traits can be resurrected, but only if a relatively small time has elapsed since their deaths. The so-called major transitions in evolutions illustrate the awesome power of path dependence in biological evolution. I shall explain how the apparent contradiction between the role of historical contingency and evolutionary convergence can be resolved by looking at engineering constraints and the details of the convergent traits. Finally, I conclude with a short summary of what we have seen.

Evolutionary units, adaptive landscapes, and the flux of mutations

Units of evolution must multiply, show heredity, produce variants, and among the traits that are inherited some should affect the chance of the units' survival and/or fecundity (see Maynard Smith, 1986). If there is enough hereditary variability in the system, there is opportunity for cumulative selection, which is the building up of traits that are adaptive, increasing the fitness of the unit carrying them. One could argue that path dependence is, after all not interesting if we arrive at selection equilibrium. This argument is similar to the one made by Bassanini and Dosi in relation to thermodynamics and irreversibility: there is path dependence in *how* we reach the equilibrium, but not *where* the equilibrium is. When we reach it, history is erased.

A crucial question concerns the nature of the 'adaptive landscape' of populations. If there are more than one nearly optimal (equally fit) solutions to the problem, well separated in parameter space, then even the 'end result' will be highly historically contingent, and this could be apparent even at a coarse scale. If the landscape is, in contrast, Fujiyama-like, then selection equilibrium may in fact erase much of history. Whether this holds at a finer scale depends on the nature of inheritance. If there are many traits, then at a fine scale history will still be preserved, owing to the cumulative nature of selection (see section on phylogeny below). If there are few traits, this may not be apparent. Consider the evolution of the eye. Squids and humans have surprisingly similar eyes, at least on a large scale. And we also know that the outcome is, at least in some regards, perfect. The human eye, when adapted to complete darkness, can detect the arrival of a single photon (I do not know about a similar experiment on squids, but I suspect the result would be the same). When we look at the eyes more closely, however, we see that the *details* are different. Interestingly, the vertebrate eye is, in some sense, clumsily 'designed': the fibres in the retina conducting the signals to the brain are wired so that they go *between* the light receptor and the light. In the case of the squid evolution got it right: the receptors are facing the light and the fibres are facing the dark.

It is a common experience of biologists that evolution produces different outcomes in replicate runs. This can be seen *in vitro* (such as with bacterial populations; Lenski and Travisano, 1994) and *in silico* (where computer programs competing for CPU time are free to mutate and to die out or survive; Yedid and Bell, 2002). Theoretical analysis shows that this depends on the relation between population size N and the rate of beneficial mutations u (Johnson et al., 1995; Wahl and Krakauer, 2000). If $Nu > 1$, then beneficial mutations occur together and can be simultaneously selected for (in sexual populations), whereas if $Nu \ll 1$ then replicate populations *cannot* evolve in parallel because each will accumulate a different set of mutations, and even for an overlapping subset the order of incorporation will be different. Note that more complex organisms tend to be bigger, and they typically have a smaller population size; hence adaptive evolution will be more path-dependent for them.

Forms of heredity

The very idea of inheritance is strongly bound to path dependence. In the case of biological systems it is eminently true that the thermodynamic state variables grossly underdetermine the system. If one exposes a population to a series of environments, and if ultimately the initial environment is restored, then (if the timescale is right) it is very unlikely that the population will be, revert into, the initial state. Apart from stochastic effects, this will be due to the action of natural selection on heritable variation.

Limited and unlimited heredity

The nature of inheritance is crucial for cumulative selection and, hence, path dependence. The capacity of a hereditary system crucially influences the scope of evolution. It is convenient to distinguish between limited and unlimited hereditary potential (Maynard Smith and Szathmáry, 1995). We have limited heredity when

number of individuals ($>$ or α) *number of potentially heritable states*;

and we have unlimited heredity when

number of individuals $\ll$ *number of potentially heritable states*.

Of course, this is an operational definition and also depends on the size of the population, but FAPP (*f*or *a*ll *p*ractical *p*urposes) it is a useful guide. It is unlimited heredity that allows ongoing evolution and the cumulative build-up of complex adaptations. Unlimited heredity for genetic sequences means that the number of possible sequences is much larger than the number of individuals in the given system. Imagine RNA molecules of the size 200 (roughly three times the size of transfer RNA). This allows for $4^{200} \approx 10^{120}$ possible sequences. One could not build a museum with the available material in the observable Universe in which just one copy of each sequence would be shown (cf. Eigen, 1971). Alternatively, one may consider the sequence space (Maynard Smith, 1970) of all possible genomes up to a length of 10^9 (the genome size of humans). In this multidimensional space each point represents a polynucleotide sequence (a string made of the building blocks of DNA), being one mutation step away from all its immediate neighbours. Imagine, further, that initially the whole space is dark. Then, sequence the genomes of all the creatures, great and small, that have ever lived on this planet, and light a bulb in the corresponding part of sequence space. *After lighting a bulb for every individual, the space will remain virtually dark.* This is comparable to the emptiness of physical space at the scale of the Universe (Jastrow, 1971). In contrast, the importance of limited heredity lies in the fact that usually it is the first to arrive in evolution and it evolves into unlimited heredity subsequently. This seems to be true for chemical systems and DNA-based heredity, epigenetic inheritance and language. I shall briefly discuss them in turn.

Chemical systems

In chemical systems inheritance rests on the existence of replicators (see Dawkins, 1976 for the introduction of this concept), and in turn replication rests on autocatalysis (when a substance catalysis its own formation at the expense of raw materials, cf. Orgel, 1992). The simplest self-replicator *of biological relevance* I know of is glycolaldehyde (a small molecule with two carbon-containing groups, $H_4C_2O_2$), the autocatalytic seed of the formose 'reaction' (Figure 10.1), discovered by Butlerow in 1861. There are a few highly relevant questions we may ask about this simple replicator: (1) does it have heredity? (2) Is there a possibility of hereditary variation? (3) What are the conditions of its spontaneous propagation? (4) In what way does such a replicator differ from more conventional ones (e.g. genes)? I discuss these issues in turn.

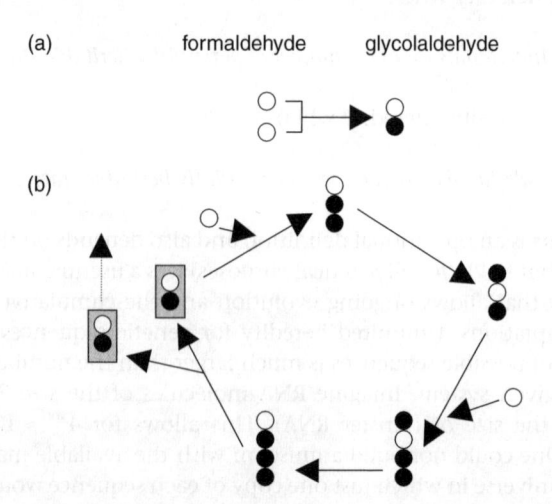

Figure 10.1 The formose 'reaction', which is, in fact, a complex network of autocatalytic sugar formation. (a) The 'spontaneous generation' of the autocatalytic seed is a very slow process; and (b) the autocatalytic core of the network. Each circle represents a group with one carbon atom

Whether one can have heredity in such systems or not is an open question, both theoretically and empirically. To be sure, there are other autocatalytic cycles of small organic molecules (such as the Calvin cycle and the reductive citric acid cycle, fixing carbon dioxide in plants and some bacteria, respectively) that could have played an early role even in chemical evolution. Wächtershäuser (1998, 1992) suggested that archaic versions of the citric acid cycle could have existed and propagated on pyrite surfaces. This suggestion is open to experimental test. We surely do not know of replicatable alternatives of glycolaldehyde in the formose reaction. Most changes in the chemical identity of the cycle intermediates will only be transient fluctuation or will just drain the system. Even if heredity is possible for such cycles, hereditary variation will be very rare, closer to what biologists call 'macromutations' (Wächtershäuser, 1992). Wächtershäuser hypothesizes that present-day intermediary metabolism is a result of the fixation of such macromutations. He has developed the methodology of 'biochemical retrodiction' whereby one could infer, looking at the topology of the network, the path evolution had taken 'before enzymes and templates'.

Epigenetic inheritance

Epigenetic inheritance is a kind of inheritance in biology that does not rely on differences between DNA sequences (see Jablonka and Lamb, 1995 for review); it affects mainly how the available genes are being used by the organism. I would like to refer to two types here: the so-called steady-state systems and the chromatin marking mechanism. An example of the former is that when the product of gene A activates the transcription of gene B, and the product of gene B enhances the transcription of gene A. Note that once such a system is turned on by an external inducer, it remains active by itself. Such systems occur even in bacteria, albeit with limited heredity (only a few cell states can be propagated). Note that the cell states can be passed on during reproduction of the cell. Cells of complex multicellular organisms (like us) rest on many different, inducible and inheritable cell states. Although steady-state systems of gene activation play a role also here, it is the chromatin marking that dramatically increases the hereditary capacity of the epigenetic system. Chromatin marks are permanently attached labels to genes that influence their expression. They also can be passed on during cell division (Figure 10.2). When a differentiated liver cell divides, and the liver is healthy, the offspring cells will also be liver cells rather than, say, neurons.

An intriguing question is whether such states can be passed on to the next generation of *organisms* rather than cells. This would be a Lamarckian dimension of inheritance, no doubt. The answer is that it *does* happen. Some of the epigenetic changes in certain cell lines do not become RESET, as it were, during the genesis of gametes or other reproductive units. The phenomenon seems to be more widespread in plants than in animals, the reason being that in the former there is no early segregation of the germ cells from

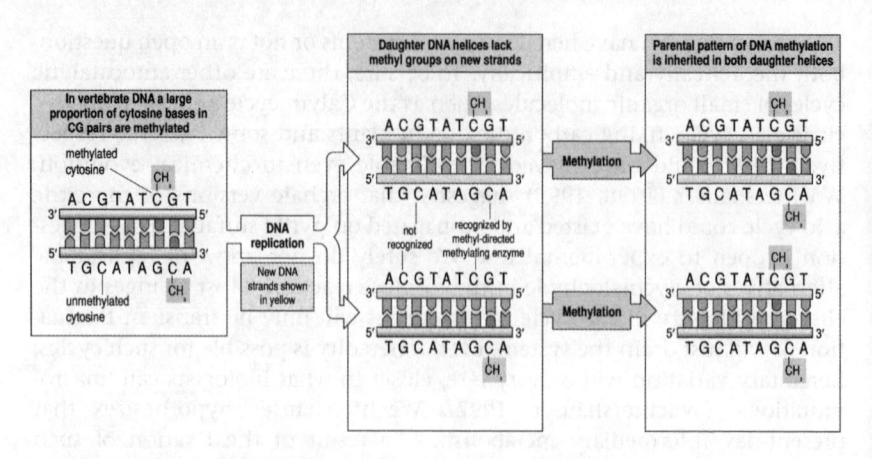

Figure 10.2 DNA methylation as a chromatin marking system*

* Such systems are used for epigenetic inheritance in diverse organisms. Cytosines in so-called GC islands in sequences that affect the transcription of genes may become methylated. The methylation mark inactivates the regulated gene. Note how this new system hitches a ride on conventional genetic inheritance based on DNA.

the rest (Jablonka and Lamb, 1995). More cell divisions in the differentiated state seem to result in an accumulation of less erasable epigenetic marks.

Memes and cultural inheritance

Learning is obviously a mechanism of path dependence for the individual, at a level higher than epigenesis (although sometimes the two phenomena are closely related: some learning results in marked epigenetic changes in brain structure, cf. Changeux, 1983). It can lead to population path dependency when social learning emerges, i.e. when animals are learning from each other (Table 10.1). Associated with the importance of social learning is the idea of memes; a shorthand for culturally transmitted items such as words, habits, fashion, scientific laws and the like: anything that can be passed on in cultural evolution with reasonable accuracy as a distinguishable unit (Dawkins, 1976). Although resentment to the idea of memes is considerable, it cannot be denied that the analogy originally made by Dawkins with genes is beguiling, to the extent that memes seemed then the only rivals to genes with – as one would put it nowadays – unlimited hereditary potential, allowing for the possibility of cumulative cultural evolution. Note that unlimited memetic heredity requires language with complex syntax. Before that, an intermediate stage is 'protolanguage'; essentially a limited set of words without syntax (Bickerton, 1990). Only natural language as we know it has the potential to convey a potentially unlimited number of different meanings.

Yet there is a crucial difference between genes and memes, also bearing on the issue of path dependence. It is well known that cultural evolution is

Table 10.1 Examples of social learning (from Szathmáry, 2002b)

Type of learning	Description
Stimulus enhancement	B learns from A to what to orient behaviour
Observational learning	B learns to what circumstances a behaviour should be a response
Imitation	B learns from A some part of the form of a behaviour
Goal emulation	B learns from A the goal to purpose

much faster than genetic evolution. Why? One crucial element of an answer seems to be the method of transmission. If memes are replicators, they replicate in very peculiar way: in sharp contrast to genes, replication (in a broad sense) of memes passes through the phenotype, as it were (Maynard Smith and Szathmáry, 1999). In the case of genes DNA, structure is directly copied, the expressed state of a gene (i.e. the protein) is not 'reverse translated' into proteins. This is not true for memes. Memes are recreated anew for each brain based on environmental input and recruitment of memes already stored in the brain. The process is more similar to prion 'replication' whereby the *conformation* of a protein rather than its sequence is passed on (Szathmáry, 2000). Aunger (2002) in his recent book analyses and refers to this type of replication as conversion. I have used the terms 'phenotypic replicator' or 'imitator' to emphasize the difference between gene and meme replication. In sum, during meme replication there is no copying of the underlying neural structure, only the information is passed on during this peculiar process, without material overlap (Figure 10.3).

The mode of meme replication entails a much faster rate of evolution, of course, hence a higher degree of historical influence. No wonder culture is so varied across time and space in human populations. There is a very strong Lamarckian component due to passing through the level of expression. Moreover, memes sharing the same phenotype in one dimension may turn out to be utterly different in another, partly due to the inertia involved in the process of conversion. It is perhaps fair to say that Newton's second law means a different thing to almost every individual. Testing in physics exams ensures a higher degree of, but surely not complete, convergence of phenotypes. The good side of it is the innovative capacities that are partly due to the 'un-pruned' parts of the memes.

It is perhaps instructive to consider a molecular analogy with meme replication further (Szathmáry, 2000). Is it simply reverse translation? It if were, it would not be conversion. Suppose there are two individuals A and B. Take protein X from individual A. Suppose you want to enable individual B to develop a molecule with the same phenotypic effect (enzymatic function,

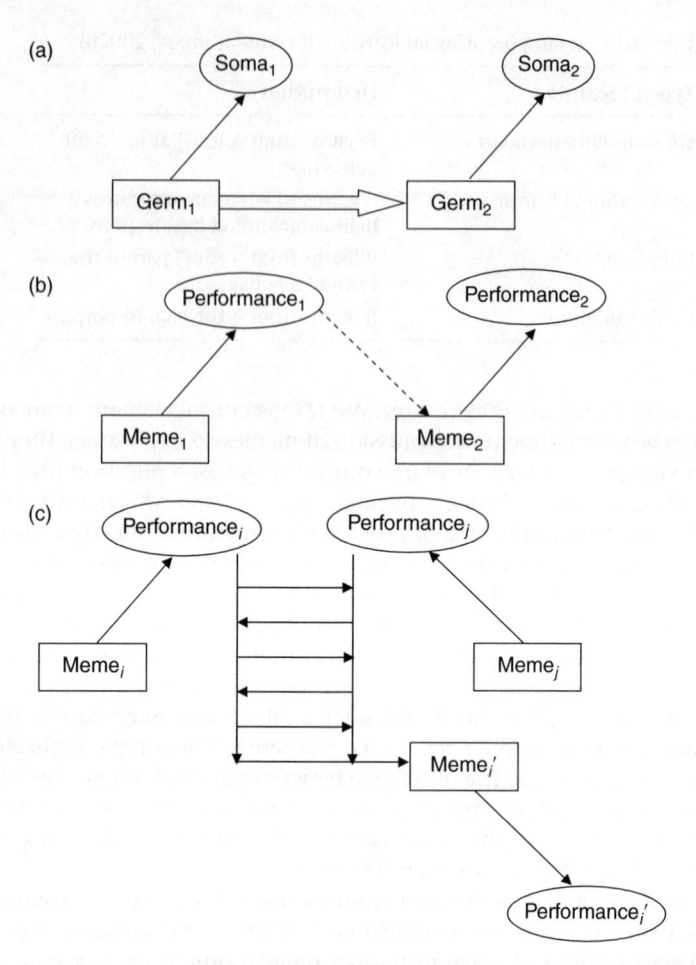

Figure 10.3 Memes and Lamarckian inheritance. (a) The Weissmanist segregation of soma and germ line; (b) transfer of memes passes through the performance level, which is mostly absent in the molecular world; and (c) in most cases a meme becomes multiplied by the interactions of two memes*

* There is interaction between memes i and j happening at the performance level, until meme i ('pre-existing knowledge') is transformed into meme i'. Such a process happens during teaching. The intervention of the performance level results in high variability (after Szathmáry, 2002a).

for example). If gene transfer from A to B is not allowed, one then must have (i) some generative mechanism for proteins in B, and (ii) some method for the assessment of phenotype. This comes very close to an immune system in B. The crucial difference is that the task now is to produce 'antibody' Y, in individual B, that shares crucial phenotypic properties with 'antigen' X, from individual A. Although both molecules would have sequences, they would most unlikely be close to one another in protein space (Maynard Smith, 1970). In all probability the other (not directly selected for, but existing) effects of the two proteins would differ. This is why cultural heredity is bound to be inexact, and cultural evolution is faster than biological evolution.

'The past as they key to the present': molecular phylogeny and paleomolecular resurrection

Due to the chemical nature of traditional inheritance, resting on DNA replication, phylogenetic reconstruction of evolution using molecular sequences is, fortunately, possible. The method is widely used nowadays for historical reconstruction as well as for testing evolutionary hypotheses. It is a means to follow the paths evolution has chosen. Recently, however, a new twist to the story has been added. It is called paleomolecular reconstruction (e.g. Benner, 2002). The main idea is that a good phylogeny should be able to retrodict ancestral, but functional, states of the molecule on question. Some attempts at such resurrection have met with considerable success; the recent work of Zhang and Rosenberg (2002) is a case in point.

They have analysed two molecules from eosinophilic lymphocytes. Since their names do not reveal much about their function (poorly understood anyway), let us call them by their acronyms: EDN and ECP. These are paralogs, i.e. they arose in an act of gene duplication about 30 million years ago in the lineage leading to humans and Old World monkeys. They seem to be related to a digestive ribonuclease (an enzyme decomposing RNA) in a certain group of mammals (including ox, deer, antelope, etc.). Zhang and Rosenberg wanted to find out about the *present-day* function of EDN and ECP in Old World monkeys. *In vitro* studies indicated that ECP kills bacteria and EDN inactivates retroviruses. Phylogenetic analyses have suggested that these proteins are likely to have been under positive selection around the time of gene duplication. The experimentally reconstructed last common ancestor of these proteins has a very low ribonuclease and antiviral activity (Figure 10.4). Two amino acid replacements in the EDN protein, relative to the ancestor, confer a high activity. Thus it is likely that indeed the positively selected function is the one that modifies *in vitro* behaviour relative to that of the ancestral state. There is no better example to show the awesome power of history in biology.

Major transitions in evolution

Table 10.2 gives a list of what lately have been called the 'major transitions in evolution' (Maynard Smith and Szathmáry, 1995). There are a few remarkable features of this table. Some major transitions in evolution (such as the origin of multicellular organisms or that of social animals) occurred a

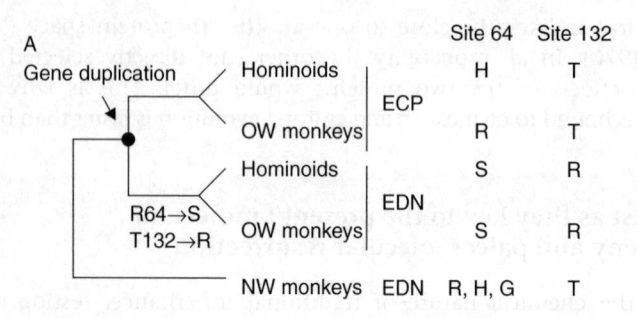

```
                                        Site 64   Site 132
A
Gene duplication  ┌── Hominoids          H         T
          ↓      ┌┤       ECP
          ●      └── OW monkeys          R         T
          │      ┌── Hominoids           S         R
          │      ┤        EDN
  R64→S   │      └── OW monkeys          S         R
  T132→R  │
          └──── NW monkeys  EDN    R, H, G         T
```

Figure 10.4 Schematic evolution of ECP and EDN proteins*

* Numbers indicate amino acid positions in the polypeptide change. Amino acids designated by their one-letter code: H = histidine, T = threonine, R = arginine, S = serine, G = glycine (after Zhang & Rosenberg, 2002).

Table 10.2 The major transitions in evolution

Before	After
Replicating molecules	Populations of molecules in protocells
Independently replicating genes	Chromosomes
RNA as gene and enzyme	DNA genes, protein enzymes
Bacterial cells (prokaryotes)	Cells with nuclei and organelles (eukaryotes)
Asexual clones	Sexual populations
Single-celled organisms	Animals, plants, and fungi
Solitary individuals	Colonies with non-reproductive castes
Pre-linguistic societies	Human societies with language

Reproduced from J. Maynard Smith and E. Szathmáry (1999) *The Origins of Life*. Oxford University Press.

number of times, whereas others (the origin of the genetic code, or language) seem to have been unique events. One must be cautious with the word 'unique', however. Due to a lack of the 'true' phylogeny of all extinct and extant organisms, one can give it only an operational definition. If all the extant and fossil species which possess traits arising from a particular transition share a last common ancestor after that transition, then the transition is said to be unique. Obviously, it is quite possible that there had been independent 'trials', as it were, but we do not have comparative or fossil evidence for them (Szathmáry, 2002b).

Contingent irreversibility

Once a transition has occurred, in many cases there seems to be no way back. But there are exceptions. For example, there are insects whose solitary state is secondary: all their living relatives are social. But, in contrast, there is no mitochondrial cancer; i.e. mitochondria, our energy-producing cell organelles, cannot reproduce out of control. This can be understood, given the fact that most mitochondrial genes had been lost in evolution; a fraction had been moved to the cell nucleus; and very few genes remain in the organelle. Emphatically, all genes necessary for the division of this organelle have moved to the nucleus, hence the latter is in complete control of mitochondrial division. We can thus appreciate contingent irreversibility as a key mechanism for 'locking-in' the result of a transition. It is not the case that reversal would be *logically* impossible, it is just far too demanding on the side of the requisite heritable variation: the number of simultaneous, chance genetic changes enabling the reversal is so large (their joint probability is so small) that *for all practical purposes* we can assume that they will not happen.

Truly unique transitions

Despite the fact that we cannot be certain about the existence of truly unique transitions, it is plausible that some of the transitions nonetheless belong to this category. The origin of the eukaryotic cell (cells with a well-formed cell nucleus, such as our cells; Cavalier-Smith, 1987) and the emergence of language (Bickerton, 1990) are good candidates. Uniqueness of a transition in such cases means that there is some sense in which it is 'difficult'. This difficulty can be due to either of two reasons, or a combination of the two; namely, a unique transition may be limited by genetic variation or by natural selection. A variation-limited transition is difficult if the set of the requisite genetic variations is very unlikely to arise. Limitation by selection means that the right conditions for the spread of the appropriate genetic variation is very special and unlikely. A sub-category within selection-limited transitions is the case of 'pre-emption': although the first transition by itself is not so difficult, it modifies the conditions to such an extent that a second, independent trial becomes virtually impossible.

Repeated evolution

If some of the transitions have been truly unique, this raises the question whether evolution would repeat itself essentially in the same way. As we have seen, De Duve (2002) thinks that all the transitions were inevitable. Let me recapitulate in part my criticism (Szathmáry, 2002c) on this matter. In chapter 12 ('The Arrow of Evolution'), central to the main scientific thesis, he distinguishes between two directions of evolution: horizontal and vertical, roughly coinciding with micro- and macroevolution, respectively. The claim is that the first is littered with contingency, whereas the latter is self-guided along certain trends towards increasing complexity. The main reason is, he thinks, that in microevolution the population can respond by many alternative mutations to the same selective challenge, whereas in macroevolution the range of genetic changes is severely limited. De Duve says that 'it is self-evident that the universe was pregnant with life and the biosphere with man', and maintains that claims to the contrary are logically fallacious. 'The universe has given life and mind. Consequently, it must have had them, potentially, ever since the Big Bang.' I think De Duve confuses (if I may borrow a phrase from Jacob) the possible with the actual. Any married couple could be said to have had the potential to give birth to their actual children, but knowing the chance elements in marriage, in meiosis and in fertilization, they can be said to have had in their potential, for all practical purposes, a countless number of different children. Having *those* and *such* children that they actually have was by no means inevitable. The genetic constitution of the actual children will critically determine many aspects of family life. I share the *belief* with De Duve that life had to originate in some more or less straightforward manner, but lacking a convincing scenario, we cannot calculate probabilities, and *cannot know*, therefore, whether this belief is true or not. I am inclined to use contingency exactly opposing De Duve. The point that with increasing complexity of the organism the fraction of permissible genetic changes (not opposed by negative selection) decreases does not reduce, but increases contingency in the *set* of possible lineages: recall the condition that for complex organisms $Nu \ll 1$ is likely to hold. Although alternatives of future evolution are reduced in number, alternatives of existing present forms are more markedly delineated from each other, due the constraints of previous turns. To use Lewis Wolpert's example, one cannot evolve angels with wings from humans, because of the past evolution of the relevant developmental mechanisms. This increased contingency would lead to inevitability only in the extreme case when, from, say, the level of bacteria upwards the number of options at major steps were reduced to a minimum, preferably one. In any other case previous turns exclude many feasible alternatives in the future. The fact that there are no asexual gymnosperms is possibly due to the condition that the egg delivers the mitochondrion and the pollen brings the plastids. Lack of parthenogenesis in mammals is partly explained by the existence of genomic imprinting. It is not unimaginable to figure out what genetic changes would be needed to allow for the appearance of asexual variants; it is just practically impossible (highly unlikely) that the genetic systems could produce them at once. Given 1000 Earth-like planets with the same initial conditions, and granted 8 billion years (to be on the generous side), how many of them would evolve eukaryotes? Vertebrates? Primates? Humans? De Duve's tacit position is that the majority would have man. If this case could be proven, it would be the most important discovery in evolution, more important than the idea of natural selection. But De Duve fails to prove his case.

Irreversibility in evolution and Dollo's law

In the 19th century Dollo stated that an anatomical feature lost in evolution would not reappear, even if the selection pressure favouring its first appearance returns. The statement is, of course, about contingent irreversibility at a smaller scale than the major transitions. Back mutations and recombination could in principle resurrect many features, but this will be statistically unlikely. Marshall *et al.* (1994) offer an interesting re-analysis of this issue, concentrating on the evolutionary resurrection of genes.

A gene can become silenced due to a change in the regulatory sites allowing for its transcription. Another change later may restore the transcription. Will the gene still be of use then? If yes, its resurrection will be successful. In order to assess the odds one must calculate the probability that a gene retains a functional sequence:

$$P(retain\ function) = P(survive\ point\ mutations)\ \Box\ P(survive\ frameshifts).$$

The above expression will depend on the elapsed time since silencing, the mutation and frameshift rate, and the degree of neutrality. Analysis of existing proteins gives a good estimate of *P(retain function)*. It turns out that genes can be easily resurrected after a few million years, but then the chance drops sharply.

If speciation is fast in a clade (a group of organisms descending from a common ancestor), such as in some salamanders, then the same gene can be silenced and inactivated a number of times independently in the different species. Salamanders are particularly interesting because they have the phenomenon of adult 'larvae', such as in the axolotl that fail to mature fully (retain gills, for example). Such species are called 'neotenic'. Artificial input of the hormone thyroxine turns the axolotl into a usual adult form, meaning that the genes for metamorphosis are still functional. Other salamander species are permanently neotenic as they fail to respond to exogenous thyroxine. The clade of Mexican ambystomatid salamanders includes metamorphosing, permanently and facultatively neotenic species. Their phylogenetic tree (Figure 10.5) testifies to the existence of 'flickering' developmental traits. Note that the crucial condition that the whole radiation took place in the last 10–12 million years, thus on the average there has been more than one speciation event per one million year.

Evidence for long-term reversals (such as the inducible teeth of hens) remains highly controversial. There are some good examples, however, such as the re-appearance of the second molar (M2) tooth in *Lynx*. Closer inspection revealed that this 'atavistic' change resulted from a simple change in the developmental mechanism that had been maintained by selection throughout and possibly did not require the resurrection of any gene.

Convergent evolution

If biological evolution is so critically path-dependent, then explaining convergent phenomena, such as the cephalopode/vertebrate eye, poses a special

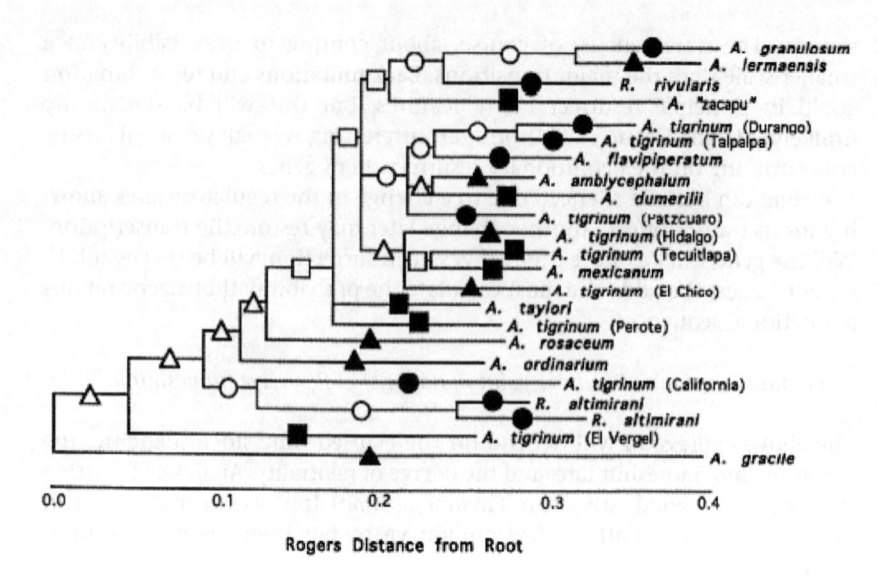

Figure 10.5 The radiation of Mexican salamanders*

* The trait of metamorphosis is flickering in this set of species. Solid symbols indicate extant developmental modes: circles, metamorphosing; squares, neotenic; triangles, facultative. Open symbols refer to inferred ancestral developmental modes (after Marshall *et al.*, 1994).

problem. In his case the genes, specifying strikingly similar organs, had to be created *de novo* rather than inherited from a common ancestor and then resurrected. Bearing in mind that there are some 40 known different ways of making an eye, this convergence must be explained by a set of special features shared in the evolution of these two lineages.

Why is it that independent evolutionary paths can converge on analogous solutions? The reason is that in some cases there is a limited set of alternative 'engineering' solutions to the same problem (Maynard Smith, 1986) that evolution may indeed discover time and time anew. These independent hits must be seen in the right context, however. The convergence to the same basic eye structure in cephalopods and vertebrates is striking, but one should not forget that some kind of eye evolved about 40 times independently (Dawkins, 1996). Depending on the niches of populations, the space of engineering solutions may be severely or loosely constrained.

The evolution of colour vision gives an excellent example (Pichaud *et al.*, 1999). Vision is the processing of information carried by light. Thus the nature of light is expected to constrain what an animal can do with it. Direct light is rich in short wavelengths whereas reflected light is more in the green/yellow range. It is revealing that primitive arthropods are indeed sensitive to UV and green wavelengths. It is this two-pigment system that evolved into more complex forms.

It is known that humans use rods for light detection and cones for colour detection. The insect retina seems to have a convergent solution: short and long visual fibres seem analogous in this respect to rods and cones, respectively. Opsins for UV, blue, and green have evolved independently in insects and vertebrates. Also, the one photoreceptor: one pigment gene regulation pattern has also been independently arrived at in these lineages. Whereas the optical structure of the eye varies widely across taxa, there is remarkable convergence in the way colour vision is organized.

Concluding remarks

All known individuals execute a genetic programme which drives them through a series of irreversible changes. Without some type of epigenetic effects or learning this by itself does not amount to path dependence on the individual level. Learning, on the other hand, is able to generate different paths even under the same gross environmental circumstances. Social learning, by accumulating cultural change, can by itself differentiate between otherwise identical genetic populations. Most of the cultural differences (including language) are due to such progressive differentiation between local human populations.

The very existence of inheritance ensures that the thermodynamic state variables do not fully determine the state of biological populations. When heredity is able to produce and propagate enough variants, cumulative selection can lead to progressive genetic differentiation between replicate biological populations. Selection may or may not be guided by multiple dynamical attractors. When there are multiple attractors, the strong form of path dependence is apparent. But even with a single domain of attraction there is likely to be a conspicuous historical effect: if population size is small enough (a condition likely to be met by complex organisms with large physical size), then a different set of adaptive mutations will be fixed: the trajectories in genotype space of replicate populations will not cross (the weak form of path dependence).

Macroevolution, especially the major transitions, is littered with path dependence. Language, for example, evolved only once (in our species). Had we had different basic means of communication (such as a two-dimensional screen on our forehead), syntax would surely have evolved in a markedly different way. It will be fascinating to see how life on other Earth-like planets evolved into various forms.

References

Aunger, R. (2002) *The Electric Meme: A New Theory of How We Think*. Free Press.
Barrow, John D. & Tippler, Frank J. (1988) *The Anthropic Cosmological Principle*. Oxford University Press.

Bassanini, A. & Dosi, G. (1999) 'When and how chance and human will can twist the arms of Clio'. *LEM Working Paper Series 1999/05*. Pisa.
Benner, S. A. (2002) 'The past is the key to the present: resurrection of ancient proteins from eosinophils'. *Proc. Natl. Acad. Sci. USA* 99, 4760–1.
Bickerton, D. (1990) *Language and Species*. The University of Chicago Press.
Cavalier-Smith, T. (1987) 'The origin of eukaryotic and archaebacterial cells'. *Ann. N. Y. Acad. Sci.* 503, 17–54.
Changeux, J.-P. (1983) *L'Homme Neuronal*. Librairie Arthème Fayard, Paris.
Dawkins, R. (1976) *The Selfish Gene*. Oxford University Press.
Dawkins, R. (1996) *Climbing Mount Improbable*. W. W. Norton.
De Duve, C. (2002) *Life Evolving: Molecules, Mind, and Meaning*. Oxford University Press.
Eigen, M. (1971) 'Self-organization of matter and the evolution of biological macromolecules'. *Naturwissenschaften* 58, 465–523.
Jablonka, E. & Lamb, M. J. (1995) *Epigenetic Inheritance and Evolution*. Oxford University Press.
Jastrow, R. (1971) *Red Giants and White Dwarfs*. Harper and Row, New York.
Johnson, P. A., Lenski, R. E. & Hoppensteadt, F. C. (1995) 'Theoretical analysis of divergence in mean fitness between genetically identical populations'. *Proc. R. Soc. Lond B* 259, 125–30.
Lenski, R. E. & Travisano, M. (1994) 'Dynamics of adaptation and diversification: a 10,000 generation experiment with bacterial populations'. *Proc. Natl. Acad. Sci. USA* 91, 6808–14.
Marshall, C. R., Raff, E. C. & Raff, R. A. (1994) 'Dollow's law and the death and resurrection of genes'. *Proc. Natl. Acad. Sci. USA* 91, 12283–7.
Maynard Smith, J. (1970) 'Natural selection and the concept of a protein space'. *Nature* 225, 563–4.
Maynard Smith, J. (1982) *Evolution and the Theory of Games*. Cambridge University Press.
Maynard Smith, J. (1986) *The Problems of Biology*. Oxford University Press.
Maynard Smith, J. & Szathmáry, E. (1995) *The Major Transitions in Evolution*. Freeman, Oxford.
Maynard Smith, J. & Szathmáry, E. (1999) *The Origins of Life*. Oxford University Press.
Orgel, L. E. (1992) Molecular replication. *Nature* 358, 203–9.
Pichaud, F., Briscoe, A. & Desplan, C. (1999) 'Evolution of color vision'. *Curr. Op. Neurobiol.* 9, 622–7.
Smith, T. F. & Morowitz, H. J. (1982) 'Between history and physics'. *J. Mol. Evol.* 18, 265–82.
Smolin, L. (1999) *The Life of the Cosmos*. Oxford University Press.
Szathmáry, E. (2000) 'The evolution of replicators'. *Phil. Trans. R. Soc. Lond. B.* 355, 1669–76.
Szathmáry, E. (2002a) 'Units of evolution and units of life'. In: Pályi, G., Zucchi, L. & Caglioti, L. (eds) *Fundamentals of Life*, pp. 181–95. Elsevier, Paris.
Szathmáry, E. (2002b) 'Cultural processes: the latest major transition in evolution'. In: Lynn Nadel (ed.), *Encyclopedia of Cognitive Science*. Macmillan Reference, London.
Szathmáry, E. (2002c) 'The gospel of inevitability: was the Universe destined to lead to the evolution of humans?' *Nature* 419, 779–80.
Wahl, L. M. & Krakauer, D. C. (2000) 'Models of experimental evolution: the role of genetic chance and selective necessity'. *Genetics* 156, 1437–48.
Wächtershäuser, G. (1988) 'Before enzymes and templates: theory of surface metabolism'. *Microbobiol. Rev.* 52, 452–84.

Wächtershäuser, G. (1992) 'Groundworks for an evolutionary biochemistry: the iron–sulfur world'. *Prog. Biophys. Molec. Biol.* 58, 85–201.
Yedid, G. & Bell, G. (2002) 'Macroevolution simulated with autonomously replicating computer programs'. *Nature* 420, 810–12.
Zhang, J. & Rosenberg, H. F. (2002) 'Complementary advantageous substitutions in the evolution of an antiviral RNase of higher primates'. *Proc. Natl. Acad. Sci. USA* 99, 5486–91.

Part IV
Institutional Inertia

11
The New Institutional Economics: Can It Deliver for Change and Development?

*Jeffrey B. Nugent**

The New Institutional Economics (NIE) has developed in order to find answers to the following kinds of questions that could not be answered (at least until very recently) by traditional economics (either neoclassical or Marxian):

(1) Why are institutions so different from one time and place to another?
(2) Why are they the way they are?
(3) How and to what extent do they explain differences in productivity?
(4) Why and how do inefficient institutions get locked in?
(5) Of all the institutional differences that exist, which ones really matter?

Clearly, these questions suggest the possibility that the relationships may be two-way, i.e. environmental conditions can influence institutions but at the same time institutions can affect economic performance and thereby environmental conditions over time.

Traditional economics generally assumed institutions to be given. Neoclassical economics derived many interesting propositions based on the assumption of a set of complete product and factor markets without explaining how these markets arose. Likewise, Marxian economics derived various behavioral propositions based on the assumption of classes and their objectives without explaining where the classes came from.

The impact of the NIE on economics or indeed social science in general has been sufficiently large as to make it difficult in recent years to draw the line between what is NIE and what is not. This is because, by bringing the need to explain institutions to the foreground of economics and social science, the NIE has elicited new explanations from economists (e.g. agency theory Jensen and Holmstrom, and public choice theory) and other social scientists of many persuasions, in many cases incorporating some of the insights of the NIE into otherwise more traditional economic models and methodologies. Analytically, it has long been deemed useful to distinguish between the demand for institutions and the supply of institutions (Nabli and Nugent 1989; Ruttan 1989). The former derives from information and transaction costs while the latter focuses on problems of collective action and credible commitments.

Before proceeding to the demand for institutions in section II and the supply of institutions in section III, section I that follows defines institutions and identifies some relevant concepts used in subsequent sections. Section IV identifies problems arising in applications of NIE to change and development. Section V illustrates some ways in which some of these difficulties may be overcome. Section VI concludes.

Some definitions and concepts

By 'institution' is meant a humanly devised set of rules that can affect or even govern behavior, both economic and political. The rules can be either formal or informal. Constitutions and law codes are important examples of formal institutions while social norms and conventions are examples of informal ones. Markets are institutions that may have either formal or informal rules. Contracts are also institutions, perhaps the simplest institutions to analyze since the rules may be quite explicit and the number of involved parties may be as few as two. In practice, the usefulness of contracts in affecting or regulating behavior depends on other institutions like law courts, penalties for violating contractual terms, the character of the relevant markets and other environmental conditions.

A distinction is made between (individual) 'institutional arrangements' and 'institutional structure'. The former is the set of rules that govern behavior in a particular domain whereas the latter is the totality of institutional arrangements in an economy at a particular point in time. An important reason for why some institutional arrangements like contracts are interrelated with others like legal institutions is that contracts are 'incomplete'. They do not spell out all the contingencies either because the costs of doing so are too high due to the large number of relevant contingencies or because of 'bounded rationality' in that people do not want to deal with such detailed contingencies even if they had been identified.

Typically, institutions serve one or more functions. This is why they are often so important or even indispensable. Primary among these functions is often the desire to economize, i.e. to make one or more agents better off without making others worse off. There may be many different ways of achieving this economizing function, such as to improve the efficiency of production, to make individual or group behavior more predictable, thereby reducing risk and uncertainty. Reducing risk, with no change in efficiency, can be the dominant function in some situations. The risk reduction and efficiency increasing functions are likely to be interdependent in that, by making behavior more predictable, people are less likely to make mistakes in reacting to the various signals they receive.

Another way to reduce risk and/or increase efficiency may be to obtain more accurate and up-to-date information, e.g. on the prices at which similar transactions are based. Typically, there are significant 'informational asymmetries' among different agents, meaning that different agents engaged in economic activities differ in the quality and quantity of the information to which they have access. For example, a seller may have better information about the product or service he is selling than the buyer. Informational asymmetries give rise to two distinct problems, 'adverse selection' and 'moral hazard'. In the presence of information asymmetries, e.g. between a seller and a buyer of insurance, adverse selection arises when given a fixed premium announced by the seller, the buyers of insurance (when such purchase is voluntary) are primarily those who represent greater risks since those with lower risks may not find it beneficial to buy the insurance. On the other hand, moral hazard arises when, once insured, the insured individuals take greater risks than they would have in the absence of insurance. Both of these problems tend to undermine the economic viability of insurance. In some situations, adverse selection is the more serious problem but in other cases it may be moral hazard.

There may exist institutional means of reducing these informational deficiencies and asymmetries by offering incentives to the participants to reveal full information. Nevertheless, even so, informational asymmetries are endemic. The presence and direction of such asymmetries may disadvantage one party or group relative to another. If so, it may allow one party to practice 'opportunistic behavior' (cheating, shirking on effort, fudging on the quantity or quality of the item to be sold or the information delivered) vis-à-vis the other. But knowing this, the disadvantaged party may change his behavior or react to this situation in a certain way. For example, the disadvantaged buyer of a used car typically reacts to this situation by suspecting that the seller wants to sell a product that is of poor quality ('a lemon'). As a result, he may offer only a low price or even nothing at all for it, thereby tending to undermine the very existence of such a market (resulting in 'market failure').[1] Similar informational asymmetries crop up in factor markets, where it may be difficult to observe the effort of a worker or that of a debtor to repay a loan. Workers and debtors have better information on their respective efforts than their employer and creditor counterparts. Naturally, if the unobserved effort is linked to measurable output or performance, this problem can be reduced. But, real world conditions often accentuate these difficulties in that random shocks like rainfall or injury tend to weaken the relationship between effort and performance.

Since different institutional arrangements can have very different distributional implications, redistribution (rather than pure economizing) constitutes another important function of institutions. If there is little competition among alternative institutional arrangements and any one individual or group is better able to impose its favored rules on others, the distributional function may dominate over the economizing one. Institutions may help to preserve power imbalances. Relative power is therefore an important determinant of institutions. When power is more equally distributed and competition among alternative institutional arrangements is stronger, the economizing function of institutions may dominate.

A particular institutional arrangement may function better when it is embedded in a more appropriate institutional structure. Hence, the efficacy of any institutional arrangement depends on how well it copes with opportunism on the part of the different relevant agents, but this in turn depends on other institutional arrangements in the institutional structure. Sometimes the elements of the institutional structure that matter most are very subtle. As a result, a very subtle change in either environmental conditions or in a particular rule within a particular institutional arrangement can make a big difference in the functioning of an institutional structure.

Some NIE practitioners ('functionalists') tend to assume that there exists sufficient competition – actual or potential – among alternative institutional arrangements, that the institutions that exist should always be efficient. But, in practice, the assumption may not be justified. Indeed, even if a given institutional arrangement is efficient (i.e. it economizes better than the available alternatives), over time as circumstances change it may no longer be so. Institutional rigidities and inertia may impede changes. Inefficient institutions may get locked in. In the extreme, inefficient institutions may persist even when no one benefits from them. More likely, they may persist when only some agents benefit from them and/or when governments or other powerful agents or groups fear that the emergence of more efficient institutional arrangements would undermine their power.

The demand for institutions

As noted above, institutions can reduce costs and risks of various kinds and increase efficiency but they require costs to establish and operate them. These costs, broadly defined as transaction costs, include those of organizing, negotiating, communicating, maintaining, monitoring, and enforcing the rules of an institutional arrangement. These transaction costs also include the costs arising by any party's failure to abide by the rules. Some of these costs, *ex ante* costs, occur before the transaction affected by the institution takes place whereas others, the *ex post* transaction costs, occur after it. The latter include the costs of adjusting to opportunistic behavior and of dispute resolution.

Contracts are an example of the kinds of rules that try to reduce the transaction costs that arise from the fact that different inputs are 'owned' or provided by different individuals, groups or firms. In many contexts, there is open competition between different forms of contract. In the context in which owners of land or capital need to contract with owners of labor, there are typically three main contracts to choose from: wage contracts, rent contracts and share contracts (the latter being a mix, wherein the owners of land or capital agree to share the proceeds with the owners of the labor). Contracts serve to illustrate the nature of the transaction costs and how they differ from one type of contract to another.

In a 'fixed' wage contract the worker receives a fixed compensation per hour or day; in a fixed rent contract the worker pays the owner of the assets (capital or land) a fixed rent, in a share contract the parties agree to share the output in a pre-specified way. Each of these forms of contract is subject to opportunistic behavior of one or more kinds. With a fixed wage contract, the worker lacks incentive to supply effort. With a fixed rent contract, the worker is the residual claimant and hence has incentive to supply effort but also an incentive to over- or misuse the assets at the expense of the capital or landowner. With a share contract, there is some but considerably less incentive for each of these forms of opportunistic behavior, but also an incentive by one of the parties to underreport the output (or over-report the inputs if inputs are also shared). These contracts also differ in terms of who bears the various sources of risk. For example, with respect to production risk, the owner bears all the risk in a fixed wage contract, the worker in a fixed rent contract and the risk is shared in a share contract. Each of these contracts can be modified in a way that may mitigate one form of opportunism but often at the expense of others and or raising supervision or contracting costs.[2]

Transaction cost economics suggests that environmental conditions determine which form of opportunism is likely to be most important and therefore allows one to predict which form of contract or other institutional arrangement will be chosen. For example, in labor-intensive activities or activities in which labor inputs are especially hard to monitor where labor-shirking should be the most important form of opportunism, fixed rent contracts may be the transaction cost-minimizing ones. But, where capital-intensity is high and productive activities vulnerable to asset misuse, wage contracts may be the transaction cost-minimizing ones. With conditions intermediate between the two and possibly with neither of the parties to the contract able to bear all the risk, share contracts may be the best alternative.

An especially important and common problem in contracting arises from asset specificity in a particular transactional relationship. For example, one firm (A) buys a specialized part from another firm (B) that requires a certain fixed asset for its production. That asset is useful in producing the part for firm A but is of little use in producing anything else. Once B makes that investment, firm A can exploit this situation by lowering the price it is willing to offer firm B to the marginal variable cost of the product, exclusive of the fixed cost. But, knowing this, firm B may under-invest in the specific asset, raising the unit costs for firm A. This problem is known as the 'holdup problem'. The relationship between the two firms is clearly vulnerable to exploitation and opportunism by one or both parties.

One solution to the holdup problem is for firm A to make a long-term commitment to firm B to buy the product at a price no lower than that offered by other firms or the estimated average total unit cost or at a pre-specified fixed price. This solution, however, can be undermined by the fact that the long-term contract at a pre-specified price may give the supplier an incentive to fulfill the contract by supplying at low quality or when the firm's commitment to long-term purchase is either not credible or too costly to enforce. Another solution is 'vertical integration' wherein Firm A buys out Firm B (or B buys out A) to assure that the efficient investments are made, thereby avoiding the need for complicated contracts that are costly to monitor and enforce. Yet, this solution may lead to monopoly and unnecessarily large size and hence additional costs.

The hold-up problem is symptomatic of a whole set of similar problems involving government, firms and households known as 'time inconsistency problems'. For example, a government may want to reduce the expected rate of inflation and thereby increase the demand for money and reduce the rates of inflation and currency depreciation. Suppose it does so by declaring that from now on it will practice tight monetary and fiscal policy. But, firms and households realize that in the short run, that government has an incentive to accelerate the money supply so as to collect the inflation tax, increase employment, output and tax revenues. As a result, promises of fiscal responsibility and monetary tightness are often not credible. An institutional innovation may be needed. For example, a government may be willing to 'tie its own hands' or make itself a 'hostage', such as by introducing a rule wherein control of the money supply is delegated to an independent and conservative Central Bank Board. Or it could put a clause in the constitution of the country (or make a treaty or monetary union with another group of countries) spelling out the penalties were the commitment not honored.

The supply of institutions

While the demand for institutional arrangements may be a necessary condition for its emergence, it is not a sufficient one. An important reason for this is that institutions have a public goods character. For example, a new law or regulation in the country applies to everyone and may generate benefits for many. But, therein lies the 'collective action' problem.[3]

This problem arises because, if everyone in a group benefits from the collective effort, none may have sufficient incentive to undertake the costs of this effort. Each has an incentive to 'free ride' on the efforts of others in the group. Yet, if enough choose to free-ride, collective action may fail, implying that the new institutional arrangement may not come into existence.[4]

The theory of collective action provides a number of rather intuitive hypotheses about conditions in which the free-riding problem may be more likely solved. Some of these have to do with the characteristics of the groups whose interest it is to bring about the new arrangement. Small groups whose members have known each other for a long time, among whom communication costs are low, with homogeneous backgrounds, easily identifiable to one another, and who trust each other may be able to overcome free-riding problems more easily than others. Also, the more threatening is the result of inaction, the more likely the group will be able to overcome the freerider problem. Somewhat less intuitive are hypotheses that (1) the greater the heterogeneity of goals, and (2) the greater the wealth inequality among group members, the more likely the group will be able to overcome free-riding and hence succeed in collective action.

While most collective action theory consists of static hypotheses of this sort, there are some of a more dynamic nature, such as that the actions of one group may be affected by the prior or foreseen actions of other groups and/or by the extent to which one can 'exit' from the group. For example, when exit from the group is possible, people dissatisfied with the status quo, instead of pushing for institutional change by exercising 'voice', may simply exit, weakening the potential for collective action (Hirschman 1970). Yet, those for whom the exit option is most attractive may have the greatest influence within that group in designing a new institutional arrangement.

Groups with characteristics that are not in themselves conducive to success in collective action may still be able to succeed by making use of selective incentives (defined as private goods that can be used to induce people to join in the collective action who might otherwise want to free-ride). Even without selective incentives, group members may be more inclined to participate in the collective action if they have a positive taste for 'solidarity' or for participating. Also, political entrepreneurs may also be willing to mobilize otherwise latent groups into collective action because doing so may further their own objectives. Because of the coercive powers of the nation state, a common means of achieving collective action is to get the state to do it. As a result, politically powerful political entrepreneurs can play major roles in collective action.

The state itself is a set of institutional arrangements, suggesting that its own rules and modus operandi can be treated by NIE. Some view the state as the resultant of collective action by different groups in a society, each exercising a different degree of leverage over it. Another contrasting view is of the 'hard' state, one that is not very permeable to pressures from civil society. Except in situations where constitutions are openly violated or suspended, constitutions provide some rules for how decisions of the state are made, and also for how the constitution can be changed (Brennan and Buchanan 1987).

In well-functioning democracies, the voters may be able to exercise influence over the composition of the legislative and executive branches of government and thereby over various government policies. Political economy modelers, therefore, typically characterize decisions as being made either directly by the median voters or with the median voter acting as a principal exercising some degree of control over the bureaucrat-agents who may be 'rent-seekers', i.e. trying to satisfy their own private interests and objectives. The rules of election and means of remunerating government officials can play an important role in determining how and why certain decisions are made. In some situations, where political parties are strong, government decisions may be largely the result of intra- and inter-party bargaining or vote-trading. In other less democratic countries, the head of state typically has greater power. With coercive powers of the state at his disposal, heads of state may be able to design the institutional framework to maximize their own interests, subject to the constraint of not doing something that would lead them to be overthrown.

As a result, it is by no means necessary that governments act in the social interest. Even more, because of the distortions that can result from variations in collective action across different groups, inefficiencies are endemic. Not only does free-riding prevent desirable institutional arrangements from appearing but the transaction costs of institutional change often produce institutional inertia. Because of the importance of institutional inertia, the importance of institutional precedent, network externalities (where the benefits to one user of the network depend on how many other users there are), relative stability in the strength of various interest groups, institutions and institutional changes are also likely to be 'path dependent'. The search and experimentation with alternative institutions is also likely to be local, making minor modifications in what is familiar. In other words, institutional history is likely to matter greatly. With so much inertia, even institutional arrangements that were optimal at one point in time may not be so at all subsequent times.

Shortcomings of NIE as a paradigm of change

From the breadth of the above ingredients of NIE, it should be clear that the NIE may be able to explain many different types of institutions, and in particular to explain why institutions may be different in different environmental conditions. When there is great demand for change and also the conditions for success in collective action for such a change are satisfied, the NIE is perfectly capable of explaining change. At the same time, for the following reasons, NIE cannot always be counted on to accurately predict institutional change.

First, since most institutions have a multiplicity of different rules, it is often difficult to know which specific rules make the most difference in explaining differences in outcomes. Different NIE practitioners may come to very different conclusions even if they use the same hypotheses.

Second, most NIE is rather static, implying that, despite useful concepts like path dependence, institutional inertia, and dynamic group interactions, the dynamics of NIE are not well developed. And yet dynamics may be more crucial in explaining change than statics.

Third, although much progress has been made over the years in devising innovative measures, typically both transaction costs and strength of collective action are easily measured without reference to the outcomes which involve the hypotheses to be tested.

Fourth, since many organizations, states included, are not very transparent about the way in which decisions are arrived at, it is often difficult to obtain the information needed to operationalize the analysis.

Fifth, because of the interdependencies among different institutional arrangements within a given institutional structure and because both demand and supply factors are relevant, it is often difficult to isolate the most relevant element(s) in institutional change.

Sixth, because of the slow pace of institutional change, many other non-institutional variables change virtually simultaneously with institutional changes, making it difficult to control for other factors and to determine the direction of causality.

Seventh, because of the dependence of such explanations on many unobserved details about current or past environmental conditions and institutions, there is a tendency to fall into the functionalist trap of assuming that what exists is efficient and to use the details available to come up with ex post rationalizations.[5]

Eighth, of special relevance to institutional dynamics are: (1) that, unlike other species, selection mechanisms of humans are informed by perceptions and beliefs about the consequences of actions, implying that culture and ideology may matter and (2) that economies of scope, complementarities and network externalities constrain institutional change relative to what the fundamentals would suggest (North, 2000).

Given these methodological difficulties confronting the explanation of institutional and other societal changes, definitive proofs of the superior ability of NIE approaches to explain change will be hard to come by. Nevertheless, in the section that follows we present some examples where NIE would seem to offer new and potentially important alternative explanations.

Examples of NIE explanations of important institutional changes

To examine the potential usefulness of NIE for explaining change, examples are deliberately chosen from a variety of historical as well as geographic contexts. In each example, the NIE explanation is contrasted with one or more traditional explanations. While some evidence is provided, in no case do we wish to suggest that the NIE explanation is definitely better than alternatives. Indeed, what NIE explanations do is to open up vast new areas of social science research and re-energize old ones.

The emergence of private property rights and their distribution

Historians and economic historians may disagree over many things but they rarely do so about the importance of well-defined private property rights for long-term economic development. So too, in the contemporary world the degree of clarity in the definition and enforcement of private property rights across countries goes far in explaining differences in levels of economic development. Not only that, but many individual factors believed to contribute to economic growth and development, such as domestic and foreign investment, technology transfer, efficiency in resource use, and well developed markets for land, capital, products and services of all kinds, have been found to be positively related to property rights.

Like any other institution there is both a demand side and a supply side for property rights. The demand for property rights over a given resource is triggered by increasing scarcity of that resource. Hence, when land is abundant and not fully utilized, there is little demand for developing a property rights system for this resource. But, when it becomes more fully utilized, e.g. as a result of a discovery of a valuable mineral or underground water, or a sharp rise in the price of an agricultural export that induces the expansion of its cultivation, people may be induced to make claims of private property (Libecap, 1978, 1989). There are several different functions that the possession of property rights can serve. First, by protecting the user's future claim on the land, the possessor of the land can be induced to make greater investments on the land than he would if his future possession were not assured. Second, the rights to rent or sell facilitate the development of a market for land such that competitive pressures can usually be counted on to contribute to its more efficient use. Third, titles to land facilitate the development of credit markets and thereby the application of capital and other complementary inputs to the land.

Yet, even with favorable demand conditions, the supply of such an institution is by no means automatic. In particular, the costs of establishing claims, defining such rights and then setting up a system of enforcing these rights may be high, especially in developing countries. To the extent that private property rights establish exclusive rights over a scarce resource, there are likely to be conflicts among different individuals or groups in the way these rights are allocated. A further problem arises from historical precedence in that, prior to the establishment of private property rights, informal rules may have applied in such a way that all members of the community may have had non-exclusive use rights to the resource. Hence, the establishment of exclusive rights to certain individuals, especially if they should be outsiders, may make many other individuals net losers from the establishment of such rights. In such situations, the high cost of conflict may rule out the adoption of private property rights. On the other hand, if the formal system of property rights were to reinforce pre-existing informal ones (de Soto 2000), no such conflicts may arise. Yet, formal legal institutions often not only fail to recognize these claims but also deliberately suppress them, greatly increasing the likelihood of costly conflicts.

Especially where markets for land or other forms of property are not well developed, the issue of property rights goes beyond mere existence. Indeed, the depth or comprehensiveness of the rights, such as whether or not they apply to current use, to underground resources, to sale and rent, are all features that may matter a great deal. Likewise, the size distribution of landed property may matter greatly for productive efficiency. Even more, because any society's political economy and institutions may depend on the size and inequality in property holdings, a given society's growth dynamics may well depend heavily on the character and distribution of private property rights.

Two interesting experiences that are useful to compare in this respect are North America and Latin America beginning in the early 19th century. At that time, these regions were at roughly the same level of income per capita. After that, their growth rates diverged substantially accumulating to their very different levels of development at present. What explains the divergence?

Economic historians Engerman and Sokoloff (1997, 2000) have used a traditional, again technological, explanation for these differences. In particular, Engerman and Sokoloff note that the crops that can be grown in the tropics of Latin America such as sugar, coffee, cocoa, bananas, cotton and tobacco, are different from those grown in North America, mainly grains, and vegetables. Among the ways in which they differ is that the former are allegedly subject to economies of scale while the latter are not. These authors allege that this made it economic for early settlers to organize agricultural production in Latin America (and the Southern part of the United States) on large-scale plantations but in the rest of North America on small farms.

Yet, they argue that the difference in technology led to different institutional trajectories over time. In Latin America, to make plantations economically viable and their exports competitive internationally, it led to institutions to suppress the organization and mobility of labor. By contrast, the small farm orientation in North America encouraged the development of secure property rights. These rights in turn fostered investments on the land and the ensuing shortage of labor encouraged the development of labor-saving innovations and encouraged the extension of private property to intellectual property. These innovations and the expansion of non-agricultural activities, in turn, stimulated educational investments.

There may be a great deal of truth in the argument, but there are also some pitfalls in the analysis. First, detailed studies of productive technologies for tropical agriculture are generally not supportive of economies of scale in production (Binswanger and Rozensweig 1986). Second, the character of technological change may well depend on many factors beyond those identified by Engerman and Sokoloff, including whether it is developed by private or public institutions, and on relative factor endowments (Hayami and Ruttan 1971; Ruttan 2000). Finally, many other differences between North America and Latin America could have been responsible for the observed difference in growth rates.

In selecting appropriate cases for institutional comparison, therefore, it may be better to concentrate on countries in which other differences can be better controlled. To that end, Nugent and Robinson (2000) focus on differences among countries growing the same principal crop, namely, coffee in the same region, Central America. They contrast Guatemala and El Salvador with Costa Rica and Colombia. All are coffee exporters with common terrain, climate and legal and colonial backgrounds as former colonies of Spain. Yet, Costa Rica and Colombia developed coffee quite early thanks to the rapid development of private property rights for mostly smallholders whereas Guatemala and El Salvador developed private property rights only somewhat later and when they did so, it was largely in the form of large plantations.

Nugent and Robinson explain the earlier development and smaller-scale orientation of the property rights in Costa Rica and Colombia in terms of a schism among the elites of these two countries. The thesis is that, when the elite fission, the ensuing political competition may require the different factions of the elite to offer something tangible and credible in order to attract people who could then be used in common defense. While offers of employment, the provision of credit or promises of future public goods production could also be used to lure people to an area to work and fight, since these can easily be rescinded, such commitments may not be credible. The provision of private property rights over well-defined parcels of land authenticated with formal registered titles clearly carries greater credibility. They also attempt to refute the general validity of the aforementioned alternative explanations such as technology, geography and cultural-legal backgrounds. The authors argue that the earlier development of property rights for smallholders in Costa Rica and Colombia led these countries to become more democratic earlier and to invest more in education.

By contrast, Guatemala was dominated by a conservative alliance of church and monopolistic merchants that did not fission. They did not allow political or other competition. These groups deliberately delayed agricultural development they could not control. When they did go into coffee, they did so on large estates that became competitive only by taking advantage of their monopsony power over labor, accounting also for why these countries adopted very regressive labor legislation.

If this theory and evidence should hold up to closer scrutiny, because of the greater extent to which it is able to control for other potential explanatory variables, these comparisons could provide a much more persuasive NIE explanation for the role of property rights and farm size in explaining inter-country differences in growth trajectories than the aforementioned ones contrasting North America with Latin America. In this case, the NIE explanation draws upon political economy.

Other institutional change applications

Numerous other applications of NIE to change and development could be cited. We cite only a few more.

Note that in both of the preceding applications it was not easy to distinguish between exogenous and endogenous variables. Acemoglu, Johnson and Robinson (2001b) provide a nice example of how property rights and other macro-level institutions can be treated as more clearly determined by strictly exogenous influences, namely geographic and health conditions, than the other way around. They show that even the same colonial power introduced into its different colonies quite different institutional orientations in different health environments. Where relatively healthy conditions prevailed and hence where people from the home country might like to settle, they brought with them property rights and other institutions favorable to development. But, in unhealthy conditions, they left the prevailing inferior institutions remain, following a divide and rule strategy for minimizing the cost of governing the colony. They substantiate the hypothesized long-run effects of these differences in colonial institutions by showing a positive correlation between those initial conditions and both the quality of current institutions and current levels of development.

In a subsequent but related paper, Acemoglu, Johnson and Robinson (2001a) argue in favor of political economy of institutional change and against the geography argument. Since geography is constant, it would suggest continuity over time in growth rates whereas the political economy-NIE view would suggest that differentials in growth rates could be reversed after the institutional changes. They show that this is what happened. The high population density, more urbanized and developed areas were left with their existing inferior institutions and/or had new extractive regimes superimposed on them. By contrast, the areas of lower population density, urbanization and levels of development were seen as more attractive to colonial settlers since the settlers could take advantage of land and other resources with less need for conflict with indigenous inhabitants. They, therefore, had greater incentive to bring their superior institutions to these areas. As a result, the previously less developed, less urbanized areas outgrew the more developed, more urbanized ones over the intervening 500 years because of their superior institutions which in time accomplished 'a reversal of fortunes'.

Still another NIE explanation of long-term differences in growth rates and current levels of development is that of La Porta *et al.* (1998). These authors show that countries differ in their legal traditions, Great Britain and its former colonies having common law, France and Spain and their former colonies having civil law, Germany and the Scandinavian countries having somewhat different versions of common law and socialist societies still another system of law. They also show that emanating from these basic differences are quite different rules for banks and other financial markets, those derived from common law backgrounds often being more favorable to financial market development than the civil law backgrounds. Since financial markets, like property rights, are deemed to be conducive to growth and development, this NIE explanation could be capable of explaining important differences in levels of development between these different countries.

Still other applications of NIE are to rapid and radical change. An example, is the fall of the Iron curtain and the subsequent changes in the institutions of Central and Eastern European countries, sudden changes in culture and norms, and various revolutions all of which were both rapid and very surprising. In these cases, the underlying phenomena are characterized by multiple equilibria. NIE practitioners have identified various mechanisms and concepts, such as informational cascades, preference falsification, self-reinforcing changes, bandwagon effects, thresholds, and domino and herd effects, whereby societies stuck in one equilibrium for a long time may then break out of it and gravitate quickly to a new and very different equilibrium. For interesting applications to the East European and other radical changes, see Kuran (1987, 1991) and Bikchandani *et al.* (1992).

Conclusion

Even the few applications mentioned above illustrates that NIE-based hypotheses can provide explanations of important institutional and other differences across countries capable of competing with other non-NIE based explanations. Numerous elements in the NIE, such as the transaction costs of change, institutional inertia, path dependence, collective action failures, constitutional constraints, the interdependencies among different elements in the institutional structure, network externalities and the fact that agents will search for institutions locally and not globally, are consistent with institutional change being slow and path dependent. Moreover, those with greater power than others, by earning higher rents than others under existing rules may have the means to block even highly desirable changes in the status quo.

While it is certainly less clear that NIE can succeed in explaining institutional change and its effects of a more radical and sudden nature, the above applications to Eastern Europe and various revolutions and cultural changes illustrate that it is possible. Moreover, even the granting of private property rights to smallholders as was done in Costa Rica in the 1830s and 1840s was a radical change for that region at that time. Admittedly, however, such applications are much easier to explain *ex post* than to predict. Time will tell how well these explanations are likely to stand up relative to others.

Yet, although still a new paradigm, at this point at least NIE appears to be quite capable of competing with other theories and paradigms in explaining change economic and social change of different kinds. This is not to say that it has been proven to be better than other alternatives. Moreover, as suggested by the methodological difficulties identified above, it may be hard to tell whether it is better or not.

While much of what has been said above would seem to suggest that the institutional and other changes that NIE can explain are those of the type that lead to improvements in social welfare, as North (2000) has stressed, failures are more common than successes. Institutional and other failures deserve more attention.

Notes

* The author expresses his appreciation to some of his past collaborators Mustapha Nabli and Justin Yifu Lin for work drawn on in the introductory sections of this chapter and to the useful comments and suggestions of John Harriss, Raghavendra Gadagkar, Ivo Welch, the workshop directors, several other participants at the workshop and an anonymous reader.
1. For further references on transaction and information costs, see Williamson (1975, 1985), Akerlof (1970).
2. For example, the wage contracts can be structured in a way that would reward a worker with a bonus or a 'promotion' to a higher paying job if positive information about the worker's effort were received and with dismissal, penalties or demotion if negative information about that effort were received. If the only thing that the employer can do to a lazy worker is to dismiss that worker, it may be useful to make the wage offers high relative to other available alternatives (an efficiency wage).
3. Some analysts exclude collective action theory and other supply side aspects of institutions from the NIE proper, preferring instead to view it as part of the separate field of public economics or public choice theory. However, the international society that promotes NIE, the International Society for the New Institutional Economics (ISNIE), typically includes collective action analyses in its conferences and publications.
4. On collective action theory, see, especially Olson (1965, 1982), Hardin (1982).
5. See Nabli and Nugent (1989) for some practical methodological measures for how to reduce falling into the functionalist trap.

References

Acemoglu, Daron, Simon Johnson and James A. Robinson 2001a. 'Reversal of fortune: geography and institutions in the making of the modern world income distribution', NBER Working Paper 8460.

Acemoglu, Daron, Simon Johnson and James A. Robinson, 2001b. 'The colonial origins of comparative development: an empirical investigation', *American Economic Review* 91(5) December 2001, 1369–1401.

Akerlof, George 1970. 'The market for lemons: qualitative uncertainty and the market mechanism', *Quarterly Journal of Economics* 84 (August), 488–500.

Binswanger, Hans P. and Mark Rosenzweig 1986. 'Material and production relations in agriculture', *Journal of Development Studies* 22, 503–39.

Birchandani, Sushil, David Hirshleifer and Ivo Welsh 1992. 'A theory of fads, fashion, custom, and cultural change as informational cascades', *Journal of Political Economy* 100 (October 1992), 992–1026.

Brennan, Geoffrey and James Buchanan 1985. *The Reason of Rules: Constitutional Political Economy*. Cambridge: Cambridge University Press.
De Soto, Hernando 2000. *The Mystery of Capital*. New York: Basic Books.
Engerman, Stanley L. and Kenneth Sokoloff 1997. 'Factor endowments, institutions, and differential growth paths among New World economies: a view from economic Historians of the United States', in S.H. Haber, ed., *How Latin America Fell Behind*. Stanford, CA: Stanford University Press.
Engerman, Stanley L. and Kenneth L. Sokoloff 2000. 'Institutions, factor endowments, and paths of development in the new world', *Journal of Economic Perspectives*, 3, 217–32.
Hardin, Russell 1982. *Collective Action*. Washington, DC: Resources for the Future.
Hayami, Y and Vernon W. Ruttan 1971. *Agricultural Development: In International Perspective*. Baltimore: Johns Hopkins University Press.
Hirschman, Albert O. 1970. *Exit, Voice and Loyalty: Responses to Decline in Firms, Organization and States*. Cambridge: Cambridge University Press.
Jensen, M.C. and W.H. Meckling 1976. 'Theory of the firm: managerial behavior, agency costs, and ownership structure', *Journal of Financial Economics* 3 (4), 305–60.
Kuran, Timur 1987. 'Preference falsification, policy continuity and collective conservatism', *Economic Journal* 97, 642–65.
Kuran, Timur 1991. 'Now out of never: the element of surprise in the East European revolution of 1989', *World Politics* 44 (October), 7–48.
La Porta, Rafael, Florencio Lopez-de-Silanes, Andrei Shleifer and Robert W. Vishny 1998. 'Law and finance', *Journal of Political Economy* 106, 1113–55.
Libecap, Gary 1978. 'Economic variables and the development of the law: the case of Western mineral rights', *Journal of Economic History* 38, 338–62.
Libecap, Gary 1989. *Contracting for Property Rights*. New York: Cambridge University Press.
Nabli, Mustapha K. and Jeffrey B. Nugent 1989. 'Collective action, institutions and development', in M. Nabli and J.B. Nugent, eds, *The New Institutional Economics and Development: Theory and Applications to Tunisia*. Amsterdam: North-Holland, 80–137.
North, Douglass C. 1990. *Institutions, Institutional Change and Economic Performance*. Cambridge: Cambridge University Press.
North, Douglass C. 2000. 'Institutions and the performance of economies over time', Paper presented at the Global Development Network, Tokyo.
Nugent, Jeffrey B. and James A Robinson 2000. 'Are endowments fate?' (Working Paper.) Los Angeles: University of Southern California.
Olson, Mancur 1965. *The Logic of Collective Action*. Cambridge, MA: Harvard University Press.
Olson, Mancur 1982. *The Rise and Fall of Nations: The Political Economy of Growth, Stagflation and Social Rigidities*. New York: Yale University Press.
Ruttan, Vernon W. 1989. 'Institutional innovation and agricultural development', *World Development* 17 (9), 1375–87.
Ruttan, Vernon W. 2001. *Technology, Growth and Development: An Induced Innovation Perspective*. New York: Oxford University Press.
Williamson, Oliver E. 1975. *Markets and Hierarchies: Analysis and Antitrust Implications*. New York: Free Press.
Williamson, Oliver E. 1985. *The Economic Institutions of Capitalism*. New York: Free Press.

12
Institutions, Politics and Culture: A Case for 'Old' Institutionalism in the Study of Historical Change

John Harriss

> Such terms as 'values' and 'culture' are not popular with economists, who prefer to deal with quantifiable (more precisely definable) factors. Still, life being what it is, one must talk about these things ...
>
> *David Landes* (1998: 215)

Institutions are, as Jeffrey Nugent explains (in his paper in this volume), humanly devised rules that affect behaviour, constraining certain actions, providing incentives for others, and thereby making social life more or less predictable.[1] They are, as Geoffrey Hodgson puts it, 'the stuff of socio-economic reality' (Hodgson, 2001: 302). It is in a way rather curious that so fundamental an aspect of social life should not have been a more important focus of study in the social sciences (outside anthropology) until relatively recently. This reflects the facts that, as Nugent says, mainstream, choice-theoretic economics has not previously problematised institutions, such even as those that are necessary for markets to function – and the expansionary pre-eminence of this kind of economics amongst the social sciences. Hodgson, however, reminds us that the dominance of this particular style of economics, with its pretensions to universality, has overlain and led to the forgetting of a rich tradition of thought about institutions, associated both with the German historical school and with American scholars such as Veblen and John Commons (Hodson, 2001). I shall pick up some of his arguments in this short essay.

Nugent distinguishes between the 'demand' for institutions and their 'supply', pointing out that the former involves in particular problems due to informational asymmetries, and the latter problems of collective action. He then shows us how it is possible to explain, parsimoniously, within the framework of neo-classical economics, why particular institutions are the way they are and why they differ from each other, and how they influence productivity, taking great care as he does so to point out the dangers of making tautological, functionalist assumptions: (on the lines of the following: these are the institutions that exist; they must therefore reduce transactions costs; therefore they exist – and they must be efficient).

Institutional theorists recognise that it is perfectly possible for a society to get 'locked in' to an inefficient set of institutions because of the interests of power-holders in their reproduction.[2] An example from some of my own work would be the existence of socially inefficient agrarian institutions, such as those that obtained in Eastern India, and which made usurious money-lending and speculative trading in foodgrains privately profitable for a small class of landowners to the extent that, for a long time, there was little or no incentive for them to make productive investments in agriculture, and certainly not those that required collective action, as in the organisation of irrigation. I think it can be shown that the institutional arrangements that under-pinned this kind of rural economy were socially inefficient; but they supported and were supported by the power of the landowning oligarchy which thus had a strong interest in their reproduction. This is one way, at least, whereby 'historical path dependence' may arise.[3]

How good is the kind of institutional theory that Nugent describes when it comes to explaining change in institutions? How valuable or effective is it, therefore, in theorising change in human societies? Nugent is explicit about the limitations of the NIE: most of it is rather static, he says, and 'because of the interdependencies among different institutional arrangements within a given institutional structure, it is often difficult to isolate the most relevant institutional change for hypothesis testing'. The examples that he gives of NIE explanations of important institutional changes are all interesting, but each of them confirms the modesty of his claims for the NIE as a way of explaining historical change. NIE explanations provide an interesting gloss on current understandings of the emergence of the factory system of production during the Industrial Revolution: the factor of the danger of asset misuse helps to explain why capital owners hired workers in, rather than hiring machines out; the demand for skilled labour probably made the tying of workers into relatively long-term contracts advantageous; there were probably advantages in terms of knowledge and information sharing and the building of trust and cooperation, when workers were brought together in factories. But Nugent says that he would not argue 'that evidence exists to suggest that the traditional economies of scale argument for the rise of the factory system is entirely dominated by these transaction cost and NIE considerations'. The second of his cases, about the emergence and distribution of private property rights involves an interesting study of property rights and coffee production in Central America and the contrasts between Guatemala and El Salvador on the one hand, and Costa Rica and Colombia on the other. The argument is that the latter pair of countries 'developed coffee quite early thanks to the rapid development of private property rights for mostly smallholders', and the reason for this is said to be 'elite schism' or in other words the fragmentation and consequent development of competition amongst the elites of the two countries – whereas Guatemala on the other hand 'was dominated by a conservative alliance of church and monopolistic merchants that did not fission'. The argument is an interesting one. The application of the NIE in this case, however, does not in itself explain change. It serves to highlight the importance of power considerations and of politics, but it does not in itself explain them at all. Rather it raises interesting questions about the political context which have to be answered through some other, historical [and political] analysis. Thinking of E.H. Carr's metaphor of fishing, in his classic study of the nature of history, the NIE is a useful net that directs our attention to particular facts that then need to be explained historically (Carr, 1961). It is not in itself a theory of historical change. I believe that John Toye's judgement that the NIE has no theory of history ('The main weakness of the NIE as a grand theory of socio-economic development is that it is empty'[4]) is substantially correct (Toye, 1995: 64).

I refer back to my East Indian example again. As I said, I think it can be shown that an inefficient set of agrarian institutions persisted over a long period; and it can be shown that this in turn explains the long run stagnation in the agriculture of Bengal. This stagnation came to an end in the early mid-1980s, and since that time the rate of growth of agricultural output in West Bengal has been amongst the highest, perhaps the highest, in the country. The explanation of this historic change is of course, complex, and exactly as Nugent says 'it is difficult to isolate the most relevant institutional change for hypothesis testing'. The precise role of the modest agrarian reforms implemented by the Marxist-led government of the state remains controversial; but they were certainly instrumental in changing the socially inefficient institutions that I have referred to. Institutional change here followed from the rise to state power of a (moderate) left wing political party (Harriss, 1993; Mohan Rao, 1995).[5] It is not too long a jump, then, to argue that the relationships between social classes, and the nature of power structures, which themselves have to be analysed historically, are of particular significance in explaining change or alternatively the lack of it over long periods of time. If we wish to explain the different historical trajectories of Guatemala and Costa Rica, for example, amongst the cases referred to by Nugent, we will also have to take account of the specifics of class relationships, for it is these which appear to underlie the institutional differences that are his focus. As Pranab Bardhan has said 'The history of evolution of institutional arrangements and of the structure of property rights often reflects the changing relative bargaining power of different social groups'; and he points out that 'North [who won the Nobel Prize for Economics in 1993, for his contributions to institutional economics], unlike some other transaction cost theorists, comes close to the viewpoint traditionally associated with Marxist historians' (Bardhan, 2001: 261). North's work, indeed, can be seen as reflecting a constant tension between his commitment to the framework of choice-theoretic economics and his awareness of the limitations which it imposes when it comes to the analysis of change. This is reflected in his admission that there is 'much to learn' from 'the "old economic historian", the institutionalists of Vebel and C.E. Ayres' persuasion, or the Marxist';[6] and in his concern, as Lazonick has pointed out, to graft onto 'mainstream economics a theory of *political* change' (North, 1978: 974).

If I may extend my point. I have long been interested in the differences between the historical trajectories of the major Indian states, in terms of rates of economic growth and of levels of development. Contrary to the theoretical presuppositions of many economists, in regard to both growth and human development, the Indian states have continued to diverge rather than to converge (Rao et al., 1999). A large number of different indices show that in many respects the states of the Hindi Heartland in the North, notably Bihar and Uttar Pradesh, lag far behind those of the South and the West. It is a much longer story than I can conceivably do justice to here: but there is a lot of evidence to suggest that a major part of the explanation has to do with the persistence of hierarchical social relationships, and of the fairly extreme social fragmentation associated with them, in the Hindi Heartland, while these have been more or less successfully challenged by the political mobilisations for over more than a century of lower caste and class people in the South and the West. I can offer an institutional explanation for different patterns of change, if you will, but one which focuses on the persistence or not of what I have referred to as 'hierarchy', or what Francine Frankel and M.S.A. Rao call the (traditional) 'dominance' of upper caste/class people who exercise authority that is sanctioned by religious beliefs. With a group of co-authors, these writers have shown how the particular political histories of the major states reflect the workings out of the persistence or not of this upper caste dominance, which is of course linked with the history of lower caste/class mobilisations.[7] Now, in terms of this framework, it becomes difficult to explain how and why two of the states of the Hindi Heartland, Rajasthan and Madhya Pradesh, should have started to grow much more vigorously and to improve levels of human development, clearly distancing themselves from Uttar Pradesh and Bihar. The answer, I think, still lies in political factors, in this case having to do with the nature of party political competition in the these two states, by comparison with Bihar and Uttar Pradesh, and with political leadership. In Madhya Pradesh, in particular, a reforming Chief Minister, with a definite vision of development for the state, has created a kind of a local version of the 'developmental state'.[8]

The question is whether new institutional economists, and notably North, have succeeded in 'grafting a theory of political change onto mainstream economics'. For North, 'a dynamic model of economic change entails as an integral part of that model analysis of the polity'. But it is not at all clear that the NIE actually has a theory of how and why polities differ. It offers no explanation of the fact that the same economic institutions can have very different consequences in distinct contexts. As Robert Bates has argued, this shows 'the necessity of embedding the new institutionalism within the study of politics', for the differences observed – for example between the outcomes of the establishment of coffee marketing boards in Kenya and Tanzania – have to do with the political context (Harriss, Hunter and Lewis, 1995). Ultimately this means studying institutions historically and so integrating theory building and the study of reality.

I have described my analysis of different patterns of change across the various major Indian states as positing an 'institutional' explanation – but, *pace* North's attempts at grafting together an analysis of political change with choice-theoretic economics, I do not think that it is one that fits within within the frame of the 'new institutional economics' outlined by Jeffrey Nugent. This is described as 'new' because, unlike the older traditions of thought about institutions in the German historical school and in early American institutionalism, it operates with the same basic assumptions, about scarcity and individual choice, as mainstream neo-classical economics.[9] The institutions to which my analysis of political regimes across Indian states refers are those of caste, and they involve ideas about authority rooted in religious belief. They have to do, then, with what is commonly referred to as 'culture'. This is one of the most awkward words in the English language, not least because it is polysemic. Here I am referring to culture in the sense of 'the (historically specific) habits of thought and behaviour of a particular group of people', or of 'the ideas, values and symbols – more generally, "meanings" – in terms of which a particular group of people act'.[10] 'Culture', in this sense, is quite often used as a kind of a residual in explanations for social change, or the lack of it, to account for what appears to be 'irrational', or in other words what is not readily explained in terms of the basic model of utility maximisation. The NIE engages with the problem of culture, as it does with politics, but with difficulty. Douglass North argues that 'culture defines the way individuals process and utilise information and hence may affect the way informal constraints get specified', which at least adds to the factors involved in explaining the nature of institutions in any particular case but leaves culture as exogenous to explanation (North, 1990: 42). It remains a residual.

Taking serious account of those aspects of social life and experience that are labelled in English as 'culture' (in the particular sense just described) starts to expose the limitations of the universalising pretensions of neo-classical economics, which depend in part upon quite simplistic assumptions about the preferences that individuals are supposed to be maximising, and upon a simplified notion of human rationality. Even rather cursory empirical examination of human behaviour shows that people very often act habitually – that is, in ways which are characteristic of their 'culture' – and that preferences too are culturally specific. Of course these preferences and actions may be subjected to rational thought by the social actors themselves,[11] but they are very often not. The strength of the 'old' institutionalism is that it does not treat culture as an awkward (though sometimes convenient) residual, but rather makes it central in analysis. My own analysis of variation in the patterns of change between the Indian states is, it follows, much more in line with the 'old' institutionalism than it is with the NIE.

The 'old' institutionalism has been criticised as being 'descriptive' and lacking in the formal rigour of mainstream economics and its off-shoot in the NIE, but as Hodgson has argued there was more to it than this for scholars from the German historical school, and the Americans like Veblen and Commons, at least *sought* to tackle the problem of historical specificity, and the serious limitations of attempts at producing universal theory in the face of the sheer complexity of society and the historical variation between different 'societies' (Harriss, Hunter & Lewis, 1995: 4–5). In doing so they did not retreat into empiricism, but aimed rather to develop 'middle range' theory, or a particular historiography, based – in Hodgson's own exploration of the tradition of the 'old' institutionalism – on certain general propositions concerning the importance for understanding of socio-economic systems of 'the laws ... that dominate the production and distribution of vital goods and services. Such laws would concern property rights, contracts, markets, corporations, employment and taxation'. These legal rules and contracts, it is held, are always and necessarily 'embedded in deep, informal social strata, often involving such factors as trust, duty and obligation (so that) a formal contract always takes on the particular hue of the informal social culture in which it is embedded'. Further, it is clear that 'The emergence of law, including property rights, is never purely and simply a matter of spontaneous development from individual interactions (but rather) is an outcome of a power struggle between citizens and the state'. Politics and power, as I argued earlier, thus become of central significance in this approach.[12]

Let me illustrate the argument further, referring again to India.[13] We may recognise that 'shared habits of thought and behaviour' (i.e. a particular culture) associated with caste are central to what it means to be Indian – though we should also recognise both that caste is an important aspect of a dominant ideology (that of orthodox Brahminical Hinduism, resisted by subordinated groups over more than two millennia[14]), and that it is an historical phenomenon. Notably, caste underwent significant restructuring under colonial rule. It is not true that caste is a colonial creation, but there is no doubt that the meaning of caste was changed by the ways in which the colonial rulers used it to classify the population (Dirks, 2001). All cultures, though by definition they involve enduring habits, are both contested within the field of power, and they are all the time being reflexively reworked or reinvented.

The values and practices of caste have tended to create relatively tight, closed social networks, so that Indian society, it has been said, is pronouncedly segmented or 'cellular'.[15] This in turn has important implications for economic action. As we know, from some of the new institutional economists indeed, most transactions involve uncertainty, arising from any one actor's incomplete knowledge about the future actions of others with whom s/he is transacting. Trust is one way of coping with this uncertainty – uncertainty that is occasioned by the freedom of others – that is never entirely removed, as Hodgson has argued (see above), even in the presence of legal rules and contracts. But where does trust come from? One important source of or basis for trust is the sharing of key characteristics with others, or from knowledge of them in particular social networks (predictability comes with familiarity). Caste relations, involving both shared characteristics and particular social networks, are an important source of trust in Indian society – and it may be said of certain caste communities that they constitute an economic organisation. A South Indian caste community, for instance, the Nattukottai Chettiars, has functioned very much like a bank, and Nattukottai Chettiars have transacted vast sums of money across long distances relying on the specific trust to which their caste relationships have given rise (West Rudner, 1994). But there is a significant difference between such *specific trust* based on particular shared characteristics or social networks, or that which depends upon personalised transactions, and *generalised trust* running through society as a whole (beyond such networks/relations as caste). Generalised trust can be shown to be desirable for an effectively and efficiently functioning market economy (Platteau, 1994). As David Landes has put it '(The) ideal society would ... be honest [or, in other words, "generalised trust" would prevail]. Such honesty would be enforced by law, but ideally the law would not be needed. People would believe that honesty is right (also that it pays) and would live and act accordingly' (Landes, 1998: 218). I think that it can also be shown that the very strength of the specific trust that is generated in caste relationships stands counterposed to such generalised trust or morality, and that this has constrained India's economic development.

For example, the private sector of the Indian economy has been dominated for a long time by a small number of powerful family business groups, which have been secretive and non-transparent, and have relied heavily on personalised, family and kinship networks – on 'specific trust', therefore – resisting the professionalisation of management. Now, in the context of India's increased integration into the global economy, these great family firms are finding themselves disadvantaged, and they are having to open themselves up more to scrutiny, in order to attract investors. New institutions of corporate governance are being introduced, or are sought to be introduced, substantially because of pressures from one fraction of the business elite. In a sense, the contest is now on, with different champions on either side from within the business world, between 'traditional', informal institutions, linked to family, caste and kinship, and formal institutions of corporate governance involving laws and codes of practice.

The case shows up key points which together help to support the argument that I am making for an approach deriving from the 'old' institutionalism in the analysis of social change, as against the static nature of the NIE. First it shows up the inter-relations of formal institutions with the 'deep informal social strata' in which they are embedded, and hence the importance of those historically specific 'shared habits of thought and behaviour' (or culture). These are not at all easily or satisfactorily explained in the would-be universal theory of mainstream economics – and they remain exogenous in the NIE. Yet they may be central to understanding what is happening! David Landes, after all, concludes his magisterial history of economic development over the last millennium by saying that 'If we learn anything from the history of economic development, it is that culture makes all the difference' (though he also points out that 'culture does not stand alone ... monocausal explanations will not work' [Landes, 1998: 516]).[16] Second – though I have only been able to suggest the argument here, and in my earlier examples – explaining institutional change, and hence social change, requires that we take account of power, as Hodgson implies and as Pranab Bardhan explicitly stated in the commentary that I cited earlier. Power is missing from the NIE. Whether or not the rules of corporate governance in India will be changed in such a way as to be effective will depend upon the outcome of a power struggle between different fractions of Indian business and their political supporters, and on 'deeper' changes in habits of thought and behaviour. The two are inter-related and the outcomes cannot be predicted. Change in human societies can only be satisfactorily explained when these historically specific factors are taken into consideration, as they are in an approach based on the 'old' institutionalism – while they are not in the NIE.

Notes

1. Note that one of the leading figures of new institutional economics, Douglass North, emphasises constraints on behaviour. With Nugent I believe it is important also to recognise the incentive effects of institutions.
2. Compare Douglass North (1990: 99): '... unproductive paths (can) persist. The increasing returns characteristic of an initial set of institutions that provide disincentives to productive activity will create organizations and interest groups with a stake in the existing constraints'. And see also Ha-Joon Chang's (2002: 117) historical review of 'Institutions and Economic Development' from which he concludes that 'in many cases institutions were not accepted ... because of the resistance from those who would (at least in the short run), lose out from the introduction of such institutions'.
3. My own work on this is Harriss (1982). The paper refers in part to Amit Bhaduri's (1973) classic paper. An authoritative study which substantiates my argument is James Boyce's (1987).
4. Douglass North (one of the leading exponents of the NIE), Lance Davis and Calla Smorodin (1971) conceded in their work on American economic growth that their 'model is not dynamic, and we know very little about the path from one comparative static equilibrium to another'. It is a moot point as to whether North has been able to develop a dynamic theory in his subsequent work, as I explain later in the main text.

5. For a contrasting view, see Lieten (1992).
6. For an elaboration of the points I have raised here see the critical discussion of North's work by William Lazonick (1991: 310–18).
7. Francine Frankel and M.S.A. Rao (1989) define 'dominance' as follows: 'the exercise of authority in society by groups who achieved socio-economic superiority and claimed legitimacy for their commands in terms of superior ritual status'. My development of the Frankel–Rao analysis is in Harriss (1999).
8. On the recent growth performance of Rajasthan and Madhya Pradesh see some passing commentary by Lloyd and Susanne Rudolphs (2001). On the theory of the 'developmental state', elaborated for Japan and other states in East Asia see Gordon White (1988).
9. Douglass North has written of NIE that it 'builds on, modifies and extends neo-classical theory'. See Harriss, Hunter & Lewis (1995: 17).
10. I am not implying that these two definitions of culture have absolutely the same meaning (see Hodgson, 2001), but both assert the historical specificity of cultural patterns. As Plateau (1994: 534) has argued 'Ultimately, the cultural endowment of a society plays a determining role in shaping its specific growth trajectory, and history therefore matters.'
11. Amartya Sen (1999) shows this in his commentary on identity politics.
12. Hodgson (2001: 301, 304, 312) notes the continuities with Marx's approach, but argues that 'the analysis goes further than Marx, by grounding property relations in shared habits and by also emphasising the concept of culture' (2001: 309).
13. The following discussion draws on Harriss (2002).
14. See A.K. Ramanujan's (1973) introduction to his translations of Veerasaivite poems, many of which expressly ridicule and repudiate caste.
15. This idea appears in some of Marx's writings on India; in Barrington Moore's (1966) great classic *The Social Origins of Dictatorship and Democracy*, and most expressly in Satish Saberwal's (1996) *The Crisis of India*.
16. The point about culture 'not standing alone' is an extremely important one, in the light of the current vitality of cultural determinism – reflected, for example, in Francis Fukuyama's book on trust (1995). On this point, as in other ways, Landes follows Max Weber, who of course also argued that culture does not stand alone in his classic *The Protestant Ethic and the Spirit of Capitalism*.

Bibliography

Bardhan, Pranab. (2001) 'Institutional impediments to development'. In: Kahkonen, S. & Olsen M. eds *A New Institutional Approach to Economic Development*. Delhi: Vistaar Publications.

Bhaduri, Amit. (1973) 'A study in agricultural backwardness under semi-feudalism'. *Economic Journal* 83 (329): 120–37.

Boyce, James. (1987) *'Agrarian Impasse in Bengal: Institutional Constraints on Technological Change'*. Oxford: Oxford University Press.

Carr, E.H. (1961) *'What Is History?'* London: Penguin Books.

Chang, Ha-Joon. (2002) *Kicking Away the Ladder: Development Strategy in Historical Perspective*. London: Anthem Press.

Davis, L.G., North, D.C. & Smorodin, C. (1971) *Institutional Change and American Economic Growth*. Cambridge: Cambridge University Press.

Dirks, Nick. (2001) *Castes of Mind: Colonialism and the Making of Modern India*. Princeton, NJ: Princeton University Press.

Frankel, Francine & Rao, M.S.A. eds (1989) *Dominance and State Power in Modern India: Decline of a Social Order, Volume* 1. Delhi: Oxford University Press.
Fukuyama, Francis. (1995) *Trust: the Social Virtues and the Creation of Prosperity*. New York: The Free Press.
Harriss, John. (1982) 'Making out on limited resources; or, what happened to semi-feudalism in a Bengal district'. CRESSIDA *Transactions* II (16–76).
——. (1993) 'What is happening in rural West Bengal? agrarian reform, growth and distribution'. *Economic and Political Weekly* 28 (24): 1237–47.
——. (1999) 'Comparing political regimes across Indian states: a preliminary essay'. *Economic and Political Weekly* 34 (48): 3367–77.
——. (2002). 'On trust, and trust in Indian business'. Working Paper Series for the Development Studies Institute No. 35, London School of Economics.
Harriss, John, Hunter, Janet & Lewis, Colin, M. (1995) 'Introduction: development and significance of NIE', in Harriss *et al. The New Institutional Economic and Third World Development*. London: Routledge.
Hodgson, Geoffrey. (2001) *How Economics Forgot History: the Problem of Historical Specificity in Social Science*. London: Routledge.
Landes, David. (1998) *The Wealth and Poverty of Nations*. New York: W.W. Norton.
Lazonick, William. (1991) *Business Organization and the Myth of the Market Economy*. Cambridge: Cambridge University Press.
Leiten, G.K. (1992) *Continuity and Change in Rural West Bengal*. Delhi: Sage.
Mohan Rao, J.M. (1995) 'Agrarian forces and relations in West Bengal'. *Economic and Political Weekly* 30 (30) 29 July.
Moore, Barrington. (1966) *The Social Origins of Dictatorship and Democracy*. New York: Beacon Press.
North, Douglass. (1978) 'Structure and performance: the task of economic history'. *Journal of Economic Literature* 16 (3): 963–78.
——. (1990) *Institutions, Institutional Change and Economic Performance*. Cambridge: Cambridge University Press.
Platteau. (1994) 'Behind the market stage where real societies exist'. *Journal of Development Studies* 30 (3): 533–78.
Ramanujan, A.K. (1973) *Speaking of Siva*. Harmondsworth: Penguin Books.
Rao, M.G., Shand, R. & Kalirajan, K. (1999) 'Convergence of incomes across Indian states: a divergent view'. *Economic and Political Weekly* 34 (27): 769–78.
Rudolphs, Lloyd & Susanne. (2001) 'Iconisation of Chandrababu: sharing sovereignty in India's federal market economy'. *Economic and Political Weekly* 36 (18): 1541–52.
Saberwal, Satish. (1996) *The Crisis of India*. Delhi: Oxford University Press.
Sen, Amartya. (1999) *Reason Before Identity*. Delhi: Oxford University Press.
Toye, John. (1995) 'The new institutional economics and its implications for development theory'. In: Harriss, J., Hunter, J. & Lewis, C. eds. *The New Institutional Economics and Third World Development*. London: Routledge.
West Rudner, David. (1994) *Caste and Capitalism in Colonial India: the Nattukottai Chettiars*. Berkeley: University of California Press.
White, Gordon ed. (1988) *Developmental States in East Asia*. London: Macmillan.

13
Exporting Metaphors, Concepts and Methods from the Natural Sciences to the Social Sciences and *vice versa*

Raghavendra Gadagkar

How could a biologist react to Jeffrey Nugent's excellent chapter in this volume? Since I have studied insect societies it should come as no surprise that I will be looking for parallels, real or apparent, in the world of social insects. Relative to our understanding of human economic institutions, our understanding of insect societies is woefully inadequate. But the little that we do know about them convinces me that there are fascinating parallels waiting to be explored to the mutual benefit of insect sociology and human economics. Let me briefly describe three of the many possible examples.

The honey bee dance language

Honey bees live in populous colonies consisting of tens of thousands of individuals. Each colony consists of a single queen, a small number of males (drones) while the rest of the colony consists of nearly sterile female worker bees. Because the drones do not contribute to colony labour and the queen merely lays eggs, all domestic duties are the responsibility of the workers. Worker bees have elaborate adaptations to undertake various tasks required for the welfare of the colony. After spending about half their life (which totals about 40 days) working inside the nest, worker bees fly out of their nests in search of food (nectar and pollen), which they bring back to the nest. To aid in this process of stocking up the nest with nectar and pollen, forager bees recruit other bees, which are idle in the nest (Gould J. L. and Gould C. G., 1988). Recruitment is not achieved by leading naive bees to new sources of food but, as the Austrian zoologist Karl von Frisch discovered in the 1940s, it is done by providing naive bees with abstract information about the distance and direction of the food sources discovered by the forager bees (Frisch, 1967). The information transfer is accomplished by means of a dance language that is a unique form of symbolic communication not witnessed in any other non-human animal. When the food is within about 100 metres form the nest, returning foragers perform a 'round' dance which conveys no specific information about the location of the food. But, having been alerted to the presence of food near their nest and having a good idea of what they should be looking for, which they get by smelling the dancing bee and by partaking drops of regurgitated nectar provided by the dancer, dance followers are able to successfully locate the source of food. When the food source is at greater distances, returning forager bees perform a 'waggle' dance, which additionally encodes information about direction and distance as well. Many different aspects of the honey bee dance language are under intense scientific investigation by dozens, perhaps hundreds of researchers (Seeley, 1995).

Many of the questions being investigated bear a remarkable resemblance to the kinds of questions that Nugent and others ask about human institutions: How are the dances of different bees different from one another? How are the dances related to the quality and quantity of food found by the dancer? How are the characteristics of the dances related to success in recruitment? How do bees in the nest respond to multiple dancers, advertising different sources of food, perhaps of differing quality and quantity and of differing value to the nest? How do bees balance short-term and long-term needs of the nest? How do bees deal with competition from neighbouring nests, which must also depend on the same sources of food? Is the recruitment system based on the dance language more efficient in some environments and less so in others? Can we find or breed bees, which recruit/get recruited more efficiently than other bees? The parallels with human institutions are hard to miss. I cannot believe that there isn't much to be gained by cross-fertilization of ideas between those of us who study insect or other animal social institutions and those of you who study human institutions.

Ant agriculture

Human agriculture which is believed to have originated some 10,000 years ago has rightly been considered the most important development in the history of our species. Virtually all the plants we consume today are derived from cultivars that have been bred and modified by humans for thousands of years. There has also been extensive exchange of cultivated crops from one part of the globe to another. The impact of agriculture on the further development of human societies has been profound – high rates of population growth, urbanisation and economic surpluses – all of which were pre-requisites for the development of modern civilization – were made possible with the advent of agriculture. Impressive as all this is, our achievements are surely humbled by the lowly ants, which appear to have invented agriculture, and as we shall see below – a fairly sophisticated type of agriculture – almost 50 million years before we did. Three different groups of insects practice the habit of culturing and eating fungi. They are, ants belonging to the tribe Attini, macrotermitine termites and certain wood-boring beetles. While the beetles in this group are few and not of comparable importance, the fungus growing ants in the new world and fungus growing termites in the old world are ecologically very dominant. With a few exceptions, all fungus-growing ants are leaf cutters – they cut pieces of leaves, bring them to the nest and use them as substrata to grow fungi. The ants derive their nutrition only from the fungi so grown and not from the leaves themselves. There are some 200 species of ants that do not know any other form of life style other than fungus farming. Because of their ecological dominance and their insatiable hunger for leaves, leafcutter ants are major pests in the new world. These ants can devastate forests and agricultural fields alike – they may maintain ten or more colonies per hectare and a million or more individuals per colony. Where they occur, leafcutter ants consume more vegetation than any other group of animals. Like in the humans, the advent of agriculture appears to have significantly affected the evolution of leafcutter ants. Today the leafcutter ants are among the most advanced and sophisticated social insects. As may be imagined, the process of fungus cultivation is a complicated business. In the field, leaves are cut to a size that is most convenient for an ant to carry them back. In the nest, the leaf fragments are further cut into pieces 1–2 mm in diameter. Then the ants apply some oral secretions to the leaves and inoculate the fragments by plucking tufts of fungal mycelia from their garden. The ants maintain a pure culture of the fungus of their choice and prevent bacteria and other fungi from contaminating their pure cultures. On the other hand, growing pure cultures of some of these fungi in the laboratory has proved difficult or impossible for us humans. How ants manage to achieve this remarkable feat remains poorly understood. Not surprisingly, they manure their fungus gardens with their own faecal pellets. When a colony is to be founded, the new queen receives a 'dowry' from her mother's nest – a tuft of mycelia carried in her mandibles! Thus these ants appear to have asexually propagated certain species of fungi for millions of years (Hölldobler and Wilson, 1990).

What kinds of fungi do these ants cultivate? Do all ants cultivate the same type of fungi? As in the case of human beings, have there been multiple, independent events of cultivating wild species? Like humans, do the ants exchange cultivars among themselves? Until recently it was not easy to answer any of these questions. Today, with the advent of powerful DNA technology, answers to many of these questions are being attempted (Mueller *et al.*, 1998). Recent studies suggest that there have been at least five independent origins of fungal cultivation by ants, rather than a single event as was supposed previously. Even more interesting, recent results suggest that ants occasionally exchange fungal cultivars among themselves because different nests of the same species sometimes contain different cultivars. Whether the ants deliberately borrow fungal cultivars from their neighbours or whether the horizontal transfer occurs accidentally is however not known. How do ants deal with pests and parasites of their agriculture? What has been the impact of agriculture (including perhaps the economic surpluses thus generated) on the evolution of the ants themselves? These and other similar questions are now engaging the attention of researchers (Currie et al., 1999). But again the parallels with human institutions are uncanny. Remarkably similar studies have been made using pretty much the same techniques and posing the same sorts of questions but concerning human domestication of plants and animals. For example, sixteen highly regarded wine grapes of northeastern France, including 'Chardonnay', 'Gamay noir', 'Aligoté', and 'Melon', all bred and cultivated since the middle ages, have now been shown to have DNA markers consistent with the possibility that they are all the progeny of a single pair of parents, 'Pinot' and 'Gouais blanc' (Bowers et al., 1999). In contrast, analysis of DNA from today's domestic horses and samples from archaeological sites suggest that the horse was repeatedly domesticated over an extended period of time throughout the Eurasian range (Lister et al., 1998; Vila et al., 2001). The parallels between ant agriculture and human agriculture and other human institutions go far beyond questions of origin and evolution. Indeed they concern economics in the most direct manner. In the context of the leaf cutting, fungus-growing ants, e.g. what are the economic principles governing leaf harvesting and fungus production? There are issues concerning the optimum numbers of harvesting individuals, optimum sizes of the cut leaves, selection of patches of vegetation to be harvested in terms of species of the plant, distance from the nest, long-term, possibly sustainable use of resources and effective competition with neighbours who are all attempting to optimize the same variables. There has been some work along these lines but it barely scratches the surface. It is crucial for biologists to pay attention to parallel situations in humans and their study by economists and crucial for economists dealing hitherto only with *Homo sapiens* not to underestimate the potential sophistication of the adaptations and capabilities of the ants.

The division and organization of labour

Division of labour is a fundamental characteristic of insect societies. First there is the division into reproductive and non-reproductive labour. In almost all insect societies, only one or a small number of fertile individuals are responsible for reproduction while the remaining individuals in the colony (referred to as workers) are engaged only in non-reproductive tasks. In most social insects (i.e. ants, bees and wasps), colonies are headed by one or a small number of fertile females referred to as queens and there are no kings (except in the termites) as the males usually mate and die (Wilson, 1971). Apart from such reproductive division of labour, workers in many insect societies divide non-reproductive tasks among themselves by certain individuals specializing in certain tasks. Thus the colony's needs are fulfilled only because of the coordination between several individuals accomplishing sub-components of larger tasks (Oster and Wilson, 1978). For example, a returning nectar forager in a honey bee colony, has to wait in a queue to be unloaded and, depending on how quickly or how slowly she is unloaded, she will decide whether to bring more nectar or perhaps something else (pollen or water) that may be in greater demand in the colony. In addition to the mechanisms that help the insects achieve effective division of labour, there are questions concerning the organization of work – how does an individual insect that plays only a small part in a complex task, know what to do when? This problem seems to have been solved in at least two different ways. In relatively primitive insects societies, such as those of many species of bees and wasps, where colony sizes are small (less than one hundred individuals), the queen is the most active and physically dominant individual in the colony. She constantly interacts with her workers and by means of these physical interactions, she suppresses any attempts by the workers to take on reproductive roles and also regulates their non-reproductive tasks – more or less, telling that what to do when. The queen in such primitive, small societies has been called a central pacemaker. If the queen is removed, the workers stop working for the welfare of the colony and may start reproducing. If the queen is retuned soon, they abandon attempts at reproduction and get back to work. Centralized control by a single leader is thus the solution that these societies have come up with. The relatively more advanced societies, such as those of honey bees, many ants and termites, where colony sizes can be very large (thousands if not millions of individuals), have arrived at a different solution. Here it is of course impossible for a single pacemaker to control her colony by physical interaction with all the workers. A fairly sophisticated form of decentralized control thus replaces centralized control in the large, complex societies. Control of worker reproduction is still largely centralized and is achieved by the queen using a chemical weapon to inhibit worker reproduction. But even this may be thought of as somewhat decentralized. There are two reasons for this. First, it appears that the chemical weapon produced by the queen is more likely to be an honest signal that the workers find in their own interest to obey. Second, workers in some situations actually assist the queen by policing errant workers. However, the regulation of non-reproductive activities of the workers is entirely decentralized and self-organized. The principles of self-organization, using distributed intelligence and simple local rules to produce complex global patterns, are just beginning to be studied and the questions that biologists can potentially ask are rich and varied (Camazine et al., 2001). In my own research for example, I study two primitive wasp societies (Gadagkar, 2001). One of them, *Ropalidia cyathiformis* is relatively more primitive than the other, *Ropalidia marginata*. As expected, *R. cyathiformis* has a queen who is physically active and dominant and centrally regulates both worker reproduction as well as their non-reproductive activities. An *R. marginata* queen on the other hand, is behaviourally docile and inactive and appears to use a chemical to regulate worker reproduction, leaving it to the workers themselves to self-organize their foraging in a decentralized manner. I believe that a comparative study of this pair of contrasting species will help understand the evolutionary transition from centralized control of worker activity to decentralized control and indeed, the evolutionary transition from relatively primitive to relatively advanced social organization.

Returning to the theme of this essay, the parallels with human institutions and their modes of organization and sources of efficiency and inefficiency are obvious. How do centralized and decentralized modes of regulation differ in their ability to promote efficiency under different conditions – conditions relating to group size, modes of communication and perhaps most important, the levels of motivation of the individuals to strive for the common good. The last point is perhaps better understood in the context of insect societies where there is a clear correlation between decentralized regulation and high 'motivation' among the individuals to work for the common good; interestingly, this high 'motivation' is created by the queen by foreclosing nearly all selfish option for the workers.

Exporting metaphors, concepts and methods

Now I wish to make some general remarks about exporting metaphors, concepts and methods from the natural sciences to the social sciences and *vice versa*. I would like to argue that exporting methods is easy and even essential, exporting concepts is desirable and exporting metaphors is difficult and may even be dangerous. Let me begin with that which I consider easy and desirable.

Exporting methods

One area in which I would like to see a significant unification of the natural and social sciences concerns research methodology. There are two fundamental differences in studying humans on the one hand and other animal or plant species on the other. One is that humans speak while other species don't. Thus, research on human systems can be based on what the humans involved have to say about the matter while research on other species has to necessarily depend on our using observation and experiment to find our facts. The second is that in the social sciences the researchers and the objects of research usually belong to the same species while that is not usually the case in the natural sciences. I believe that these two special features of research in the social sciences make it all the more important that, even though there is a simple option of 'asking' your subjects what they think of the matter, social scientists should increasingly adopt observations and experiments to gather their data. I would like to see many more studies of human institutions that depend on data gathered by direct observations using observational and sampling methodology so commonly used in studying animals. The most striking example of the pitfalls of not using direct observations and depending too much on what the human subjects have to say about themselves is the 'Fateful Hoaxing of Margaret Mead' in her *Coming of Age in Samoa* (Mead, 2001 [1928]; Freedman, 1983, 1998). Supplementing what humans have to say about themselves with what a relatively objective outsider can observe and infer for himself will surely reduce the probability of such errors.

Exporting concepts

Exporting concepts is not always necessary but can be useful and never particularly harmful. My impression is that here natural sciences, particularly the study of insect societies, has much to gain by importing concepts from human sociology, psychology and economics. Division of labour, the adaptive significance of castes, market economy, the relationship between demand and supply, group benefits versus individual interests, egalitarian versus despotic systems of control, private property *versus* common property, are some of the many concepts that come to mind as having potential application in the study of both human economic institutions and insect societies.

Exporting metaphors

Metaphors play a useful role in both human and animal research. But metaphors can be potentially dangerous if misunderstood and that danger is especially great when metaphors are exchanged across disciplines. The reason for this is that metaphors are specially defined for a particular context and are usually understood by the parishioners of a field and easily misunderstood by outsiders who are unaware of the restricted usage. Borrowing metaphors for one field into another without understanding such restricted meanings has caused endless debate and controversy. In the study of insect and other animal societies, it is quite common to use such terms that originally belong to human institutions as royal, king, queen, police, soldier, army, caste, labour, selfish, altruistic, nepotistic, egalitarian, despotic, democratic, revolt, loyal, rebel, rape ... the list is long. When used in the context of insect and other animal societies, these terms are defined in very specific ways that strip them of much of the connotation that is inevitably tied to their usage in the human context. Consider some of the more controversial ones – selfish, altruistic and nepotistic. A selfish act by an insect is defined as any interaction that increases its own genetic fitness at the cost of the fitness of the recipient of that interaction. Conversely, altruism is defined as any interaction that reduces the genetic fitness of the actor while increasing that of the recipient. Nepotism is defined as any behaviour that enhances the fitness of genetic relatives of the actor. There is no implication of any associated moral or ethical values or of conscious awareness on the part of the actors. But when a biologist says that natural selection usually favours selfish or nepotistic behaviour, it is immediately interpreted by those outside the field to mean that biologists defend and condone and even encourage selfishness and nepotism. What is the solution to this problem? There are two. One is to insist that biologists should coin new terms (more jargon?) that have no chance whatsoever of being associated with the value system of humans. The other is to respect the restricted, metaphoric use of terms by biologists and not borrow metaphors 'illegitimately' from one field to another. I certainly prefer the latter solution. I don't believe that inventing new words would help either the cause of animal behaviour or of human ethics. Instead, it would make communication of science to non-specialists, and, indeed, even communication among specialists, more difficult and it would not necessarily prevent anybody from justifying any kind of human behaviour they wish to encourage (Gadagkar, 1997).

Exporting metaphors, concepts and methods between the natural and social sciences is useful if done with care but can be dangerous, especially with metaphors, if done without adequate understanding of their original function in the field from which they are being imported.

References

Bowers, J., Boursiquot, J-M., this, P., Chu, K., Johansson, H. and Meredith, C. 1999. 'Historical genetics: the parentage of chardonnay, gamay, and other wine grapes of northeastern France'. *Science*, 285: 1562–5.
Camazine, S., Deneubourg, J-L., Franks, N. R., Sneyd, J., Theraulaz, G. and Bonabeau, E. 2001. *Self-Organization in Biological Systems*. Princeton: Princeton University Press.
Currie, C. R., Scott, J. A., Summerbell, R. C. and Malloch, D. 1999. 'Fungus-growing ants use antibiotic-producing bacteria to control garden parasites'. *Nature* 398: 701–4.
Freedman, D. 1983. *Margaret Mead and Samoa: The Making and Unmaking of an Anthropological Myth*. Cambridge, Mass.: Harvard University Press.
Freedman, D. 1998. *The Fateful Hoaxing of Margaret Mead: A Historical Analysis of Her Samoan Research*. Boulder, Colorado: Westview Press.
Frisch, K. von. 1967. *The Dance Language and Orientation of Bees*. Cambridge, Mass.: Harvard University Press.
Gadagkar, R. 1997. *Survival Strategies – Cooperation and Conflict in Animal Societies*. Cambridge, Mass.: Harvard University Press.
Gadagkar, R. 2001. *The Social Biology of* Ropalidia marginata *– Toward Understanding the Evolution of Eusociality*. Cambridge, Mass.: Harvard University Press.
Gould, J. L. and Gould, C. G. 1988. *The Honey Bee*. New York: Scientific American Library, Freeman.
Hölldobler, B. and Wilson, E. O. 1990. *The Ants*. Cambridge, Mass.: Harvard University Press.
Lister, A. M., Kadwell, M. Kaagan, L. M., Jordan, W. C., Richards, M. and Stanley, H. F. 1998. 'Ancient and Modern DNA from a variety of sources in a study of horse domestication', *Ancient Biomolecules* 2: 267–80.
Mead, M. 2001 (originally published in 1928). *Coming of Age in Samoa: A Psychological Study of Primitive Youth for Western Civilisation*. Harper Perennial.

Mueller, U. G., Rehner, S. A. and Schultz, T. R. 1998. 'The evolution of agriculture in Ants'. *Science* 281: 2034–8.
Oster, G. F. and Wilson, E. O. 1978. *Caste and Ecology in the Social Insects*. Princeton: Princeton University Press.
Seeley, T. D. 1995. *The Wisdom of the Hive: The Social Physiology of Honey Bee Colonies*. Cambridge, Mass.: Harvard University Press.
Vila, C., Leonard, J. A., Götherström, A., Marklund, S., Sandberg, K., Liden, K., Wayne, R. K. and Ellegren, H. 2001. 'Widespread origins of domestic horse lineages'. *Science* 291: 474–7.
Wilson, E. O. 1971. *The Insect Societies*. Cambridge, Mass.: Harvard University Press.

Part V
The Multilinear Modernization of Societies

Part V

14
Multiple Modernities in the Framework of a Comparative Evolutionary Perspective
Samuel N. Eisenstadt

Introduction

In this chapter I would like to analyze the implications of the concept of multiple identities from the point of view of paradigms of social and cultural change, especially the evolutionary one. The notion of 'multiple modernities' denotes a certain view of the contemporary world – indeed of the history and characteristics of the modern era – that goes against the view long prevalent in scholarly and general discourse (Eisenstadt, 2000, 2002a, 2002b; Roniger and Waisman, 2002). It goes against the view of the 'classical' theories of modernization and of the convergence of industrial societies prevalent in the 1950s, and indeed against the classical sociological analyses of Marx, Durkheim, and (to a large extent) even of Weber, at least in one reading of his work. They all assumed, even if only implicitly, that the cultural program of modernity as it developed in modern Europe and the basic institutional constellations that emerged there would ultimately take over in all modernizing and modern societies; with the expansion of modernity, they would prevail throughout the world (Eisenstadt, 2002a).

All these scholars, with the partial exception of Weber, and above all the classical theories of modernization including Parsons, were closely identified with a certain evolutionary – above all Spencerian paradigm, perhaps one which assumed that organizations and societies, like organisms, develop from simple to more differential ones and that the more differentiated ones entailed greater adaptive capacities. But the reality that emerged proved to be radically different. Indeed, the developments in the contemporary era did not bear out this assumption of the 'convergence' of individual societies and have attested to the great diversity of modern societies, even of societies similar in terms of economic development, like the major industrial capitalist societies – the European ones, the US and Japan. Far-reaching variability developed even within the West – within Europe itself, and above all between Europe and the Americas – the US, Latin America, or rather Latin Americas. The actual developments in most modern and modernizing societies indicated that the various institutional arenas – the economic, the political and that of family, while indeed characterized by a genuine trend to differentiation – exhibit continually relatively autonomous dimensions that come together in different ways in different societies and in different periods of their development. The same was even more true with respect to the relation between the cultural and structural dimensions of modernity. A very strong – even if implicit – assumption of the studies of modernization, namely that the cultural dimensions or aspects of modernization – the basic cultural premises of Western modernity – are inherently and necessarily interwoven with the structural institutional ones, became highly questionable.

While the different dimensions of the original Western project have indeed constituted the crucial starting and continual reference points for the processes that developed among different societies throughout the world, the developments in these societies have gone far beyond the original 'Western' or European model of modernity.

Modernity has indeed spread to most of the world, but did not give rise to a single civilization, or to one institutional pattern, but rather to the development of several modern civilizational patterns, i.e. of societies or civilizations which share common characteristics, but which yet tend to develop different even if cognate ideological and institutional dynamics. Moreover, far-reaching changes which go beyond the original premises of modernity have been taking place also in Western societies. In this paper I shall examine some of the major assumptions of the notion of multiple modernities; especially the view of modernity as a distinct and yet multifaceted civilization; the basic premises and institutional characteristics of this civilization; at the same time, second, I shall analyze the roots of the tendency to the development, within the framework of this civilization, of multiple versions thereof, of multiple modernities. In the last part of this chapter I shall analyze the implications of the concept of multiple modernities for the analysis of theories of social change, especially evolutionary ones.

* * *

It is a central assumption of the notion of multiple modernities that these different modernities do share some common characteristics but that at the same time there develop great differences between them – not just local variations, but indeed differences with respect to the core characteristics of modernity. This double characteristic of modern societies, which constitutes the very core of the notion of multiple modernities, can best be seen with respect to a crucial aspect or component of modernity – namely that of protest.

One of the most important characteristics of modernity is indeed the centrality of protest. Symbols of protest – equality and freedom, justice and autonomy, solidarity and identity – which can be found in the margins, peripheries, or in movements of protest in all human societies became central components of the modern project of human emancipation. The incorporation of such themes of protest into the center heralded the radical transformation of various popular and/or sectarian utopian visions from peripheral or subterranean views into central components of the political and cultural program, and became also the ideological bases of the legitimation of modern regimes – as can be seen in the trilogy of the French Revolution – liberté, egalité, fraternité. These themes were promulgated by numerous social and political activists, above all although not exclusively by the social movements often leading to regime change. These movements, which constitute a basic component of the modern political process, appeal to the typical modern ideals of equality and justice. However, the discourses of justice and the political mechanisms of regime change differ among different modern societies according to different cultural contexts, thus attesting to the heterogeneity within the modern project; or, in other words, to the continual development of multiple modernities. Sombart's old question 'Why is there no socialism in the United States?' formulated in the first decade of the 20th century is perhaps the first recognition of such variability of the characteristic movements of protest in different modern societies – a variability which became even more visible when moving to other countries – Japan, India or Muslim societies (Sombart 1976). In all of these societies there developed modern institutional and ideological patterns and movements of protest which while sharing these basic mode orientations yet differed greatly from the 'original' European ones and from each other.

It is central to the analysis of continually changing multiple modernities that such distinctive patterns of modernity, different in many radical ways from the 'original' European ones, crystallized not only in non-Western societies, in societies that developed in the framework of the various great civilizations – Muslim, Indian, Buddhist, or Confucian – under the impact of European expansion and in their ensuing confrontation with the European program of modernity. They evolved also – indeed first of all – within the framework of the Western expansion in societies in which seemingly purely Western institutional frameworks developed in the Americas. Whereas it was sometimes assumed that European patterns of modern development were repeated in the Americas, it is now clear that North America, Canada, and Latin America developed from the start in distinctive ways. Indeed, throughout the Americas we can trace the crystallization of new civilizations, and not just, as Louis Hartz claimed, of 'fragments' of Europe (Hartz, 1964). In these Western institutional and cultural frameworks, derived and brought over from Europe, there developed not just local variations of the European model or models, but radically new institutional and ideological patterns. It is quite possible that this was the first crystallization of new civilizations since that of the great 'Axial' civilizations and also the last to date. The crystallization of different modernities in the Americas attests that even within the broad framework of Western civilization – however defined – there developed not just one but multiple cultural programs and institutional patterns of modernity (Eisenstadt, 2002c).

The basic characteristics of modernity – modernity as a distinct civilization

This view of multiple modernities entails certain assumptions about the nature of modernity. The first is that modernity is to be viewed as a distinct civilization, with distinct institutional and cultural characteristics. According to this view, the core of modernity is the crystallization and development of mode or modes of interpretation of the world, or, to follow Cornelius Castoriadis' terminology, of a distinct social 'imaginaire,' indeed of the ontological vision, of a distinct cultural program, combined with the development of a set or sets of new institutional formations – the central core of both being, as we shall see later in more detail, an unprecedented openness and uncertainty (Eisenstadt, 2001).

The second such assumption is that this civilization, the distinct cultural program with its institutional implications, which crystallized first in Western Europe and then expanded to other parts of Europe, to the Americas and later on throughout the world, gave rise to continually changing cultural and institutional patterns that constituted different responses to the challenges and possibilities inherent in the core characteristics of the distinct civilizational premises of modernity.

* * *

The modern project, the cultural and political program of modernity as it developed first in the West, in Western and Central Europe, entailed distinct ideological as well as institutional premises. It entailed several sharp shifts in the conception of human agency, of its autonomy, and of its place in the flow of time. It entailed a conception of the future in which various possibilities that can be realized by autonomous human agency – or by the march of history – are open. The core of this program has been that the premises and legitimation of the social, ontological, and political order were no longer taken for granted; there developed a very intensive reflexivity around the basic ontological premises as well as around the bases of social and political order of authority – a reflexivity which was shared even by the most radical critics of this program, who in principle denied the legitimacy of such reflexivity.

The core of this cultural program has perhaps been most successfully formulated by Weber. To follow James D. Faubian's exposition of Weber's conception of modernity:

> Weber finds the existential threshold of modernity in a certain deconstruction: of what he speaks of as the 'ethical postulate that the

world is a God-ordained, and hence somehow meaningfully and ethically oriented cosmos.' ...

... What he asserts – what in any event might be extrapolated from his assertions – is that the threshold of modernity has its epiphany precisely as the legitimacy of the postulate of a divinely preordained and fated cosmos has its decline; that modernity emerges, that one or another modernity can emerge, only as the legitimacy of the postulated cosmos ceases to be taken for granted and beyond reproach. Countermoderns reject that reproach, believe in spite of it. ...

... One can extract two theses: Whatever else they may be, modernities in all their variety are responses to the same existential problematic. The second: whatever else they may be, modernities in all their variety are precisely those responses that leave the problematic in question intact, that formulate visions of life and practice neither beyond nor in denial of it but rather within it, even in deference to it ... . (Faubion, 1993: 113–15)

These ideological developments culminated in what is probably the core characteristic of the modern project – namely, to follow Claude Lefort's nomenclature, the loss of the markers of certainty – and the concomitant continual search for the restoration of some such markers – a search which would never be fully realized (Lefort, 1988). This search cannot be fully realized because as all such responses leave this problematic intact. The reflexivity that developed in the program of modernity focused not only, as in the axial civilizations, on the possibility of different interpretations of the transcendental visions and basic ontological conceptions prevalent in a society or societies, but came to question the very givenness of such visions and of the institutional patterns related to them. It gave rise to awareness that there evolved many such visions and patterns and that such visions and conceptions can indeed be contested (Eisenstadt, 1952, 1986).

Such awareness was closely connected with two central components of the modern project, emphasized in the early studies of modernization by Dan Lerner and later by Alex Inkeles. The first such component is the recognition, among those becoming and being modernized – as illustrated by the famous story in Lerner's book about the grocer and the shepherd – of the possibility of undertaking a great variety of roles beyond any fixed or ascriptive ones, and the concomitant receptivity to different messages which promulgate such open possibilities and visions. Second, there is the recognition of the possibility of belonging to wider translocal, possibly also changing, communities (Lerner, 1958; Inkeles and Smith, 1974).

Concomitantly, closely related to such awareness and central to this cultural program there developed an emphasis on the autonomy of man; his or her – but in the initial formulation of this program certainly 'his' – emancipation from the fetters of traditional political and cultural authority and the continuous expansion of the realm of personal and institutional freedom and activity. Such autonomy entailed several dimensions: first, exploration of nature and its laws; and second, active construction, mastery of nature, possibly including human nature and society. In parallel, this program entailed a very strong emphasis on the autonomous participation of members of society in the constitution of social and political order and on autonomous access by all members of the society to these orders and their centers.

Out of the conjunctions of these conceptions there developed a belief that society could be actively formed by conscious human activity. Two basic complementary but also potentially contradictory views of the best ways to do this developed within this program. First, the program as it crystallized above all in the Great Revolutions gave rise, perhaps for the first time in human history, to the belief that it was possible to bridge the gap between the transcendental and mundane orders, to realize through conscious human actions in the mundane orders, in social life, some of the utopian, eschatological visions; Second, there was increasing acceptance of the legitimacy of multiple individual and group goals and interests and of multiple interpretations of the common good (Eisenstadt, 1992).

* * *

The loss of markers of certainty inherent in the modern political and cultural program and search for their restoration was manifest in the major institutional arenas of modern societies – above all in the political arena and in the constitution of collectivities and collective identities.

The modern program also entailed a radical transformation of the conceptions and premises of the political order, of the constitution and definition of the political arena, and of the basic characteristics of the political process. The core of the new conceptions was the breakdown of traditional legitimation of the political order, the concomitant opening up of different possibilities for the constitution of such order, and the consequent contestation about how political order was to be constituted to no small extent by human actors (Eisenstadt, 1999b).

By virtue of all these characteristics, the modern political program combined orientations of rebellion and intellectual antinomianism with strong orientations to center-formation and institution-building, giving rise to social movements and movements of protest as a continual and central component of the political process. These conceptions were closely connected with the transformation in modern societies of the basic characteristics of the political arena and processes. The most important of these characteristics were, first, the charismatization of the center, the openness of this arena and of the political process. Second were the strong tendencies of the centers to permeate the periphery and of the periphery to impinge on the center, blurring the distinctions between center and periphery. Third was a strong emphasis on at least potentially active participation by the periphery of society, by all its members, in the political arena. Fourth was the combination of the charismatization of the center or centers with the incorporation of themes and symbols of protest mentioned already above – equality and freedom, justice and autonomy, solidarity and identity. These themes became central components of the modern project of human emancipation. It was indeed the incorporation of such themes of protest into the center that heralded the radical transformation of various sectarian utopian visions from peripheral views to central components of the political and cultural program (Eisenstadt, 1999b).

This quest of the periphery or peripheries for participation in the social, political, and cultural orders, for the incorporation of themes of protest into the center, and for the concomitant possible transformation of the center, was indeed often guided by the various attempts to reconstitute the markers of certainty the political arena grounded in utopian visions – visions promulgated above all by the major social movements that developed, as we shall see later on, as an inherent component of the modern political process.

Out of the combination of symbols and demands of protest into the central symbolic repertoire of society and their consequent transformation; of the recognition of the legitimacy of multiple interests and of visions of social order, the continuous restructuring of center-periphery relations – the reconstitution of the realm of the political has become a central component of political process and dynamics in modern societies. The various processes of structural change and dislocation which took place continually in modern societies as a result of the development of capitalism, of economic changes, urbanization, changes in the process of communication, and of the new political formations have led in modern societies not only to the promulgation by different groups of various concrete grievances and demands, but also to a growing quest for participation in the broader social and political order and in the central arenas thereof – indeed to the reconstitution thereof.

These demands for participation in the center and for the reconstruction of the realm of the political were closely connected with the crystallization of the basic characteristics of the modern political processes – the common denominator of which has been the openness thereof. While these characteristics are naturally most visible in open, democratic or pluralistic regimes, they are also inherent in autocratic and totalitarian regimes even if the latter attempt to regulate and control them in such a way as seemingly to 'close' them. The first of these aspects of the political process in modern societies, attesting to such openness, has been the emergence of a new type of 'political class' or 'classes' – and of new types of political activists – a non-ascriptive class, the recruitment to which was in principle, if not in fact, open to everybody. The second is the continual attempts of this 'class' or these 'classes' and activists to mobilize political support through open public contestations. The third is the fact that such attempts at the mobilization of such support and governance are closely related to the promulgation of different policies and their implementation. Fourth are the very strong tendencies – unparalleled in any other regimes, with the possible partial, but very partial, exception of some of the city-states of antiquity – of potential politicization of many problems and demands of various sectors of the society and of conflicts between them.

It was in close relation to these tendencies that there developed in modern societies the continual struggle about the redefinition of the realm of the political which has been borne above all by different social movements. Unlike in most other political regimes in the history of mankind, the drawing of the boundaries of the political has in itself constituted one of the major foci of open political contestation and struggle in modern societies, and it was such contestation that constituted one of the most important manifestations of the loss of markers of certainty and of the search for their restoration.

The same basic dynamics developed also with respect to the distinctive mode of constitution of the boundaries of collectivities and collective identities that developed in modern societies. The most distinct characteristic thereof, very much in line with the general core characteristics of modernity, was that such constitution was continually problematized. Collective identities were no longer taken as given or as preordained by some transcendental vision and authority, or by perennial customs. They constituted foci of contestations and struggles, often couched in highly ideological terms (Shils, 1975; Eisenstadt and Giesen, 1995; Roniger and Sznajder, 1998; Eisenstadt *et al.*, 2001; Eisenstadt, 2002b, 2002d). These contestations and struggles were focused around the basic characteristics of the constitution of modern collectivities, the most important among which were first the development of new concrete definitions of the basic components of collective identities – the civil, primordial, and universalistic and transcendental 'sacred' ones – and of the ways they were institutionalized. Second, there developed a strong tendency to absolutize them in ideological terms. Third, their civil components became increasingly important; and fourth, the construction of political boundaries and those of the cultural or 'ethnic' and national collectivities became closely connected. Fifth, territorial boundaries of such collectivities were emphasized giving rise to continual tension between their territorial and/or particularistic components and broader, potential universal ones.

Such different modes of the constitution of modern political order and collective identities were promulgated by many political activists and intellectuals, especially by the major social movements in modern societies. It was indeed one of the most distinct characteristics of the modern scene that the construction of collective boundaries and consciousness could also become a focus of distinct social movements – the national or nationalistic ones. While in many modern societies, as for instance England, France, Sweden, the crystallization of new national collectivities and identities, of different types of nation states took place without the national movements playing an important role, the potentiality of such movements existed in all modern societies. In some societies – in Central and Eastern Europe, some Asian and African, and to some extent Latin-American societies – they played a crucial role in the development of the new nation states.

A central component in the constitution of modern collective identities was the self-perception of a society and its perception by other societies as 'modern,' as the bearer of the distinct cultural and political program – and its relations from this point of view to other societies – be it those societies which claim to be – or are seen as – bearers of this program, and various 'others.'

The roots of the multiplicity of modernities

These contestations around the different political programs, constitution of collectivities as well as other aspects of life such as for instance the designation of the basic characteristics of civilized persons – constituted part of the perennial search for the restoration of markers of certainty in modern societies. But this search could never be fully realized, not only because of the internal characteristics of the cultural program of modernity, of the continual confrontation with the continually developing institutional reality, but also because the concrete contours of the different cultural and institutional continuous patterns of modernity as they crystallized in different societies have indeed been continually changing.

These institutional contours of modernities have been changing, first as a result of the internal dynamics of the technological, economic, political and cultural arenas as they developed in different societies and expanded beyond them. Second, these contours changed with the political struggles and confrontations between different states, and between different centers of political and economic power. Such confrontations developed within Europe with the crystallization of the modern European state system and became further intensified with the crystallization of 'world systems' from the sixteenth or seventeenth century on. Third, these contours changed in tandem with the shifting hegemonies in the different international systems that developed concomitantly with economic, political, technological and cultural changes (Tiryakian, 1985, 1994).

Fourth, such changes were generated by the fact that the expansion of modernity entailed confrontation between the basic premises of this program and the institutional formations that developed in Western and Northern Europe and other parts of Europe and later in the Americas and Asia: in the Islamic, Hinduist, Buddhist, Confucian and Japanese civilizations.

Fifth, such changes were rooted in the continual confrontations between on the one hand different interpretations of the basic premises of modernity as promulgated by different centers and elites, and on the other hand the concrete developments, conflicts, and displacements that accompanied the institutionalization of these premises. These confrontations activated the consciousness of the contradictions inherent in the cultural program of modernity and the potentialities conferred by its openness and reflexivity; and gave rise to the continual reinterpretation by different social actors of the major themes of this program, and of the basic premises of the civilizational visions, and of the concomitant grand narratives and myths of modernity.

* * *

These different cultural programmes and institutional patterns of modernity were not shaped as presented in some of the earlier studies of modernization as natural evolutionary potentialities of these societies – indeed, potentially of all human societies; or, as in the earlier criticisms thereof, by the natural unfolding of their respective traditions; nor just by their placement in the new international settings. Rather they were shaped by the continuous interaction between several factors, the most general being the various constellations of power, i.e. different modes of elite contestation and co-optation in different political systems and different ontological conceptions and political ideologies that influenced the nature of the emerging discourse of modernity, with various political activists, intellectuals, in conjunction above all with the social movements constituting the major actors like processes of reinterpretation and formation of new institutional patterns.

Or, in greater detail, these programmes were shaped first by basic premises of cosmic and social order, the basic 'cosmologies' that were prevalent in these societies in their 'orthodox' and 'heterodox' formulations alike as they have crystallized in these societies throughout their histories. A second shaping factor was the pattern of institutional formations that developed within these civilizations through their historical experience especially in their encounter with other societies or civilizations. The third set of factors shaping such program was the internal tensions, dynamics and contradictions that developed in these societies in conjunction with the structural-demographic, economic, and political changes attendant on the institutionalization of modern frameworks, and between these processes and the basic premises of modernity.

Fourth, the different programs of modernity were shaped by the encounter and continual interaction between the processes mentioned above, and the ways in which the different societies and civilizations were incorporated into the new international systems, and the ways in which they were placed or were able to place themselves, in these systems, to insert or become inserted into the global system. Thus indeed international constellations have also to be taken into account as influencing the mode of modernization in a particular context.

It is the combination of such factors that can explain some of the puzzling aspects of multiple modernities. Thus for instance the combination of the prevalence of pluralistic organization of centers; the relative devaluation of the political system as the major arena for the implementation of the basic ontological visions of Hindu civilization and the history of relatively long centralized modern colonial and post-colonial regimes explain the rather astonishing fact that India has developed as a modern, vibrant democracy and continued to be so (Eisenstadt, 2003a).

As against this, the prolonged semi-colonial experience combined with the decline of the Ottoman Empire; the construction, in the old provinces of the Ottoman Empire, under the auspices of the great powers of semi-independent states in the Middle East, and the tradition, in Islamic societies, of oscillation between pragmatic political attitudes and extreme religious sectarianism may explain the difficulties of development of democratic regimes and the relative fragility of such regimes that developed in the successor states of the Ottoman Empire (Eisenstadt, 2003b).

One of the most interesting illustrations of the crystallization of such distinct programs of modernity is Japan (Eisenstadt, 1996). The Japanese program of modernity ushered in by the Meiji Restoration was rooted in the non-Axial, immanentist ontologies, and it guided the crystallization of the Meiji state and later on the development of modern Japanese society, and shaped to some extent at least the specific characteristics of the major institutional formations of modern Japan. These formations were not grounded in the conceptions of principled, metaphysical individualism or in a principled confrontation between state and society as two distinct ontological entities. One of the most important such characteristics was the strong tendency to the conflation of the national community, of the state and of society. Such conflation has had several repercussions on the structuring of the ground rules of the political arena, the most important of which have been the development, first of a weak concept of the state as distinct from the broader overall, in modern terms national community (national being defined in sacral, natural and primordial terms); second, of a societal state characterized by a strong tendency to emphasize guidance rather than direct regulation and permeation of the periphery by the center; and third, a very weak development of an autonomous civil society, although needless to say elements of the latter, especially the structural, organizational components thereof (such as different organizations) have not been missing.

The specific type of civil society that developed in Japan is perhaps best illustrated by the continual construction of new social spaces which provides semi-autonomous arenas in which new types of activities, consciousness and discourse develop, which however do not impinge directly on the center.

Those participating in them do not have autonomous access to the center, and are certainly not able to challenge its premises. The relations between state and society have been rather effected in the mode of patterned pluralism, of multiple dispersed social contracts (Eisenstadt, 1996).

Accordingly, changes in the types of political regimes, or in the relative strength of different groups, have not necessarily implied changes in principles of legitimation and in the basic premises and ground rules of the social and political order.

Closely connected to these characteristics of civil society in modern Japan there has also developed a rather distinct pattern of political dynamics, especially of the impact of movements of protest on the center. The most important characteristic of this impact was the relatively weak principled ideological confrontation with the center – above all the lack of success of leaders of such confrontational movements to mobilize wide support; the concomitant quite far-ranging success in influencing, if often indirectly, the policies of the authorities and the creation of new autonomous but segregated social spaces in which activities promulgated by such movements could be implemented.

This weakness of civil society was not due to its suppression by a strong state, but rather to the continual conflation of state and civil society with the national community. While it is those close to the center – oligarchies, bureaucracies, politicians and even heads of economic organizations – who have on the whole shaped the contours of this community, yet they have not done it in a continuous confrontational response to the demands of other sectors of society.

All these characteristics of the political arena and of the relations between nation, country, state and society were very closely related to the specific strongly immanentist and particularistic ontological conceptions and their dynamics that have been prevalent in Japan throughout its history. The strong universalistic orientations inherent in Buddhism, and more latently in Confucianism, were subdued and 'nativized' in Japan. When Japan was defined as a divine nation, this meant a nation protected by the gods, being a chosen people in some sense, but not a nation carrying God's universal mission. Such transformation had far-reaching impacts on some of the basic premises and conceptions of the social order such as the Mandate of Heaven, with its implication for the conception of authority and the accountability of rulers, as well as conceptions of community. Unlike China, where in principle the emperor, even if a sacral figure, was 'under' the Mandate of Heaven, in Japan he was sacred and seen as the embodiment of the sun and could not be held accountable to anybody. Only the shoguns and other officials – in ways not clearly specified.

These specific institutional and cultural dynamics that developed in Japan were major characteristics of the elites and their coalitions in Japanese society. The common characteristic of these elites was that they were relatively non-autonomous and that their major coalitions were embedded in groups and settings (contexts) that were mainly defined in primordial, ascriptive, sacral and often hierarchical terms, and much less in terms of specialized functions or of universalistic criteria of social attributes. Linked to these characteristics of the major elites was the relative weakness of autonomous cultural elites. True, many cultural actors – priests, monks, scholars, and the like – participated in such coalitions. But with very few exceptions, their participation was based on primordial and social attributes and on criteria of achievement and social obligations according to which these coalitions were structured and not any distinct, autonomous criteria rooted in or related to the arenas of cultural specialization in which they were active. These arenas – cultural, religious, or literary – were themselves ultimately defined in primordial-sacral terms, notwithstanding the fact that many specialized activities developed within them.

It was also the combination of all these factors in combination with its specific political ecological location that does also explain the mode of the incorporation of Japan into the modern international system.

* * *

Other constellations of elites, their power relation with other sectors of their respective societies, the ontological conceptions promulgated by them; the impact of international forces and modes of incorporation into the emerging and continually changing international system, gave rise in other societies – be it indeed in the first, classical European modernity, in the Americas or in the multiple modernities that developed in the realm of Islamic, Hinduist, Buddhist civilizations, to other ideological and institutional programs of modernity. In all these processes it was the characteristics of elites, especially the extent of autonomy – i.e. of their being recruited and organized according to distinct criteria promulgated by them as against being embedded in various particularistic groups and recruited and organized according to the rules of such groups; the ontological visions promulgated by them, and the relations between such elites and between them and the broader social strata were of crucial importance. The importance of international factors can be also illustrated for instance, among others, by the fact that in Japan and Germany after World War II, different versions of the culturally specific modernities have been privileged and institutionalized.

The revaluation of the evolutionary perspective in social sciences – against evolutionary functionalism and the belief in automatic progress

The preceding analysis of multiple modernities has some important implications for the analysis of paradigms of social and cultural change, especially for a reconsideration of the evolutionary perspective in social science, which has greatly influenced 'classical' studies of modernization of the fifties – as could be seen perhaps most clearly in Talcott Parsons's work in classical evolutionary theory – a central concept in which was that of cultural and social differentiation (Parsons, 1977).

As is well known, these studies and the evolutionary perspective have come under strong attack from many quarters (Eisenstadt, 1995). There is no doubt that the original version of the evolutionary perspective, which in a way was indeed fully epitomized in the studies of modernization – which stressed the unilineal development of all societies on a universal evolutionary scale and the conflation between the differentiation of all institutional arenas and symbolic dimensions of social interaction – is not tenable.

But all these criticisms notwithstanding, the evolutionary perspective has a strong kernel of truth in it: namely, the recognition of the propensity of human action to continuous expansion. Processes of differentiation may be seen as a very important dimension of such a tendency to expansion. The core of such processes is the decoupling of the different components or dimensions of social action from the frameworks within which they are embedded and from one another. Such decoupling may develop with respect to both the structural and symbolic dimensions of social interaction and structure.

On the structural level, the major process of such 'decoupling' has been that of *structural* differentiation, best manifest in the crystallization of specific, organizationally distinct roles – such as for instance occupational ones as against their being firmly embedded in different family or local settings, and of the concomitant development of new integrative mechanisms. On the symbolic level, the process of such decoupling is manifest above all in the disembedment of the major cultural-orientations from one another – i.e. the growing autonomy of the different components of such orientations. Such decoupling is usually connected with a growing problematicization of the conception of ontological and social order, and with an increasing orientation to some reality beyond the given one and with growing reflexivity and second order thinking. Some of the most important illustrations of such decoupling can be seen in the transition from immanentist to transcendental orientations, or in the constitution of collectivities and model of legitimation of regimes, from primordial to civil and transcendental ones (Eisenstadt, 1952, 1986; Shils, 1975).

Contrary, however, to the presuppositions of classical evolutionary and structural-functional analyses, and indeed of the first studies of modernization, different dimensions of structural differentiation and disembedment of cultural orientations and a growing problematicization of the perceptions of the sources/dimensions of human existence do not always go together.

Here of special importance is the distinction between on the one hand the core of structural differentiation, of social division of labor, and on the other hand what has been called the basic elite functions – those functions or activities which are oriented to the problems generated by the very constitution of social division of labor – i.e. above all the constitution of trust, regulation of power and provision of meaning. While the processes of differentiation as they develop in different historical contexts greatly influence the range of elites and elite activities that develop in different situations, yet the concrete constellations of such elites are analytically distinct from structural differentiation. It is the continual interaction between such structural processes and the different constellation of elites as they crystallize in different historical and international settings that shapes the concrete contours of institutional formations that develop in different historical contexts. It is the degree to which the elites are autonomous or embedded in ascriptive units, or act as representatives of such units in the society, as well as the relation between different elites and broader social sectors that provides the crucial clue, as we have seen in the case of Japan, to the ways in which concrete institutional patterns are shaped.

These considerations apply to different evolutionary 'stages' – be it 'archaic' (Africa) or Imperial, or modern ones. In all such 'stages' or situations there develop, as I have shown in great detail in the *Political Systems of Empires*, relatively similar structural formations, yet it is only when combined with different constellation of elites as they crystallize in different historical and international settings that shape the concrete contours of different societies (Eisenstadt, 1963).

The same considerations apply perhaps even more forcefully to the analysis of modernization, of modern societies. Modernization does indeed entail far-reaching processes of decoupling between different dimensions of social action, of differentiation manifest in such processes as urbanization; in more intensive processes of communication or industrialization, and these processes do indeed give rise in different societies to similar problems – the similarity of which is enhanced by processes of globalization. But this does not mean that the 'answers' to these problems, i.e. the symbolic and institutional constellations that develop in different modern societies, are the same, that they converge. They do indeed vary greatly between them, shaped by the forces analyzed above, and this is indeed the core of the argument of multiple modernities.

The same applies to the more recent patterns of globalization. Contrary to the statements to be found in many of the recent studies of globalization which emphasize very much, in line with the earlier studies on the convergence of industrial societies, the homogenizing, uniform impact thereof, many recent studies, such as of Suzanne Berger, C. Dore, Michael Mann and Andreas Wimmer clearly indicate the very great institutional and cultural responses developing under the impact of the very intensive process of contemporary globalization (Berger and Dore 1996; Mann 2001; Wimmer 2001).

* * *

The development and expansion of modernity was not, indeed, contrary to the optimistic views of modernity as progress, peaceful. It bore within it very destructive possibilities, which were indeed voiced, and also often promulgated, by some of its most radical critics, who saw modernity as a morally destructive force, and emphasized the negative effects of some of its core characteristics. The crystallization of the first modernity and the development of later forms thereof were continually interwoven with internal conflicts and confrontations, rooted in the contradictions and tensions attendant on the development of capitalist systems and, in the political arena, the growing demands for democratization. They were also interwoven with international conflicts that developed in the framework of the modern state and imperialist systems. Above all they were closely interwoven with wars and genocides, repressions and exclusions constituted continual components thereof. Wars and genocide were not, of course, new in the history of mankind. But they became radically transformed and intensified, generating continuous tendencies toward specifically modern barbarism, the most important manifestation of which was the incorporation of violence, terror and war in an ideological framework, manifest first in the French Revolution, and later in the Romantic movement. This transformation emerged from the interweaving of wars with the basic constitutions of nation states, with those states becoming the most important agent – and arena – of the constitution of citizenship and symbols of collective identity; from the crystallization of the modern European state system; from European expansion beyond Europe; and from the development of the technologies of communication and of war (Wimmer, 2002).

These destructive forces, the 'traumas' of modernity which undermined the great promises thereof, emerged clearly during and after the First World War in the Armenian genocide, became even more visible in the Second World War, above all in the Holocaust, all of them shaking the naive belief in the inevitability of progress and of the conflation of modernity with progress. These destructive forces of modernity were paradoxically ignored or bracketed out from the discourse of modernity in the first two or three decades after the Second World War. Lately they have reemerged again in a most frightening way on the contemporary scene, in the new 'ethnic' conflicts in many of the former republics of Soviet Russia, in Sri Lanka, in Kosovo, and in a most terrible way in Cambodia and in African countries, such as Rwanda.

* * *

The extent to which such destructive tendencies developed in modern societies was greatly influenced by some of the modes of constitution of modern collective identities as they were borne by the different elites, by the relation between them and the broader areas as they were interwoven with the strength and flexibility of the centers, the mutual openness of elites, and their relations to broader social strata.

One of the most important aspects of the constitution of collective identities and the ways in which different themes of collective identity – primordial, such as 'ethnic' or 'national'; civil and sacral (religious or secular ones) – were interwoven in such constitution in their respective societies. In all modern societies there developed a continual tension or confrontation between these components of such identity, reconstructed in such modern terms. The mode of interweaving of these different components of collective identity, and especially the extent to which in the historical experience of those societies none of these dimensions has been totally absolutized or set up by their respective carriers against the other dimensions, or contrary-wise the extent to which there developed rather multifaceted patterns of collective identity greatly influenced the extent to which the construction as against destruction of potentialities of pluralistic and totalistic tendencies of the cultural and political program of modernity developed in these societies (Eisenstadt, 2002d).

Thus to take up some illustrations from Europe, it was insofar as the primordial components were relatively 'peacefully' interwoven in the construction of their respective collective identities with the civil and universalistic ones in multifaceted ways – that the kernels of modern barbarism and the exclusivist tendencies inherent in them were minimized.

In England, Holland, Switzerland and in the Scandinavian countries, the crystallization of modern collective identity was characterized by a relatively close interweaving – even if never bereft of tensions – of the primordial and religious components with the civil and universalistic ones, without the former being denied, allowing a relatively wide scope for pluralistic arrangements. Concomitantly in these countries there developed also relatively weak confrontations between the secular orientations of the Enlightenment – which often contained strong deistic orientations – and the strong religious orientations of various Protestant sects.

As against situations in these societies, in those societies (as was the case in Central Europe, above all in Germany and in most countries of Southern and Central Europe) in which the construction of the collective identities of the modern nation-state was connected with continual confrontations between the primordial and the civil and universalistic, and as well as between 'traditional' religious and modern universalistic components of collective identity, there developed a stronger tendency to crisis and the breakdown of the different types of constitutional arrangement. In the more authoritarian regimes, such primordial components were promulgated in 'traditional' authoritarian terms – in the more totalitarian fascist or national-socialist movements, in strong racist ones – while the absolutized universalistic orientations were promulgated by various 'leftist' Jacobin movements.

France, especially modern Republican France from the third republic on, but with strong roots in the preceding periods, constitutes a very important – probably the most important – illustration of the problems arising out of continual confrontations between Jacobin and traditional components in the legitimation of modern regimes – even within the framework of relatively continuous polity and collective identity and boundaries. The case of France illustrates that under such conditions, pluralistic tendencies and arrangements do not develop easily, giving rise to the consequent turbulence of the institutionalization of a continual constitutional democratic regime (Eisenstadt, 1999a).

The constitution of different modes of collective identity has been connected in Europe – and beyond Europe – with specific institutional conditions; the most important among them being the flexibility of the centers, the mutual openness of elites, and their relations to broader social strata. There developed in Europe, and later in other societies, a close elective affinity between the absolutizing types of collective identity and various types of absolutist regimes and rigid centers, and between the multifaceted pattern of collective identity in which the primordial, civil, and sacred components were continually interwoven with the development of relatively open and flexible centers and of mutual openings between various strata.

It was in so far as such multifaceted modes of construction of collective identities and of strong but flexible centers faltered that the two major forms of absolutizing tendencies, bearing within themselves the kernels of barbarism, of destruction, of drastic exclusion, demonization and annihilation of others – the Communist and the extreme fascist, especially the National Socialist movements and regimes – triumphed.

All these destructive potentialities and forces are inherent potentialities in the modern program, most fully manifest in the ideologization of violence, terror and wars, and the total ideological exclusivity and demonization of the excluded are not outbursts of old 'traditional' force – but outcomes of modern reconstruction, of seemingly 'traditional' forces reconstituted in modern ways. Thus indeed modernity is, to paraphrase Leszek Kolakowski's felicitious and sanguine expression – 'on endless trial' (Kolakowski, 1990).

The preceding discussion shows that the concept of multiple modernities is thus not a theory that would deterministically foresee which developments can occur in which countries. Rather, it is a heuristic concept that can, when applied to different empirical constellations in different parts of the globe, produce a series of falsifiable hypotheses with regard to the political and cultural factors that determine the course of modernization processes at specific historical junctures – and it can provide also important clues for some more general problems in the analysis of social change in human societies.

Bibliography

Berger, Suzanne & Richard Dore. (1996) *National Diversity and Global Capitalism*. Ithaca, N.Y.: Cornell University Press.
Eisenstadt, S.N. (1952) 'The axial age: the emergence of transcendental visions and the rise of clerics'. *European Journal of Sociology* 23 (2) 294–314.
——. (1963) *The Political Systems of Empires*. New York: Free Press.
—— ed. (1986) *The Origins and Diversity of Axial-Age Civilizations*. Albany, NY: SUNY Press.
——. (1992) 'Frameworks of the great revolutions: culture, social structure, history and human agency'. *International Social Science Journal* 44 (385–401).
——. (1995) 'Social division of labor, construction of centers and institutional dynamics: a reassessment of the structural–evolutionary perspective'. *Proto-Soziologie* 7 (11–22).
——. (1996) *Japanese Civilization: A Comparative View*. Chicago: The University of Chicago Press.
——. (1999a) *Fundamentalism, Sectarianism and Revolution: The Jacobin Dimension of Modernity*. Cambridge: Cambridge University Press.
——. (1999b) *Paradoxes of Democracy: Fragility, Continuity and Change*. Washington, DC: The Woodrow Wilson Center.
——. (2000) 'The reconstruction of religious arenas in the framework of "multiple modernities" '. *Millenium: Journal of International Studies* 29 (3) 591–611.
——. (2001) 'The civilizational dimension of modernity: modernity as a distinct civilization'. *International Sociology* 16 (3) 320–40.
——. (2002a) *Multiple Modernities*. New Brunswick: Transaction Publications.
——. (2002b) 'The first multiple modernities: collective identities, public spheres and political order in the Americas'. In: Roniger, Luis & Carlos H. Waisman eds. *Globality and Multiple Modernities: Comparative North American and Latin American Perspectives*. Brighton: Sussex Academic Press.
——. (2002c) 'The civilizations of the Americas: the crystallizations of distinct modernities'. *Comparative Sociology* 1 (1) 43–61.
——. (2002d) 'Cultural programs, the construction of collective identity and the continual reconstruction of primordiality'. In: Malesevic, Sinisa & Mark Haugaard eds. *Making Sense of Collectivity: Ethnicity, Nationalism and Globalization*. London: Pluto 33–38.
——. (2003a) 'The puzzle of Indian democracy'. In: *Comparative Civilizations and Multiple Modernities*. Leiden: Brill Academic Publishers 801–31.
——. (2003b) 'Civil society and public sphere: the myth of oriental despotism and political dynamics in Islamic societies'. In: *Comparative Civilizations and Multiple Modernities*. Leiden: Brill Academic Publishers 399–435.
Eisenstadt, S.N. & B. Giesen. (1995) 'The construction of collective identity'. *European Journal of Sociology* 36 (1) 72–102.
Eisenstadt, S.N., Wolfgang Schluchter & Bjorn Wittrock eds. (2001) *Public Spheres and Collective Identitites*. New Brunswick: Transaction 105–32.
Faubion, James D. (1993) *Modern Greek Lessons. A Primer in Historical Constructivism*. Princeton: Princeton Universtiy Press 113–15.
Hartz, L. (1964) *The Founding of New Societites*. New York: Brace and World.
Inkeles, A. & D.H. Smith. (1974) *Becoming Modern: Individual Change in Six Developing Countries*. Cambridge, Mass.: Harvard University Press.

Kolakowski, L. (1990) *Modernity on Endless Trial*. Chicago: The University of Chicago Press.
Lefort, C. (1988) *Democracy and Political Theory*. Cambridge: Polity Press.
Lerner, D. (1958) *The Passing of Traditional Society: Modernizing the Middle East*. Glencoe, Ill.: Free Press.
Mann, Michael. (2001) 'Globalization and September 11'. *New Left Review* 12 (Nov.–Dec.) 51–72.
Parsons, Talcott. (1977) *The Evolution of Societies*. New Jersey: Prentice Hall.
Roniger, Luis & Carlos H. Waisman eds. (2002) *Globality and Multiple Modernities: Comparative North American and Latin American Perspectives*. Brighton: Sussex Academic Press.
Roniger, Luis & Mario Sznajder eds. (1998) *Constructing Collective Identities and Shaping Public Spheres: Latin American Paths*. Brighton: Sussex Academic Press.
Shils, E. (1975) 'Primordial, personal, sacred and civil ties'. In: *Center and Periphery: Essays in Macrosociology*. Chicago: The University of Chicago Press 111–26.
Sombart, W. (1976) *Why Is There No Socialism in the United States?* New York: M. E. Sharpe.
Tiryakian, E. (1985) 'The changing centers of modernity'. In: E. Cohen, M. Lissak & U. Almagor eds. *Comparative Social Dynamics: Essays in Honor of S.N. Eisenstadt*. Boulder, CO and London: Westview 131–47.
——. (1994) 'The New Worlds and sociology – an overview'. *International Sociology* 9 (2) June 131–48.
Wimmer, A. (2001) 'Globalizations avant la lettre: a comparative view of isomorphization and heteromorphization in an inter-connecting world'. *Comparative Studies in Society and History* 43 (3) July 435–66.
——. (2002) *Nationalist Exclusion and Ethnic Conflict: Shadows of Modernity*. Cambridge: Cambridge University Press.

15
On Modernity and Wellbeing
Oded Stark

A statement that there are different types of modernity is interesting but not very useful. We would want to know which type of modernity is more supportive of, or conducive to, economic betterment, what are the conditions that yield one type of modernity as opposed to another, and whether the evolutionary path from pre-modernity to modernity is amenable to policy intervention.

The idea that there are different types of modernity – 'multiple modernities' – is not alien to economics. In economic analysis we have many dynamic systems that converge to multiple steady-state equilibria. (A steady state is a situation in which all the relevant variables completed their adjustment to exogenous changes.) We often have steady-state equilibria that are stable – perturbations around them will set in motion adjustment processes that bring us back to where we were prior to the disturbance, and equilibria that are unstable – once tinkered with, we will be thrown far off course. Do we have a similar characterization of states of modernity?

Moreover, we are also able to rank equilibria – for example a high per-capita income steady-state equilibrium as opposed to a low per-capita income steady-state equilibrium, and characterize the transition from one steady state to another. Can a similar ranking be invoked and can a characterization akin to the one we employ in economics apply to states of modernity?

Taking the view that the division of labor, specialization, and the associated needs for cooperation and coordination are major constituent elements of modernity brings us quite close to basic concepts of material development and economic growth, and hence modernity is of natural interest to economists. But in economics we would seek to know whether a particular social and organizational structure is better at coordinating, more effective in inducing cooperation, more successful at prompting and promoting trust. Note, however, that contrary to the received wisdom in contemporary social science, the proposition that people are better off in a society with trust than in a society without trust need not necessarily hold. Consider a two-players, single-shot prisoner's dilemma game with the strategies and payoffs as per Table 15.1. Both

Table 15.1

	Player F	
	C	D
Player E — C	3, 3	1, 4
Player E — D	4, 1	2, 2

Table 15.2

	Player F	
	C	D
Player E — C	3, 3	$2\frac{1}{2}, 2\frac{1}{2}$
Player E — D	$2\frac{1}{2}, 2\frac{1}{2}$	2, 2

players agree to play C which entails the highest per-capita payoff in the economy. In an economy with no trust, player E conjectures that player F will not trust him to stick to C. Player E's best response to player F's expected playing of D is to play D himself. Due to the symmetry of the game, the same reasoning applies to what player F conjectures, and so on. Thus, we end up with both players playing D. In an economy with trust, player E trusts that player F will keep his word to play C, which entices player E to choose D. Again, symmetry prompts player F to reason and act likewise, resulting in both players ending up playing D.

Interestingly, if the players were sufficiently altruistic towards each other – attaching each a weight that is a little more than 1/3 to the wellbeing of the other, and a little less than 2/3 to his own wellbeing, the economy will settle at CC. For example, if each player were to attach a weight that is a little less than 1/2 to the wellbeing of the other and a little more than 1/2 to his own wellbeing, Table 15.1 will be converted, approximately, to Table 15.2 and the economy will be at CC. If altruism is a trait, per capita income and thereby wellbeing in a society with altruistic individuals will be higher than per capita income and wellbeing in a society with trusting individuals. How does the altruism trait come to be? Which processes, institutional forms, and modes of incorporation of individuals and communities into a larger society are likely to be conducive to the evolution of altruism and cooperation? These are questions to which the multiple-modernities line of inquiry is yet to provide answers.

The argument that modernity entails expansion of the set of communities to which people belong is tantamount to stating that the onslaught of modernity brings about a substitution of a large reference group for a small reference group. This substitution raises the interesting spectre that modernity introduces new complications when it comes to the sensing of improved wellbeing. It could also explain why modern societies are characterized by both 'a culture' of political protest, and by 'a culture' of uncertainty. Suppose that an individual i whose income is 10 belongs to a small reference group in which the incomes of the other three members are 12 each. If, as for example in Stark and Wang (2000), we measure i's relative deprivation, RD(i), by the proportion of those in i's reference group who are wealthier than i times their mean excess income, we have that

$$\text{RD}(i) = \frac{3}{4} \cdot \frac{2+2+2}{3} = 1.5.$$

If, while holding i's income unchanged, i's reference group expands to include one additional member with an income of 12, then

$$\text{RD}'(i) = \frac{4}{5} \cdot \frac{2+2+2+2}{4} = 1.6.$$

Individual i may even have an income that is a little bit higher than 10, say $10 + \varepsilon$, $\varepsilon > 0$, such that

$$\text{RD}''(i) = \frac{4}{5} \cdot \frac{2-\varepsilon+2-\varepsilon+2-\varepsilon+2-\varepsilon}{4} = \frac{8-4\varepsilon}{5} = 1.6 - \frac{4}{5}\varepsilon.$$

Yet, for a small enough ε,

$$1.6 - \frac{4}{5}\varepsilon > 1.5;$$

the material gain that modernity confers may not be enough to counter the increased relative deprivation and the associated feeling of eroded wellbeing that could arise from modernity's expansion of the reference group.

Drawing the attention of economists to the concept of modernity and to the process of modernization is very tantalizing. Both concept and process raise challenging questions, some of which I have sought to pose. A thorough dialogue between sociologists and economists on these and related questions is yet to begin. The study of the linkages between modernity and economics lies at the very frontier of social science research.

Reference

Stark, Oded and Wang, You Qiang. 2000. 'A Theory of Migration as a Response to Relative Deprivation'. *German Economic Review* 1, pp. 131–43.

16
Multiplicity in Non-Linear Systems
Somdatta Sinha

Introduction

While analysing the processes of rapid and fundamental change in the contemporary world with diverse cultures, it is clear that reality of the present is different from what was prophesied based on past social theories of the evolution of modernity. Different societies are in no sense becoming identical following the cultural programme of 'western' modernity (Eisenstadt, 2000). Originator of the term 'multiple modernities' and considered a pioneer in advancing an alternative view, Shmuel N. Eisenstadt, in his paper in the workshop 'Paradigms of Change', developed the concept of 'multiple modernities' in the context of a comparative evolutionary perspective in the social sciences. He based his arguments on the central concept – in classical structural evolutionary theory – that of differentiation, cultural and social differentiation and evolution. At the same time he questioned their stress on the unilineal development of all societies. He sees the process of differentiation as a tendency for expansion of human action, and the core of such processes of differentiation as the 'decoupling of "formerly" mutually embedded activities. Such differentiation may develop with respect to both the structural and symbolic dimensions of social interaction and structure'. Eisenstadt argues that the cultural and institutional patterns constitute different responses to the challenges and possibilities inherent in the core characteristics of the distinct civilizational premises of modernity, thereby giving importance to the social processes along with the structural differentiation. Thus, he sees the history of modernity as a 'story of continual development and formation, constitution and reconstitution of a multiplicity of cultural programs of modernity and distinctively modern institutional patterns, and of different self-conceptions of societies as modern – of multiple modernities'.

Does this notion of 'multiple modernities' bear any resemblance with the ideas of evolutionary processes of change in the field of natural sciences? In this chapter, I will try to show that though the players and language of interaction in these two fields of enquiry are quite different, yet there is a definite convergence in the underlying ways of reasoning in the context of the principles of evolution of system behaviour through interaction of processes intrinsic and extrinsic to the system. I will be using only one approach – that of the nonlinear dynamical systems from physics, but approaches also experimental and theoretical exist in material sciences to analyse principles of organisation in systems/organisms and their evolution.

Differentiation and change in complex systems

Most natural systems are organised entities in space and time, and are composed of multiple variables and parameters with nonlinear interactions among them. The macro-level functional behaviour of a complex system depends not only on its internal attributes, but also on how they are connected in space and time. Traditionally, linear analysis is used in all disciplines of natural and socio-economic sciences as it is mathematically tractable. But it neglects heterogeneity among the constituent entities and assumes simplistic random interactions, and thus, is restrictive and not representative of the system behaviour. Such complexities can yield quite non-intuitive system behaviour. Understanding how variation in the internal factors and external environment influence the spatiotemporal dynamics of the system is the subject of enquiry in complex system research.

Eisenstadt uses the concept of 'differentiation' which he describes as 'the decoupling of formerly mutually embedded activities' to explain the existence of multiplicity in perception and response of different societies to internal and external reality. The verb 'differentiate' normally means – to form or mark differently (i.e. to be unlike in nature and qualities) from other such things; to change or alter; to become unlike or dissimilar; to make a distinction, etc. The important theme here is 'change from one – to – something else'. In biology, 'differentiation'; in the context of development; involves structural and functional specialisation of individual cells from one of a number of common basic cell types, which possess the competence/potency to undergo changes in different ways in response to suitable internal or external. Though differentiation is the origin of cellular diversity, it may be remembered that for the development and evolution of the whole organism, this is not sufficient, and other processes such as cell interaction, movement, growth, spatial organisation, and interaction with the internal or external environments, are crucial also (Slack, 1983). The information of the final state is embedded in the genetic map of the cells, but the spatial and temporal history and environment of the cells play a major role in the manifestation of these changes. It is becoming increasingly evident, that the same cells may take quite different paths of differentiation under the influence of noisy extra/intra-cellular environment signals (McAdams and Arkin, 1999). In the context of ecology and evolution, where controversy over definitions are common place (Dietrich, 1992), the process of 'speciation' may be considered to lead to changes in organisms from one kind to another (see Haldane, 1931). A related and interesting phenomenon is the case of 'phenotypic plasticity's' where the individuals of the same population can show a great deal of differences in their structure and/or behaviour, when exposed to different environment for several generations through gene–environment interactions (Via and Lande, 1985). In all these cases there is a precursor system (cell, organism, population) that changes over a time scale that is as varied as cell cycle time, developmental time, or evolutionary (i.e. spanning several generations) time. What resulting changes are feasible depends on the present state of the system (which incorporates its past history), the intrinsic processes (e.g. variables, functional interactions and structural plan) that make up the system, and the interaction with the environment. In Figure 16.1, I give a schematic representation of the above-mentioned scenario. The state of the system at time $t + 1$ would be dependent on both global and local processes acting at different spatial and temporal scales within the system and their interaction with the environment (Hogeweg, 2002; Levin and Pacala, 2002). In both biological and physical systems, multiple processes of wide ranging time (nanoseconds to million years) and space (atoms to ecosystems) scales interact among themselves, and what is 'observed' depends much on the window of the observation time and measurement length. Most natural systems are 'thermodynamically open', where there is a continuing exchange of influences between the system and the environment – each modifies, and can be modified by the other.

In dynamical systems theory, the notion of 'bifurcation' represents a change of state of the system. 'Bifurcation' occurs when changes in the system attributes lead to quantitative and qualitative changes in the dynamic state of the system (Strogatz, 1994). Such altered states of the system can be of different types – a stable state can yield – (a) new stable state (b) states with different dynamics such as, oscillatory, chaotic, etc. Similar bifurcations can arise in the spatial domain also, and one can study changes in form and structures. Bifurcation has been thought to underlie cell differentiation (Kaneko and Yomo, 1997; also see Kauffman, 1993), and has been implicated in different areas in biology, which deal with the structure and pattern forming processes

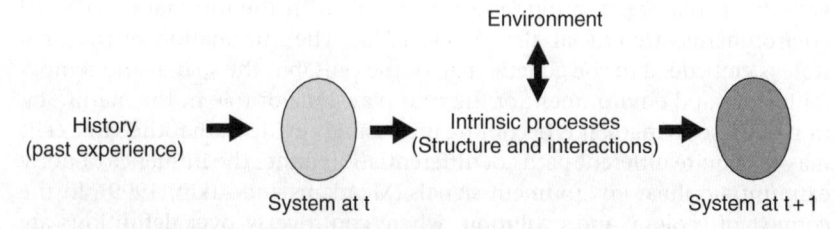

Figure 16.1 Processes involved in the evolution of a complex system

(Murray, 1989; Tyson et al., 2001). In the heart of all these are the fundamental issues of stability/constancy and change. Thus common principles exist that can be used to address problems of change across a variety of processes in different disciplines. To elaborate on some of the above points in the present context, I start with the hypothesis that a *society or a civilisation (inclusive of its cultural and political programme of modernity) is a structured system composed of nonlinear, multiply-coupled variables open to external environment*. The evolution of such a system ('society'), that constitutes a whole gamut of interacting socio-cultural processes, orders, and institutions, under changes in internal and external factors, is not easy to predict *a priori*. It can take divergent routes and lead to quite different expressions, or, it may not! For the sake of ease of explanation, it is a common practice in natural sciences to use simple models involving one or few variables as examples. Here, I assume that the difference in dynamical states, which describe the qualitative and quantitative aspects of temporal evolution, is a representative of change in the system behaviour. A large literature exists on spatiotemporal systems (including collective behaviour in multi-agent systems), where structural and temporal evolution of systems have been studied using nonlinear dynamical systems theory and other approaches (Turing, 1952; Thom, 1983; Langton, 1989; Holland, 1998; Maree and Hogeweg, 2001; Fontana, this volume).

Sources of multiplicity in nonlinear dynamical systems theory

Bifurcation as a process of change

Here I show, using two widely used model systems from physics and biology, that changes in intrinsic parameters lead systems to undergo bifurcation and induce different system behaviour.

One dimensional maps

These nonlinear discrete equations (maps) are used for modelling growth of organisms having discrete generations (annual plants, insects, etc). Two of these simple systems (May, 1976), having a single variable (population density, X) and a single parameter (growth rate, r) are given by

1(a) Logistic map: $X(t + 1) = rX(t)[1 - X(t)]$
1(b) Exponential map: $X(t + 1) = X(t)\exp\{r[1 - X(t)]\}$

Here $X(t)$ and $X(t + 1)$ are the population sizes at time t and $(t + 1)$. The plot $X(t)$ versus $X(t + 1)$ in Figure 16.2 (a and b) show that the growth functions in both models have a 'hump' shape, but they differ in details.

These two models are considered to belong to the same universality class (Devaney, 1985), and show very similar dynamical behaviour with increasing the parameter 'r' (see the bifurcation diagrams in Figure 16.2(c and d)).

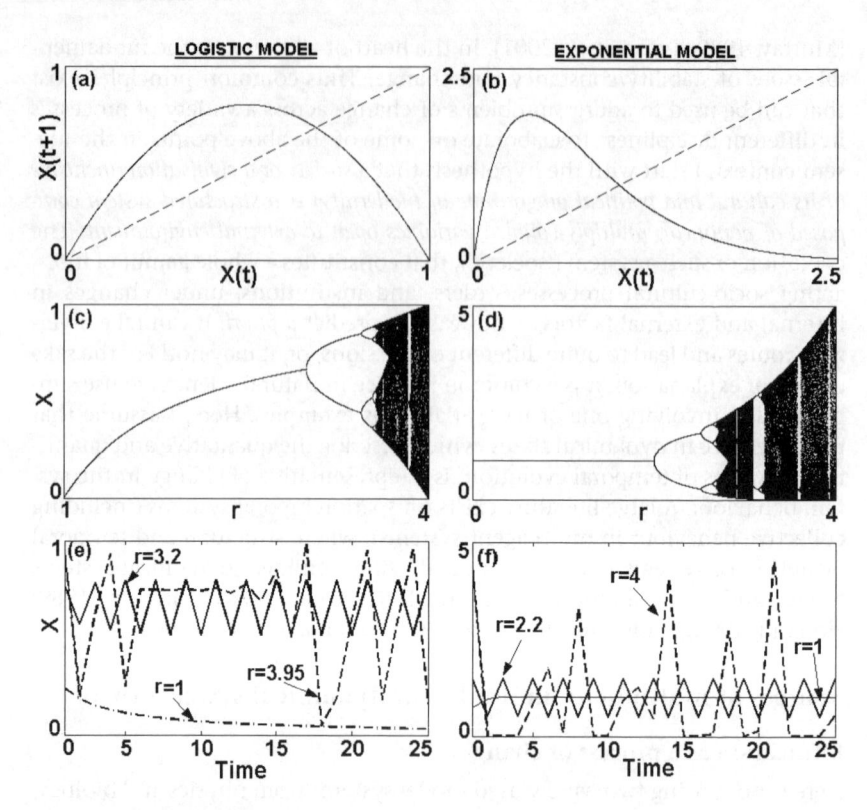

Figure 16.2 Structure and dynamical behaviour: (a) logistic; and (b) exponential maps

X shows equilibrium behaviour for a range of small values of 'r' (shown for r = 1 in Figure 16.2(e and f)). As 'r' is increased, a bifurcation occurs, giving rise to a new type of dynamical behaviour and the system exhibits periodic variation in X (r = 3.2 and r = 2.2 in Figure 16.2(e and f)). The system goes through further bifurcations for small changes in 'r' and the dynamics of X becomes more complex. Beyond a critical value of 'r', X varies erratically/chaotically (r = 3.95 and r = 4 in Figure 16.2(e and f)) taking on many values. Thus, such simple nonlinear systems even while harbouring different internal structures (growth functions) can show a similar evolutionary pattern of multiplicity in behaviour due to bifurcation in response to internal parametric variation. Also, the same system exhibits robust behavioural dynamics for a large range of low values of r, but at higher values it is quite susceptible to small variations.

The Lorenz system

A simple three variable (x, y, z) model for the flow of fluid under temperature difference (Lorenz, 1963) is given by the following three differential equations

$$dx/dt = s(y - x); \quad dy/dt = -xz + Rx - y; \quad dz/dt = xy - bz$$

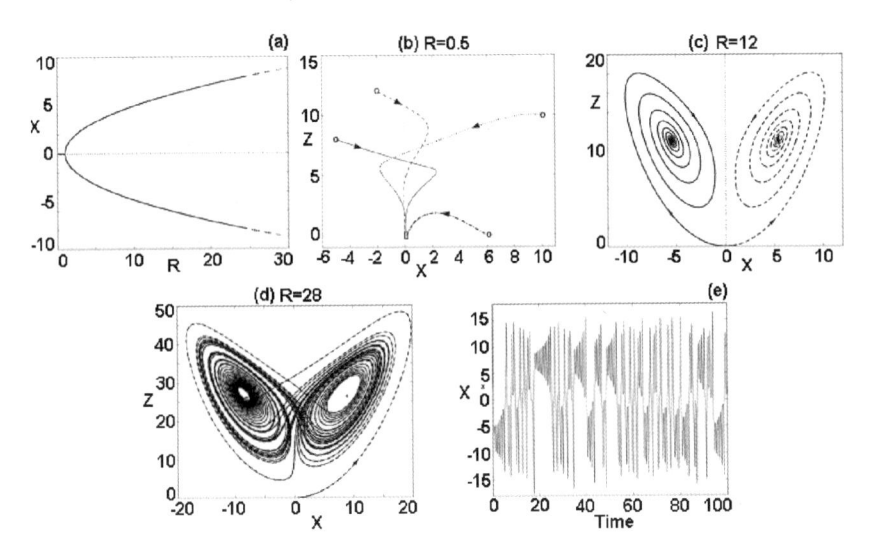

Figure 16.3 Structure and dynamics of Lorenz system

These equations tell us that, given a location in state space (decided by the variables and parameters s, R, and b), by how much each variable will change in the next small increment of time Fe(dt). Figure 16.3(a) summarises the three types of behaviour of the flow (only x variable shown here) for changing the parameter R (Hilborn, 1994). For R < 1, there is a single stable steady state (x = 0, y = 0, z = 0), and for different initial conditions, the system returns to this equilibrium point. This is shown by the four trajectories in the (x–z) phase plane in Figure 16.3b for R = 0.5. Bifurcation occurs at R = 1, and the system then possesses two new stable steady states (the two solid lines in Figure 16.3(a)). Here, the system behaviour is bistable, i.e. it can attain any of the two stable states (shown in Figure 16.2(c) for R = 12), where two trajectories, starting very close, reach the two different stable steady states. This indicates that two Lorenz systems, with similar initial conditions, having the same system attributes (parameter values) can evolve to two different final states. Further bifurcations occur as R is increased, and the system approaches a turbulent state where irregular, chaotic motion takes place around the three unstable steady states (dots in Figure 16.3(d) for R = 28). The temporal organisation of the chaotic dynamics of x is shown in Figure 16.3(e). In this state, two systems with almost identical properties and conditions can become completely uncorrelated and far apart in phase space – a phenomenon termed as 'butterfly effect' – indicating that as small a perturbation as the flutter of a butterfly's wing can cause a major long-term change in system behaviour. Chaos is also associated with instability, large changes, and visiting extreme situations/values. This example also shows that the same system can manifest a variety of behaviour and evolve to different end states with changes in its internal parameters.

(A) Logistic map

(B) Exponential map

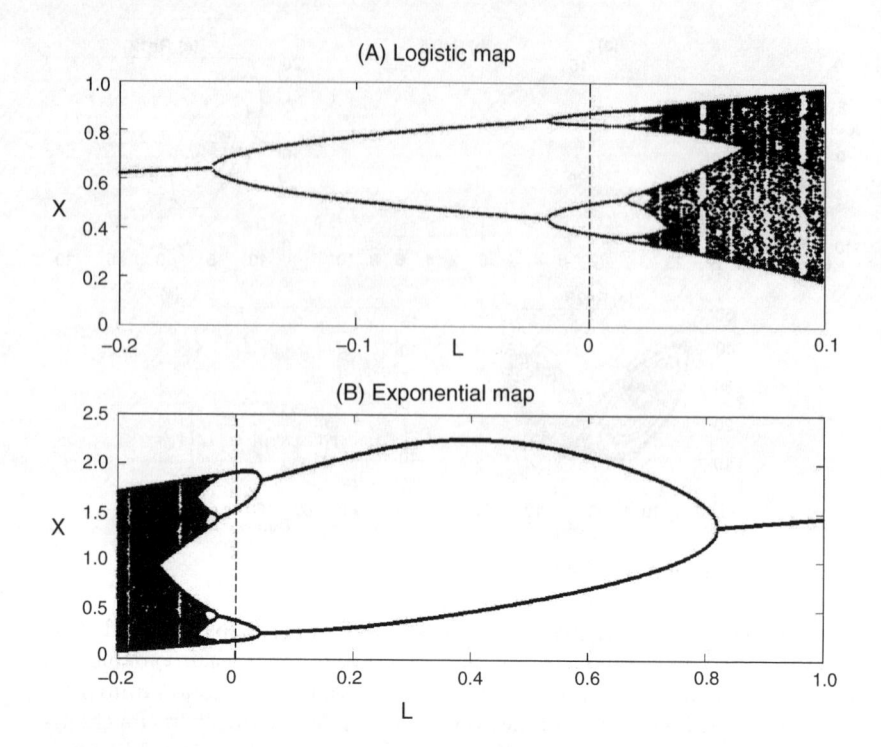

Figure 16.4 Dynamics of (a) logistic; and (b) exponential maps under external perturbation

Thus both these model systems show that bifurcation acts as a process of change to yield multiplicity in system behaviour under changes in some internal parameter, which can also depend on the initial state of the system.

Different reactions to external perturbations

Here I show that similar systems behaving similarly to changes in internal parameters may respond quite differently to external perturbations. Internal structural variations can act as *modifiers of change*.

One dimensioneal maps under external perturbation

The two simple one dimensional maps (see Section I) under a fixed external perturbation 'L' are given by

Logistic map: $X(t + 1) = rX(t)[1 - X(t)] + L$
Logistic map: $X(t + 1) = X(t)\exp\{r[1 - X(t)]\} + L$

The model parameters are chosen such that they exhibit the same dynamics (four period oscillation) when there is no perturbation (the dotted line in Figure 16.4 at L = 0). Figure 16.4(a and b) shows the bifurcation behaviour

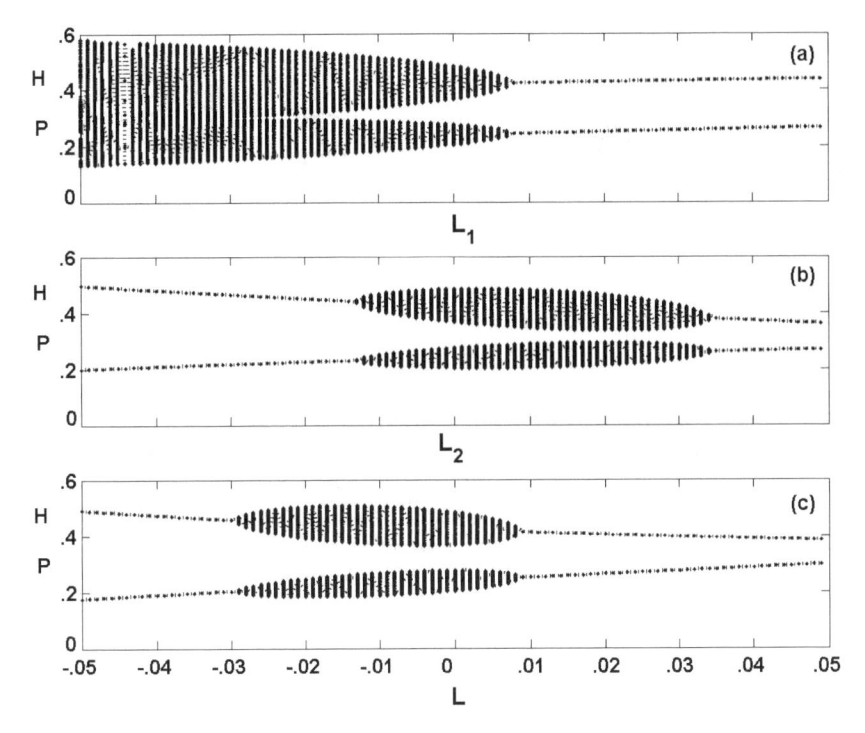

Figure 16.5 Dynamics of H and P when external perturbation is applied to (a) H, (b) P, and (c) to both H and P

of the two systems for positive and negative values of 'L'. It is clear that the dynamical response of these two similar maps to external perturbation is opposite. The Logistic map exhibits increased complexity in dynamics leading to chaos as 'L' is increased; and, regular dynamics leading to equilibrium as 'L' is decreased (Figure 16.4(a)). The Exponential map (Figure 16.4(b)) shows exactly the opposite response – positive 'L' inducing stability and negative 'L' enhancing complexity in dynamics.

This contrast in response of similar systems to the same external perturbation has been shown (Parthasarathy and Sinha, 1995; Sinha and Parthasarathy, 1996) to be due to the small difference in their internal structure (see Figure 16.2(a and b)), even though this variation does not elicit any difference in system behaviour for changes in the internal parameter 'r' (ref. Figure 16.2(c and d)). Thus, systems considered equivalent can respond very differently to some perturbations and similarly to some others. This feature highlights the need to look deeply into the differences in similar systems. Here only single maps are considered, but larger systems composed of coupled system of maps also exhibit similar results (Parekh *et al.*, 1998; Parekh and Sinha, 2002).

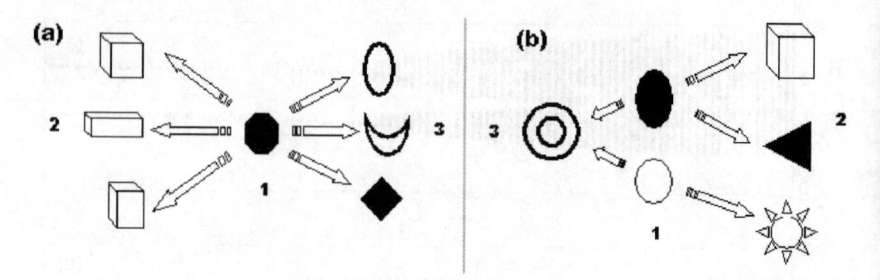

Figure 16.6 Paradigms of evolutionary change

Interaction between perturbation and system variables

It is not intuitively obvious as to how a coupled, multi-variable system will respond to external perturbation that can interact with the variables individually and collectively. In this section, I take an interacting two variable system exposed to constant external perturbation, and show that the system's response differs depending on the variable perturbed.

The discrete Host–Parasite Model (Sole *et al.*, 1992) is given by

$$H(t+1) = r.H(t)\{1-H(t)\}.\exp[-bP(t)], \qquad P(t+1) = H(t)\{1-\exp[-bP(t)]\}$$

The host species (H) grow according to the logistic map with intrinsic growth rate 'r', but their population size is modulated by the parasite (P) that grows only at the expense of the host. Parameter 'b' is the attack rate of the parasite. The H–P system exhibits quasi-periodic dynamics for parameter values $r = 4$, $b = 3.5$. When external perturbation (as used in Section II) is applied to this coupled system, it shows diverse behaviour depending on the variable perturbed. Figure 16.5(a, b and c) show the bifurcation diagrams for H (in black) and P (in grey) when external perturbation is applied to (a) H alone (L_1), (b) P alone (L_2), and (c) simultaneously to both H and P (L), respectively. It is clear from the Figures 16.5(a and b) that the system responds very differently to similar perturbation when applied to different variables – negative perturbation in H increases the amplitude of complex oscillations, whereas that in P stabilises the system dynamics. Given the differential influence of the perturbation on H and P, it is not easy to predict the system behaviour when perturbation acts on both variables. Figure 16.5(c) shows qualitative similarity to Figure 16.5(b), but with quantitative differences indicating the differential domination of the variables at different strengths of perturbation. This example shows that a generalised perturbation can induce different dynamic response in a multi-variable system depending on the nature of nonlinear interaction between the variable and the perturbation signal. One can get similar results with the three-variable Lorenz system, and the system's response to external perturbation is dependent on the specific variable perturbed (Parekh and Sinha, 2003).

Conclusion

A meaningful analysis of actual processes of change is an ongoing process in scientific enquiry, and the advent of new facts and concepts leads to new ways of analysing this process. It is only recently that the role of complexity (i.e. heterogeneity, feedback, nonlinearity, higher order organisation, etc.), and uncertainty (intrinsic and extrinsic noise and perturbations) in system structure and behaviour are being considered to have important implications. The role of historical contingency and the interfaces of intrinsic processes with history and environment in the understanding of the macroscopic structures and transitions of structure to be expected are also being considered important (Mittenthal and Baskin, 1992). The primary issue addressed here is the role of alteration of internal and external parameters/attributes on the dynamics of nonlinear, multivariable systems as shown in Figure 16.1. I have used dynamical systems theory, which tells us how the generic properties of a dynamical system depend on system attributes (parameters).

Figure 16.6 summarises the results. The figure estimates the possible final states [2 and 3] of (a) any system, and (b) similar systems, at the present state [1]. Figure 16.6(a) shows the scenario (refer to Figure 16.3(c)). where qualitatively similar but quantitatively different final states [2] can evolve from a parent system [1], indicative of 'similarity arising due to common ancestry' in biological evolution. Figure 16.6(a) also shows that relatively dissimilar systems [3] can arise from a common state [1] due to multiplicity embedded in the system that unfolds through bifurcations (refer to Figure 16.2(c) and (d)) and Figure 16.3(b, c and d)). Such divergence in the final state of the same system can also arise due to differential effects of perturbations acting on its different components (ref. Figure 16.5(a and b)). Figure 16.6(b), on the other hand, cartoons the fate of similar systems (refer to Figure 16.2(a and b)). Here two similar (but not identical) systems [1] may converge as internal constraints operating on the transformation processes in both the systems lead to similarity in behaviour [3]. Figure 16.6(b) also shows that similar systems, under the influence of external perturbations (refer to Figure 16.4(a and b)), may exhibit divergence in their final states [2]. Thus, Figures 16.1 and 16.6 summarise the paradigms of evolutionary change – patterns of constancy and variability – in the present context.

In biological evolution, organisms and their environment evolve as a composite dynamical system from which patterns of constancy and variability emerge. To describe constancy, it has been argued (Wake, 1991) that 'similarity in morphological form may arise from common ancestry (failure to evolve), from parallel evolution, from convergence, or from reversal to an apparently ancestral condition' (also see Donoghue, 1992). Diversity, of course, has been the hallmark of higher levels of organisation – structural, functional and single to multi-cellular. In recent times, the predominant approach to analyse the evolution of large networked systems – be it the world-wide-web or biochemical networks – has been through their design and the generative processes at different scales. Biologists have delineated the different levels at which systems work, and the effect of the omnipresent noise, operating at these specific levels, on the final state of the cell (McAdams and Arkin, 1997; Swain *et al.*, 2002). 'Modularity' in functional organisation and 'protocols' that prescribe allowed interfaces between modules with their role in evolution are being thought of as design principles in the new area of 'Systems Biology' (Hartwell *et al.*, 1999; Csete and Doyle, 2002). Yet, what stimuli and constraints lead to convergence or divergence in structure and function are important unanswered questions in biological evolution even today.

There are limitations in extending the arguments of biological evolution to human society. Much of human evolution (biological and sociological) may best be described as the interaction of biological and cultural evolutionary processes (Cavalli Sforza and Feldman, 1981). Both genes and ideas (or 'memes') can be transmitted across generations and the pressures of natural selection can also be generated by cultural attributes. Humans tend to modify local resource distributions, through their activities, metabolism, and choices, thereby influencing both the structure and evolution of their own and the environment (Odling-Smee *et al.*, 2003) in a short time scale.

How good are these observations from the natural sciences perspective *vis-à-vis* 'The concept of multiple modernities in the framework of comparative evolutionary perspective' by Eisenstadt? Describing or modelling processes are specific within the context of each particular field. Reducing many variables and parameters to the bare minimum for reproducing the important behavioural transitions is the rule of the game in the natural sciences. Thus, to be able to summarise a large amount of description to one or few equations and laws is respectable in physics and mathematics. Given the divergent approaches in the disciplinary fields, I have tried to bring out the underlying similarities through reasoning based on a 'grammar' of nonlinear systems theory that studies system behaviour based on its components, interactions, and environment. A social system constitutes a set of interacting socio-cultural processes, orders and institutions having different centre–periphery relations and different modes of control, and its evolution carries along internal conflicts and confrontation having destructive possibilities. For example, Eisenstadt offers several effectors that lead to continuous change of cultural and institutional patterns of modernity in different societies – (a) the internal dynamics of the technological, economic, political, and cultural arenas of each society; (b) interaction between states; and, (c) reorganisation of the structure of network of interactions within and between societies in response to expansion of modernity. That history of the multi variable interacting system plays a major role in inducing heterogeneity in response to external influences has also been emphasised by many (Daedalus, 2000). Much contradiction, continual tension, and internal conflicts have been associated (Tagore, 1961 (1908); Eisenstadt, 2000). Such phenomena are reminiscent of chaotic states where massive alteration in processes and sensitive dependence of final state to initial conditions are observed. Such final states do not obviously result in a homogeneous expression across the different societies, but diverge depending on the properties and the historical and cultural contexts of the systems.

When isolated systems are exposed to external influences, it has been argued (Subrahmanyam, 1998; Kaviraj, 2000) that the way the society, with its plurality of processes, would respond to (or, assimilate/ translate) this perturbation is completely decided by the internal processes ('strong local roots and colours') (Subrahmanyam 1998; Kaviraj 2000). I have shown that variations in the structure of systems that are unimportant to its behaviour under some changes can lead to major difference in response to other challenges. This warns us against simple categorisations (based on economics, religion, or geographical proximity) like 'Islamic world', 'Asians', 'Western Countries', 'Minority Communities', etc. as modernity has its own set of cultural and social premises in the components of each of these groups that are far from being identical. The overall treatment given in this chapter is to some extent mechanistic, but it has the prospect of being a reasonable intellectual endeavour in a trans-disciplinary framework for analysing change. It would require much dialogue to sort out the problems of setting the discourse, isolating the most important variables and parameters, and quantifying the interactions and environment. General principles that govern constancy and change in structure and behaviour of systems may be discovered from stronger interactions between natural and social sciences and from an appreciation of evolutionary constraints.

Bibliography

Cavalli Sforza, L. L. & Feldman, M. W. (1981) *Cultural Transmission and Evolution: a Quantitative Approach*. Princeton, NJ: Princeton University Press.
Csete, M. E. & Doyle, J. C. (2002) 'Reverse engineering of biological complexity'. *Science* 295 1664–9.
Daedalus (Winter, 2000) *Multiple Modernities*, 129–1.
Devaney, R. L. (1985) *An Introduction to Chaotic Dynamical Systems*. Menlo Park, CA: Addison-Wesley.
Dietrich, M. R. (1992) 'Macromutation'. In: Fox Keller, E. & Lloyd, E. A. ed. *Keywords in Evolutionary Biology*. Cambridge: Harvard University Press 194–201.
Donoghue, Michael J. (1992) 'Homology'. In: Fox Keller, E. & Lloyd, E. A. ed. *Keywords in Evolutionary Biology*. Cambridge, MA: Harvard University Press 170–9.
Eisenstadt, S. N. (2000) 'Multiple modernities'. *Daedalus* 129 (1) 1–29.
Haldane, J. B. S. (1931) *The Causes of Evolution*. Princeton: Princeton University Press (1990).
Hartwell, L. H., Hopfield, J. J., Leibler, S. & Murray, A. W. (1999) 'From molecular to modular cell biology'. *Nature* 402 (supp), c47–c52.
Hilborn, R. C. (1994) *Chaos and Nonlinear Dynamics*. Oxford: Oxford University Press.
Hogeweg, P. (2002) 'Multilevel processes in evolution and development: computational models and biological insights'. In: Lässig, M. & Valleriani, A. eds. *Statistical Physics*. Springer-Verlag 217–39.
Holland, John H. (1998) *Emergence: From Chaos to Order*. Reading, MA: Addison-Wesley Publishing Company.
Kaneko, K. & Yomo, T. (1997) 'Isologous diversification: a theory of cell differentiation'. *Bulletin of Mathematical Biology* 59 139–96.
Kauffman, S.A. (1993) *The Origins of Order* (chs. 5 and 12). New York: Oxford University Press.
Kaviraj, Sudipto. (2000) 'Modernity and politics in India'. *Daedalus*. 129 (1) 137–62.

Langton, Christopher G. (1989) *Artificial Life*. Santa Fe Institute Studies in the Sciences of Complexity, vol. 6. Redwood City, CA: Addison–Wesley.
Levin, S. A. & Pacala, S. W. (2003) 'Ecosystem dynamics'. In: K.-G. Mäler and J. Vincent, eds. *Handbook of Environmental Economics*. Amsterdam: Elsevier/North Holland 61–95.
Lorenz, E. N. (1963) 'Deterministic non-periodic flow'. *Journal of the Atmospheric Sciences* 20 (2) 130–41.
Marée, A. F. M. & Hogeweg, P. (2001). 'How amoeboids self-organize into a fruiting body: multicellular coordination in *Dictyostelium discoideum*'. *Proceedings of the National Academy of Sciences*, USA 98 3879–83.
May, R. M. (1976) 'Simple mathematical models with very complicated dynamics'. *Nature* 261 459–67.
McAdams, H. & Arkin, A. (1997) 'Stochastic mechanisms in gene expression'. *Proceedings of the National Academy of Science*, USA 94 814–19.
McAdams, H. H. & Arkin, A. (1999) 'It's a noisy business! genetic regulation at the nano-molar scale'. *Trends in Genetics* 15 65–9.
Mittenthal, J. E. & Baskin, A. R. eds. (1992) *The Principles of Organization in Organisms*. Proceedings Volume XIII Santa Fe Institute Studies in the Sciences of Complexity. Reading, MA: Addison-Wesley Publishing Company.
Murray, J. D. (1989) *Mathematical Biology*. Berlin: Springer-Verlag.
Odling-Smee, F. J., Laland, K. N. & Feldman, M. W. (2003) *Niche Construction: the Neglected Process in Evolution. Monographs in Population Biology*. Princeton University Press.
Parekh, Nita & Sinha, Somdatta. (2002) 'Controlling spatiotemporal dynamics in excitable systems'. *Physical Review E* 65 036227–1 to 9.
Parekh, Nita & Sinha, Somdatta. (2003) 'Controllability of spatiotemporal systems using constant pinnings'. *Physica A* 318 200–12.
Parekh, Nita, Parthasarathy, S. & Sinha, Somdatta. (1998) 'Global and local control of spatiotemporal chaos in coupled map lattices'. *Physical Review Letters* 81 1401–4.
Parthasarathy, S. & Sinha, Somdatta. (1995) 'Controlling chaos in unidimensional maps using constant feedback'. *Physical Review E* 51 (6) 6239–42.
Sinha, Somdatta & Parthasarathy, S. (1996) 'Unusual dynamics of extinction in a simple ecological model'. *Proceedings of the National Academy of Sciences*, USA 93 1504–8.
Slack, J. M. W. (1983) *From Egg to Embryo*. Cambridge: Cambridge University Press.
Sole, R. V., Valls, J. & Bascomte, J. (1992) 'Stability and complexity of spatially extended two-species competition'. *Journal of Theoretical Biology* 159 469–80.
Strogatz, S.H. (1994) *Nonlinear Dynamics and Chaos*. Westview Press.
Subrahmanyam, Sanjay. (1998) 'Hearing voices: vignettes of early modernity in South Asia 1400–1750'. *Daedalus* 127 (3) 75–104.
Swain, P. S., Elowitz, M. B. & Siggia, E. D. (2002) 'Intrinsic and extrinsic contributions to stochasticity in gene expression'. *Proceedings of the National Academy of Sciences*, USA 99 (20) 12795–800.
Tagore, Rabindranath. (1961) (1908) 'East and west'. In *Towards Universal Man*, a collection of essays. Asia Publishing House, New York 1961.
Thom, Rene. (1983) *Mathematical Models of Morphogenesis*. Chichester: Ellis Harwood.
Turing, A. M. (1952) 'The chemical basis of morphogenesis'. *Philosophical Transactions of the Royal Society*, London. B237 37–72.
Tyson, John J., Chen, Kathy & Novak, Bela. (2001) 'Network dynamics and cell physiology'. *Nature Reviews* (Molecular & Cellular Biology) 2 908–16.
Via, S. & Lande, R. (1985) 'Genotype–environment interaction and the evolution of phenotypic plasticity'. *Evolution* 39 505–22.
Wake, David B. (1991) 'Homoplasy: the result of natural selection, or evidence of design limitations?' *The American Naturalist* 138 (8) 543–67.

Part VI

Constellations of Contingency: Political History

17
Historical-Institutionalism in Political Science and the Problem of Change*

Ellen M. Immergut

Continuity and change in political science

For nearly a century, 'change' *per se* was not regarded as a pressing problem in political science. Scholars studying politics in the late nineteenth and early twentieth centuries were mainly concerned with the normative and procedural bases of politics, and tended to focus on constitutions – and, especially, the juridical principles embodied therein. Neither the politics of making constitutions nor the impact of these constitutions on everyday political life were particularly emphasized, although, to be sure, many writers in the tradition of Montesquieu also analysed the 'goodness-of-fit' between these formal, legal documents and the particular societies or cultures they had been drafted to govern.

The 'behavioral' revolution in political science of the 1950s and 1960s called into question the relevance of constitutions or other formal political institutions for understanding political life. Political scientists should focus their efforts on direct observation of political behavior, the new generation urged, and infer generalizations about politics directly from these raw data rather than from constitutional declarations of principle, or other subjective interpretations of political reality. Dahl's (1961) path breaking work, *Who Governs?*, for example, investigated power relations in the city of New Haven, Connecticut by directly observing local political decisions. Similarly, Truman's (1971 [1951]) equally renowned book, *The Governmental Process*, developed a theory of the political process based on empirical observation of the activities and achievements of interest groups in the United States. A great number of 'behavioralist' studies focused on public opinion and voter behavior. Here, investigators examined statistical relationships between social characteristics – such as religion, length of education or size of income – and voter preferences.

In the area of comparative politics, the concept of political development became a dominant term, as various studies sought to understand the relationship between economic growth and political democracy in terms of a unified 'modernization' model, whereby political structures became ever better adapted to the social functions (such as aggregation, integration, representation) they were to serve. (Almond and Coleman, 1960; Pye, 1965; LaPalombara, 1966)

Ironically, even though the behavioralist approach focused on dynamic themes – such as modernization, transformation and development – it can be viewed as sharing the older, formal institutional approach's neglect of the problem of change, if for very different reasons. Whereas the older approach focused on formal political institutions, but considered them largely as invariant givens – products of the sediment of time, so to speak – the newer approach assumed political institutions to be so malleable and efficient as to be virtually irrelevant. Change is part and parcel of the political process, but it is not in any way problematic. As a variety of external circumstances – ranging from economic development to technological breakthroughs, as well as new ideas – confront citizens with new problems, their demands on government – and hence, eventually, governmental policies – will change, such that responsive and effective democratic government automatically leads to renewal and change in public policies. As David Truman put it, 'The total pattern of government over a period of time presents a protean complex of criss-crossing relationships that change in strength and direction with alterations in the power and standing of interests, organized and un-organized' (1971 [1951]: 320).

This view of the political process has sometimes been called 'pluralism,' referring both to the empirical finding that a plurality of social interests – as opposed to a 'ruling class' or 'power elite' – is responsible for political outcomes, and to the normative view that, as the state is not infallible, governments should respond to societal pressures rather than attempting to effect an absolute separation between state and society, as in the classical continental doctrine.

The institutional approach

As is often the case in the study of politics, political events caused a reappraisal of the dominant paradigm. The large-scale social protests and emergence of new social movements in the 1960s and 1970s gave rise to various criticisms of the behavioralist model of politics (Katznelson, 1992, 1997; Thelen and Steinmo, 1992). If pluralist politics were so efficient, why did it take a hundred years as well as recourse to unconventional political methods – such as protest marches – for the grievances of the civil rights movement to be put on the political agenda? Indeed, comparative studies of both political disruption and the organization and activities of more conventional interest groups led to skepticism about whether the problems and preferences of individual citizens could really account for variation in the level and success of political movements over time and across nations. Scholars of social movements developed the 'resource mobilization' approach based on empirical evidence that waves of social protest could be better accounted for by variations in the ability of groups to obtain the resources necessary for political activities, as well as political opportunities, than by changes in the level of social 'discontent' or feelings of 'relative deprivation' (Tilly, 1978; Kitschelt, 1986; Zald and McCarthy, 1987; Tarrow, 1993).

Studies of more conventional interest-groups, such as trade unions and employer associations, as well as agricultural groups, found patterns of interest-group activity to be quite stable over time, causing them to question the pluralist claim that the array of organized groups and their impact on government constituted an accurate representation or vector sum of the demands and preferences of individual citizens. Instead, the 'corporatist' theories argued, many groups had been organized not 'from below' by citizens, but rather 'from above' by states. Moreover, once formed, successful groups did not simply disappear once their particular grievance was re-solved, but continued to seek out new political issues. Indeed, one could question – as Michels had pointed out several decades before, referring to the case of political parties – whether leaders of groups really represented the interests of their members, or those of themselves. Furthermore, groups with long-standing relationships to government were privileged in the competition for political influence. Consequently, the pattern of interest-*intermediation* – a term coined to emphasize that the direction of influence did not necessary go from citizen to group to state, but could also be recursive – was a critical variable in politics, shaping and perhaps distorting, the political process (Schmitter and Lehmbruch, 1979; Berger, 1981b; Maier, 1987; Lehmbruch, 2001).

Increased attention to differences in the power and organization of both societal and government actors had ramifications for the discipline's understanding of the process of political development as well. An historically informed conflict theory questioned the modernization paradigm. Scholars like Moore (1966) and Anderson (1974) showed paths of economic and political development to be diverse, and forged by intense class conflict. Wallerstein (1974), as well as writers from the 'Dependencia' school, such as Frank (1978), viewed economic development as a conflictual result of regions' quest for dominance of the international division of labor. Whereas these studies were highly influenced by Marxist theory, scholars such as Tilly (1975), Poggi (1978) and Skocpol (1979) drew on heterodox thinkers from the modernization era, such as Bendix (1969) and Rokkan (1970), to re-emphasize the central political role of the institutions of the nation-state. By the early 1980s, a large volume of literature had accumulated that questioned the 'efficiency' model of political representation and political development.

The 'new' institutionalism

March and Olsen (1984) gave a name to this general trend in political science – which, they noted, extended to other disciplines, as well – calling the stress on how political life is organized, 'the new institutionalism.' They defined the scope of the new institutionalism so broadly as to include quite different approaches. However, despite the many differences, they do indeed share a renewed interest in the role of constitutions and political institutions, and have abandoned the assumption that the 'black box' of politics is efficient. In essence, a common lesson of these various studies of politics is that similarly-situated social movements and interest-groups have responded to their situations in very different ways, and these claims on the state have been met with very different responses from governmental policy-makers. Similarly, processes of political development have been fraught with conflict, and have certainly not been characterized by a uniform process of modernization or arrival at a uniform 'end destination' of economic, political and social organization.

If one takes the defining feature of the new institutionalism to be an interest in the inefficiencies of politics, there is no contradiction in including many varied approaches under this very broad label. Furthermore, it obviates the need for adopting a common definition of the term 'institution' or even a common theoretical or methodological framework. The new institutionalism is nothing more (in the view of this author) than an interest in the distorting effects of the political process, in whatever form or stage of the process they may be found. Having solved one problem, however, we have created another, which brings us back to the main topic of this essay, namely the problem of institutional change.

The problem of change

Departing from the March and Olsen definition, it has become common to divide new institutionalist writings into three groups: rational choice or positive political theory; sociological institutionalism (or organization theory); and historical-institutionalism (see Hall and Taylor, 1996; Rothstein, 1998; Peters, 2001). Despite their different analyses of the inefficiencies of institutions, however, all share a limitation when it comes to explaining institutional change. Rational choice institutionalism refers to the effort to understand political behavior based on micro-economic models of individual choice. One important building block of many rational choice analyses is the Condorcet paradox or Arrow theorem, which demonstrates that even if individuals have coherent preferences (they prefer chicken to steak to fish, for example), there may be a problem in aggregating these individual preferences into a collective choice using majority rule (a group cannot decide which restaurant to choose). One solution to this problem is that institutional rules for political decision-making may allow for stable political outcomes by setting limits on the political process, such as granting some actors a monopoly on the setting of the political agenda, or the custom that more radical amendments are voted on first, the *status quo ante* last. Such a decision outcome has been termed 'structure-induced' equilibrium, in order to distinguish it from a 'preference-induced' equilibrium. (For good overviews see Riker, 1980; Shepsle, 1986, 1989; Weingast, 1996.) If particular rules stabilize outcomes, however, it is not clear why institutional arrangements should become unstable (Shepsle and Weingast, 1981; Colomer, 2001; Shepsle, 2001: 516–17).

Similarly, sociological institutionalism/organization theory emphasizes 'norms of appropriateness' and 'standard operating procedures' as guides to behavior (March, 1986; Powell and DiMaggio, 1991). The idea is that human beings cannot possibly process all of the information they would need to make fully rational decisions – and the cost of gathering all of this information, if it were even remotely feasible, would be prohibitive. Consequently, they (and organizations such as business enterprises) rely on short cuts, such as calling the references listed on a CV rather than exhaustively investigating a job applicant's entire employment history, or reading each and every publication produced by an aspiring professor. If many individuals rely on the same 'standard operating procedures', coordinated action is possible. Similarly, if individuals within a society or an organization follow norms of appropriate behavior, one can explain certain societal or political outcomes. Some studies, such as that of Zucker (1991) provide striking experimental evidence for how quickly a norm of behavior may be introduced and made to persist over time. Following in the footsteps of anthropologists such as Boas, Sapir and Kroeber (see Elwert, 2001), Zucker shows that norms of appropriate behavior – in this case conforming to the expectations of work colleagues – change experimental subjects' reported-perceptions of the distance that a beam of light travels; cultural frameworks screen perceptions even of the natural world. The problem, however, is how to explain changes in these norms and standard operating procedures: why does routine behavior stop being routine?

Historical institutionalism

The third new institutionalist approach, historical institutionalism, suffers from two problems at once. First, the meaning of the term 'historical' is not generally well understood. Second, the problem of institutional change is particularly vexing for this approach. Indeed, efforts to improve the coherence of the term 'historical institutionalism' have in some ways made the problem of explaining institutional change appear to be even more challenging. Why is this the case?

'Historical' institutionalism refers to a rather loose collection of writings by authors that tend to mix elements of rationalistic and constructivist explanations – or the 'calculus' versus the 'cultural' approach in Hall and Taylor's terminology (1996). This group was initially inspired by the historical studies mentioned above, which confronted modernization theory with a more conflictual approach, and led to a more serious focus on the state as holder of a monopoly on legitimate force – although newer generations have tended to abandon the quest to speak to larger questions of social change and social justice, in favor of a more limited, if more precise, research agenda (Katznelson, 1997). Thelen and Steinmo – drawing on the work of Ikenberry (1988) – define this approach in terms of an emphasis on how pre-existing institutions structure contemporary political conflicts and outcomes (1992: 2). On this view, the past influences present-day politics through a variety of mechanisms, ranging from concrete political institutions to patterns of interests associations to broadly accepted definitions of justice or even mundane ideas about the accepted way of doing things. As Thelen and Steinmo note, not only is this approach rather eclectic, but it is not necessarily distinct either in its theoretical framework or methodology from other types of new institutionalism, except possibly for its interest in explaining how distinct sets of political preferences develop, and for its rejection of absolute theories of human rationality. Indeed, a closer look at the competing 'institutionalisms,' as well as at the many 'border crossers' in this field have increased doubts about the existence of historical institutionalism as a distinct approach (Thelen, 1999: 370–1).

At the same time, the emphasis on the 'heavy hand of history,' as Ikenberry (1994) puts it, has led to a continuing focus on the problem of change – and in particular, institutional change – as a problem or 'frontier issue,' for this approach (Thelen, 2000, 2003). For, if institutions created in the past are thought to constrain future developments in some way – and particularly if they affect the preferences of political actors – it is not immediately obvious how the same account can explain both the lasting impact of institutions over time, and – at one and the same time – account for institutional change. Ironically, some efforts to better define the research agenda, theory and methodology of historical institutionalism have made the problem of explaining institutional change even more challenging.

Pierson (2000b) has suggested that models of 'path dependency' (which depend upon positive feedback loops 'locking in' particular institutional arrangements) could constitute a rigorous way to show that history matters, and that these models are applicable under conditions that are quite common for a range of political phenomena. Path dependency, in turn, is a specific case of a more general focus on the importance of 'timing' and 'sequence' in the analysis of politics, which, following discussions in the field of sociology, may be investigated with techniques such as 'event sequencing analysis' (Abbott, 1990, 1991, 1992; Pierson, 2000d). It is not unlike the concept of hysteresis discussed in Higgins' chapter (this volume) on climate change, and of course, it is identical to theoretical concept in economics, as discussed in Castaldi and Dosi's and in Nugent's chapters (this volume) on path dependency and the new institutionalist economics. In political science (and sociology), however, the application of the concept has met with some controversy. Should it be used in the restrictive economic sense of maintaining an off-equilibrium outcome? Or does it just mean that the past somehow influences the present? (See discussion in Mahoney, 2000 and Thelen, 2003.) And can one really measure equilibrium outcomes in politics (Genschel 2001)? Further, stress on path dependency, and even sequences and events, have aggravated the problems of explaining institutional change. In order to explain institutional origins and development, an analyst must account both for institutional change *and* for institutional stability. That is, a convincing account of institutional change must contain within itself its own negation, and yet somehow remain consistent. This task is particularly difficult for models that contain feedback loops, such as those emphasizing path dependency and endogenous preferences. If institutions socialize actors and thus endogenize preferences, for example, then it is difficult to explain why these actors would suddenly prefer a new set of institutions. Or, if increasing returns reinforce a particular institutional set-up, it is not clear how one can explain a switch from one particular path to another. In classical social theory, these concerns resulted in such artifices as Rousseau's *deus ex machina* 'the Legislator' or Hegel and Marx' historicist focus on contradiction (van Parijs, 1982; Boudon and Bourricaud, 1994; Offe, 1996). In contemporary theory, formulations such as 'punctuated equilibrium,' 'critical junctures' and 'exogenous shocks' are more common, but these, also, have been criticized for being incomplete, as it is not clear what causes the switch from stability to instability (Krasner, 1989; Collier, R. B. and Collier, D., 1991; Hall and Taylor, 1996; Thelen, 1999).

Finally, the emphasis on 'path dependency' has met with some resistance from historically-minded scholars (see for example, Bridges 2000). The reason for this may be that much of 'historical institutionalism' is, in essence, a defense of the idiographic in political analysis. Many historical institutionalist scholars wish to study the 'poetry' in politics, and are concerned that 'we murder to dissect,' as Wordsworth put it. It may not be possible or desirable to make historical analysis more scientific and exact, if the very purpose of the historical approach is precisely to capture the unpredictable, contingent nature of human action, which stems precisely from the self-reflective capacities of human actors. 'History' may give observers a critical vantage point, from which one may understand better the times when human agency has made a difference, and to remind ourselves of the 'road not taken,' in order to expand the range of alternatives we consider to be within the scope of the possible. Further, scholars interested in 'narrative' and 'identities' often view history not as a set of objective facts, but as an interpretation or myth, which has an impact on the present as actors re-tell and re-consider their own histories. Indeed, in the full 'constructivist' view, actors construct and are constructed by symbolic constructs such that culture and institutions are 'constitutive' of human agency (Sewell, 1985; Calhoun, 1991, 1994; Somers, 1994; Sabel and Zeitlin, 1997; Hattam, 2000; Jupille *et al.*, 2003).

The logic of historical explanation

The debate about 'path dependency' – and indeed even the larger ongoing debate about the role of qualitative research in political science – has some similarities to the discussion about the 'taming' of historical sociology. A specific model or method cannot capture the spirit of the larger enterprise, which is to pay closer attention to 'historicity' or 'temporalities' (Orren and Skowronek, 1994; Calhoun, 1996; Sewell, 1996; Katznelson, 1997). Event sequencing programs, for example, which have been developed by mathematical biologists for analyzing the genetic code, may indeed be suitable for analyzing encoded social products – like rituals, myths and texts (Abbott, 2001). But this perspective in 'events' is very different from that of Sewell, who focuses on how events can change the *context* for interpretation and action (1996: 262–3). In the new context, the same sequence can now mean something different. To be sure, there may be very exact models for specifying how such context effects may work. Fontana's chapter (this volume) on RNA-folding shows how many genetic mutations may have no effect; only when a mutation causes a non-neutral electro-static or steric change, does it have an impact on the phenotype – in this case, on molecular shape. Models such as informational or reputational cascades (e.g. Kuran, 1998) are examples of mechanisms that model context effects that could be used systematically in historical research, in the manner proposed by Hedström and Swedberg (1998). Nevertheless, as worthwhile as it is to undertake a search for such mechanisms, it is nevertheless important to stress that the *historian*'s program is definitely not one of developing 'arguments that can "travel" in some form beyond a specific time and place' (Pierson, 2000d: 73) or 'to try and "capture the impact of time in as timeless a way as possible" ' (Thelen cited in Pierson 2000a: 119). Instead, historians wish to travel through time themselves, to experience another time as experienced by those that lived it, just as anthropologists aspire to travel across societies.

What then is the relationship of historical-institutionalism to history, and how does this affect the explanation of institutional change? Just as there are a great many different research questions and methods that have inspired historical studies, the range of historical-institutionalist studies is quite broad. Nevertheless, there are some key, recurrent institutionalist themes and questions that lend themselves particularly to historical research. (For a more extended discussion, see Immergut, 1998; Mahoney and Rueschemeyer, 2003.) One is the issue of preference formation and the construction of interests. If one wants to study how institutions shape political outcomes by affecting actors' definitions of their own interests or selecting one definition of an issue at the expense of another, one needs examples of how definitions of preferences, interests and issues change over time or vary across societies. It is not simply that historical research can provide such examples, but that the very nature of the historical endeavor – delving into the world views and mind-sets of persons with alternate views on what is rational in a particular situation – is particularly attuned to the research question at hand: what factors shape people's definitions of their political situation, their political goals, and their assessments of the best course of action? In any particular case of political behavior, it is difficult to show that there is anything unusual about the preferences expressed by political actors; it is really simpler to follow the behavioralist course of assuming that behavior reveals preferences. However, if one wishes to step behind the expressed preferences, political choices or decisions, and to analyse to what extent they are artifacts of institutions, then one needs a different method. One possibility is to use a formal model; another is to rely on a broader, comparative–historical perspective. A formal model can help one to disentangle the strategic effects caused by interactions amongst actors from the effects of the preferences themselves. A comparative–historical perspective can provide a critical vantage point for relativizing preferences. Thus, the work of economic anthropologists such as Malinowski, Boas and Mauss on phenomena like the Kula Ring, the Potlatch, the spirit of the Gift – or even Polanyi's depiction of the development of the 'satanic mill' – has called the Adam Smith's Robinson Caruso Myth of man's 'natural tendency to truck, barter and exchange' into question, just as Rousseau's *Discourse on the Origins of Inequality* tried to discredit Hobbes' and Locke's myth of origin for the Leviathan. Similarly, historical studies, such as E.P. Thompson's *The Making of the English Working Class*, have shown the relationship between objective economic situation and subjective understandings of this situation, and hence, definitions of interests to be complex and contingent. Thick description is one way to try to pin down the complicated relationships between institutions, actors and interests, even though it is in the nature of thick description to provide multiple and thickly-layered interpretation, as in Clifford Geertz' famous interpretation of the Balinese cockfight.

A second important common theme is the issue of contextual causality. Most – or at least many – institutional effects are interaction effects. They can indeed be measured by statistical analysis, but only if the institutional variable is correctly specified. It is rarely the case that a formal institution is directly linked to a specific political outcome. Instead, it is the interaction of political institutions with specific electoral results or preference patterns that is significant. 'Historical' research in the sense of looking at micro-political events, using standard historical methods, but without necessarily restricting oneself to any particular historical period, is a way of going into the 'black box' of politics to understand better the interactions of actors, preferences and institutions. Since institutionalism as a whole is concerned with the distorting affects of the political process – or the 'mobilization of bias' in Schattschneider's well-known phrase – one needs methods that allow one to view these distorting effects. Consequently, Pierson warns that the causes of institutional design cannot be 'read-off' from any hypothetical functions they might have or inferred from their unintended consequences, but instead,

> one must consider dynamic processes that can highlight the implications of short time horizons, the scope of unintended consequences, the emergence of path dependence, and the efficacy or limitations of learning and competitive mechanisms. This requires *genuinely* historical research. By genuinely historical research I mean work that carefully investigates processes unfolding over time. (2000c)

Similarly, Hall (2003) stresses the importance of 'process-tracing' and hence an historical stance for institutional research. Thelen (2003) puts this into practice by showing how the meaning of institutions can change over time both as a reaction to their changing social and political context, but also as a constitutive part of this changing context.

The third area of selective affinity between institutionalist analysis and historical research is the role of contingency. Institutions do not generally exert their effects in a continuous manner. Political institutions, for example, may have a strong impact on a political decision taken at a discrete point in time. This decision may have lasting consequences, but there is no direct relationship between the institutions and the consequences. For example, the set of social programs we associate with the concept of a 'welfare state' are the result of individual political decisions taken for a variety of reasons – and there are not that many of them. Once such a law is enacted, it can have a large impact on the size of government. But although a specific political-institutional configuration may have been a crucial part of the story of the passage of the law, we would not expect a lasting and continuous relationship between the political system and size of government. And indeed, efforts to find such relationships – for example through analysis of variance methods – may be picking up mainly spurious correlations (Cutler and Johnson, 2001). Consequently, institutionalist analysis requires analysis of discrete events. These will be composites of unexplainable, random events and, possibly, some systematic institutional effects.

Change and continuity

If one takes this broader view of the 'historical' part of 'historical institutionalism,' it is clear that change is not necessarily problematic for this approach. If one does not posit that social or political phenomena are at equilibrium, there is no concern with explaining shifts away from equilibrium.

As Oakeschott states,

> And the only explanation of change relevant or possible in history is simply a complete account of change. History accounts *for* change by means *of* a full account of change. The relation *between* events is always other events, and it is established in history by a full relation *of* the events. The conception of cause is thus replaced by the exhibition of a world of events intrinsically related to one another in which no lacuna is tolerated. (cited in Roberts 1996: 39)

But this outlook does not lead one to any singular or arching model of institutional change. Nor does it eliminate the problems of causality in historical explanation. Indeed, in trying to fuse institutionalism with an historical approach, one may be embarking upon an inherently contradictory program, which does not do justice to the epistemological differences between and within both the historical and the institutionalist approaches. One way out this very complicated corner is to follow Robert's suggestion of focusing less on the theory of history and more on its practice:

> To this eminently sensible conclusion that it is wise to rely upon 'available generalizations' in guiding human conduct, one can add a second truth. Those 'available generalizations' will be more dependable, more useful, more profitable if they are based upon a right understanding of the causes of events in the past, which right understanding depends in large measurement upon a right understanding of the logic of historical explanation and of the logic of historical interpretation. (1996)

To be sure, Robert's view is controversial. Nevertheless, we may follow his lead in taking a pragmatic approach to the theory of knowledge, and asking ourselves, what is good scientific practice in the specific field of historical-institutionalism in comparative politics? This is a field of knowledge in which some objects – namely political institutions – have a good chance of working with law-like regularity, if one specifies the boundary conditions correctly. However, it is also a field where human agents interpret the workings of these institutions and adjust their behavior according to these interpretations, including acting in ways to change the institutions. Thus, it could be useful to follow the historical practice of studying these activities and events over time, as well as researching micro-political decisions, paying close attention not only to which actors were active, and what they did, but also to what they thought they were doing while they did it. Such an approach is not limited to one type of research question, model or method, but can draw freely on a variety of approaches. What gives it its coherence is an interest in studying the distortions of institutions, and a commitment to checking whether generalizations really apply to the self-understandings of real people in the real world. An example may illustrate this pragmatic approach to study the history of institutional change.

A brief empirical example

The research presented here draws on two standard methods of historical research and historical interpretation: the construction of an historical counterfactual, and recourse to written documents available in archives. The case at hand is the partial revision of the Swedish Constitution, which was voted on in 1967 and 1968, and went into force in 1970. The historical puzzle in this instance was that the previous constitution was widely viewed as favoring the Social Democratic Party. Why then would members of the Party agree to give it up?

The provision of the old Constitution that had proved advantageous to the Social Democratic Party was the bicameral legislature, with an Upper House that was indirectly elected by the County Councilors, the local politicians at the County Council or provincial level. This Upper House (or First Chamber, as it was called) was the *quid pro quo* granted to the Nobility in 1866, for their consent to the dissolution of the Parliament of Four Estates. Once property qualifications had been abandoned in 1907 and 1918 (through laws that went into effect in 1909 and 1921, respectively), and after the phenomenal electoral successes of the Social Democratic Party of the 1930s and early 1940s, the First Chamber became an important element of Social Democratic Party parliamentary strength. For while electoral results averaging 47.6% of the vote from 1936 through 1968 rendered the party relative majorities of averaging 49.9% in the directly-elected Second Chamber, the indirect elections to the First Chamber, which were based on County Council elections in which the Social Democratic Party averaged 47.6% of the vote during the same period, rendered the Party absolute majorities averaging 51.7% of the seats in the First Chamber. Although these differences in the disproportionality between votes and seats may appear small, the difference between 49% of seats and 51% is highly significant politically. It means the difference between full control of government and legislation versus shared control. Moreover, under the old Swedish Constitutions, both Chambers were equal and both were required to approve legislation. Therefore, control of the First Chamber – which was guaranteed by the Social Democratic Party's absolute majority in this chamber from 1936 to 1969 – was sufficient for the Social Democratic Party to veto any and all legislation proposed in this period. In addition, votes on the budget were taken by the united parliament (a joint session of the two chambers), in which the First Chamber majority was sufficient to give the Social Democratic Party a majority in the united parliament, and hence, control over all budgets in this period. Given the significant enhancement of parliamentary power accruing to the Social Democratic Party through the indirectly-elected First Chamber, why would

Historical-Institutionalism in Political Science 249

members of this Party vote to relinquish the First Chamber at a time that the Party held either a relative (1967) or an absolute majority (1968) in the Second Chamber, and an absolute majority in the First Chamber (1967 and 1968)? Three theoretical possibilities present themselves. First, members of the Social Democratic Party may have found their advantages of the old Constitution to be unjust and incompatible with their vision of democracy or constitutional *principles*. Second, for some not immediately apparent reason, social democratic politicians may have decided that these provisions were not longer in their *interests*. Third, these decision-makers may have *miscalculated* or *misunderstood* the effects that the new Constitution might have on their parliamentary power. (Buchanan and Vanberg, 1989; See discussion in Vanberg and Buchanan, 1989 and 1996.)

In order to be able to even discuss whether political actors may have made errors of judgment, however, one needs two very different pieces of information. First, one needs to know what the effects of their actions were; second, one needs to know why these political actors made these choices. The first question raises a general problem for the analysis of institutional change. Namely, that it is extremely difficult to assess the impact of institutional change. Of course one knows which specific constitutional rules were changed through the partial constitutional revision of 1967/68, but the impact of these changes on the relationship between parliamentary and executive power and the sharing of parliamentary power amongst the various political parties – that is, the political meaning of these changes – is not evident from the changes in rules alone. For the impact of constitutional and electoral rules on politics depends upon electoral results. This is what makes it difficult to assess the impact of changes in the rules both looking forward into the future, and backwards into the past. To this end, a quantitative historical-counterfactual was constructed: Had the Swedish Constitution not been changed, what would have been the distribution of parliamentary seats amongst the political parties, based on the actual electoral results from 1970 through 1994?

Counterfactual reasoning has long been associated with the historical method, as it is one way to try to disentangle causally important factors from unimportant ones (Weber, 1978; Roberts, 1996; Lebow, 2000; Tetlock and Lebow, 2001). As has been superficially mentioned in previous sections of this chapter, however, the very notion of causality in history is controversial. So too, is the use of counterfactual reasoning. Some scholars consider counterfactual reasoning to be at the heart of *any* notion of causality; others consider to be suspect under all circumstances. (See discussions in Lewis, 1993; King *et al.*, 1994: 10–11; Edgington, 1995.) In this case, however, a counterfactual is not only defensible, but also necessary. It is defensible, because all that has been done is to use the constitutional and electoral rules applicable in 1968 (at the time of the constitutional revision) to simulate a 'counterfactual' parliament using the electoral results obtained from 1970

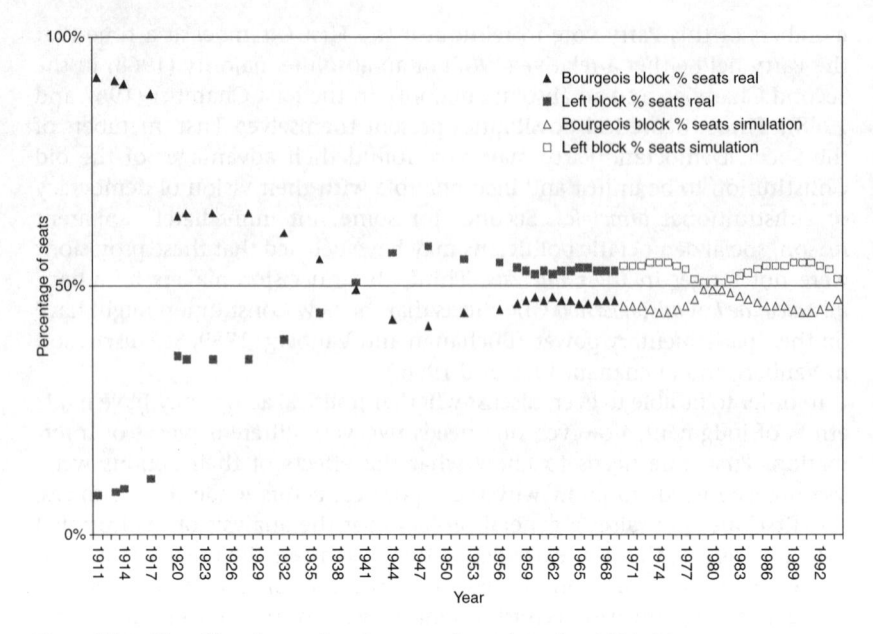

Figure 17.1 First Chamber real and counterfactual results, 1911–94

on. (For details see Immergut, 2002.) That is, both the rules and the electoral results are based on fact; indeed, as the First Chamber elections were indirect, and based on older electoral results, the electors for the first few years of First Chamber simulation were already elected when the constitutional change went into effect, and thus 'factually' existent. It is necessary, because, there is no other way to assess the magnitude of the constitutional change. As Voigt (1997: 19) points out, in analysing institutional change, one is always comparing 'the effects of a realized (unrealized) institutional arrangement with an unrealized (realized) one.' Consequently, counterfactual reasoning is central to the problem of institutional change. The results of the counterfactual are straightforward. As Figure 17.1 shows, under the old constitutional rules, the Social Democratic Party would have held a majority of seats in the First Chamber for all electoral results that it achieved in the 1970 to 1994 period.

One can note on the graph, as well, the steady rise of Social Democratic representation between 1911 and 1944, coupled with the abrupt increase caused by the constitutional reform implemented in 1921. One sees the steady absolute majority held by the Social Democratic Party in the First Chamber from 1944 through 1969, and sees that this majority continues uninterrupted until 1994 – except for a decline at the end of the 1970s. This decline is not sufficient, however, to cause a loss of the First Chamber majority. For technical reasons, the figure combines social democratic and communist seats, but it should be noted that the numbers of communist seats are estimated to be minimal – at most 2 or 3 seats compared to 79 or 80 for the Social Democrats. The simulation experiment result presented here is based on two experimental conditions: the electoral results actually obtained from 1970 to 1994, and the use of electoral alliances exactly as they had actually been entered into during the 1966 and 1968 elections. Although the electoral results from 1970 to 1994 are on average lower than those from 1944 to 1968, the simulation shows First Chamber results with quite a bit of continuity with the 1944 to 1968 period. Indeed, under the new constitution, the electoral results from 1970 to 1994 led to several changes in government, with the Social Democratic Party out of office from 1976 to 1982 and from 1991 to 1994. Under the old constitutional rules, however, the same electoral results allow the Social Democratic Party to control the First Chamber for the entire period.

A second condition of the simulation concerns the use of electoral alliances. The alliances were a method by which the smaller parties could improve the number of seats they obtained; the parties wishing to enter into an alliance were required to campaign under a common party label, but the individual parties could then divide up the seats amongst themselves after the election. Even though the use of electoral alliances had increased dramatically during the 1960s, and reached a peak in the 1966 election, the application of these alliances to the 1970 to 1994 period does not eliminate the majority held by the left bloc in the First Chamber.

This 'instrumental' counterfactual can be used to raise questions about the historical process, thus providing one approach to the problem of how and when to 'cut into' history. It shows that the First Chamber would have provided a critical source of parliamentary power for the Social Democratic Party. Such a counterfactual can structure the research, and provide plausibility tests for various hypotheses that can help in disentangling whether principles, interests or miscalculation were responsible for the decision to eliminate the First Chamber. But it cannot substitute for an investigation of the history of the constitutional reform itself. A full description of this process – which lasted from 1948 through 1968, and on into the 1970s – is beyond the scope of this essay. But a few comments can be made on the basis of the parliamentary debates, the minutes of the meetings of the social democratic party's congresses, its parliamentary group and its executive council, as well as from newspaper accounts.

Constitutional principles

The Liberal Party was the driving force putting the issue of constitutional reform on the agenda. In early 1953, both Liberal and Conservative members of parliament submitted several motions calling for elimination of the First Chamber of the Parliament. In response, five Social Democrats submitted a motion for a full government investigation of the issue of Constitutional reform. Strategically, this motion is surprising, because it widened the scope of the constitutional issue, and put it irreversibly upon the political agenda. As one of the elder statesmen of the Social Democratic Party, Per Edvin Sköld, said at the end of the reform process,

> We must not forget that it was the social democrats that breathed life into the constitutional issue. That happened because the government appointed a commission almost ten years ago. There does not seem to be great interest within the party now for constitutional reform and one can ask oneself whether the party wants a reform or not. My answer is that once we have started the ball rolling, we cannot stop it, even if the negotiations should lead to us not getting what we wanted from it[1]

Nine years earlier, in the meeting of the social democratic parliamentary group that discussed the social democratic motion for a complete investigation of constitutional reform, one of the signers of the motion, Ossian Sehlstedt, explained his motivations to the Social Democratic Parliamentary Group. On the one hand, it was important to update the constitution in line with societal changes, and to increase both the influence of the people and the effectiveness of parliament. Sehlstedt was particularly interested in the possibility of introducing a majoritarian electoral system, as in Britain. On the other, it was important to counter the fact that 'the opposition is in certain points using pure social democratic propaganda against social democracy. Without choosing to do so, now we are behind.' Gösta Nélzen concluded that the signers of the motion were united on only one point, namely that it was time 'to deprive the opposition of the initiative in this question.'[2] Thus, tactical and not just normative considerations were behind the proposal for reform, and there was certainly no groundswell of principled social democratic opinion against the First Chamber.

Constitutional interests

During the internal party debates about Constitutional reform in the early 1960s, many discussions focused on party interests, and how best to argue for these party interests with the opposition and to communicate them to the public. It was difficult to challenge the Liberal Party's arguments that 'each vote should count the same,' and that indirect elections based on outdated electoral results were counter to the principle of popular sovereignty. To solve this problem, various proposals were drafted that would allow for disproportional representation of the Social Democratic Party through various mechanisms for regional representation. Some of these proposals maintained the indirect elections, others called for regional representatives within a unicameral parliament, but all shared the common feature of giving slight overrepresentation to the Social Democratic Party. Rather than speaking of 'overrepresentation,' 'disproportionality' or 'indirect elections,' however, the Prime Minister elaborated a new political term, 'kommunala sambandet' or the 'local link' (Ruin, 1990; von Sydow, 1989: and primary sources; Stjernquist, 1996).

Up until the 1966 communal elections, the Prime Minister and the party leadership dragged their feet in the negotiations over constitutional reform. After the elections, however, in which the Party lost its majority in the County Councils, the Prime Minister decided that it was time to get the issue of constitutional reform off the agenda, and that it was necessary to enact a reform before the next parliamentary elections would be held in 1968. There seem to have been several grounds for this sudden urgency. First, the constitutional issue had been used as a campaign issue for the first time, and any delay was now viewed as a liability for the Party. Second, the loss of the County Council majority made the First Chamber seem vulnerable for the first time.[3] In these discussions, the focus was very much on the electoral alliances used by the bourgeois parties: 'It is necessary that one in the question of the electoral method take a position that can hinder in future the very disturbing electoral manipulations that have occurred in this year's electoral campaign, namely the alliance of the bourgeois parties in those areas where they stand the most to gain from cooperation, and that each party campaigns separately in those areas where they can gain by that solution. ... This cannot continue this way.' The solution chosen by the party was full proportionality. Calculations carried out showed that under a more proportional electoral method, electoral alliances made no difference to the final distribution of seats. The Prime Minister reasoned, 'It remains for us to find out whether the opposition could imagine accepting a ban on campaigning together. Perhaps instead of a prohibition one could stop the whole thing with such a [proportional] method.'[4]

Constitutional knowledge

Constitutional interests – as they were interpreted by the party leadership and in particular by the Prime Minister – can explain the timing of the constitutional reform, and its content. The 1966 elections created a sense of urgency for reform, and focused the party leadership on a fully proportional electoral formula. At the same time, however, the process contains many accidents of sequence and 'bounded' rationality. The Party leadership did commission simulation experiments almost identical to the ones whose results were presented here. However, the Social Democratic calculations focused exclusively on the Second Chamber, and did not take the First Chamber into account in any way. Although the vague term 'local link' had initially been used to refer to an electoral equivalent of the First Chamber within a new unicameral legislature, by the end of the negotiations, the term had been reduced to a common election day for national and local elections. Not only did the Social Democratic Party leadership give up the tremendous advantages of the First Chamber for the relatively paltry concession of a common election day, but many members of the party's executive council were either unsure whether the common election day would benefit the party or opposed to it altogether. As one member of the executive council put it, 'I think that had we from the beginning been confronted with the choice of finding a form for the local link our solution would not have been the common election day. We cannot, after all the years that have gone by and we have campaigned for the local link, get out of this and therefore I accept with a sigh that the only way will be the common election day.'[5]

The causes for the Swedish partial reform of 1967/68 were thus the continued pressure of the Liberal Party's campaign against the First Chamber, the fear of younger Social Democrats that the Liberals were outflanking them on the constitutional question, the general lack of interest or commitment of party base to institutional issues, and the credible threat provided by the use of electoral alliances on the part of the bourgeois parties in the 1966 election. Not just these ultimate causes, but the twists and turns of the political process, changed the array of political alternatives and the final provisions that these actors chose. Indeed, there is quite a bit of evidence that many actors were dissatisfied with the final outcome. One can find direct historical evidence relatively easily for all of these causes and the effects of sequence, in the form of direct statements by the participants in these events. But no mono-causal explanation can be fitted to these events; the sorting out of the relative importance of each of these factors remains a question of interpretation and argument, and thus of ongoing debate amongst scholars.

Conclusions

Historical research based on a variety of methods can certainly be used to investigate institutional change. Some of the regularities of institutional effects lend themselves to systematic and even quantitative analysis. But the political process itself is so shaped by contingent events and subjective perceptions that it is highly unlikely that institutional change can be modeled systematically. In any case, no single model of change or the impact of past events can do justice to the multiple levels of causality at work in historical explanation. Instead, general models, even in the form of covering laws can be used to pose questions; but there is no substitute for empirical research in finding the answers regarding a particular case.

Notes

* This essay was presented in rough form at the conference Paradigms of Change, held at the Center for Development Research in Bonn from 23 to 25 May 2002, and at the Hanse-Wissenschaftskolleg on January 30, 2003. I thank the participants in these meetings for their comments, but especially the late Georg Elwert, Reinhart Köessler, Alfred Neuwen and Andreas Wimmer. The final draft was written while in residence at the Hanse-Wissenschaftskolleg as a Fellow, an opportunity for which

I am very grateful. I thank Isabelle Schulze for help with the graphic and the references.
1. ARAB, SAP Partisytrelsens Protokoll 1964–67, 'Protokoll vid sammaträde med Social demokratiska Partistyrelsen den 27 februari 1964 i Stora Partilokalen, Riksdagshuset,' p. 3.
2. Ossian Sehlstedt, 24 February 1953, cited in von Sydow, 1989: 90 and direct citation, ARAB, SAP Riksdagsgruppensprotokoll, 'Protokoll fört vid sammanträde i stora partilokalen med socialdemokratiska riksdagsgruppen tisdagen den 24 februari 1953 kl. 18.00,' p. 3. Gösta Nélzen, ibid., p. 7.
3. Tage Erlander, ARAB, 'Protokoll partistyrelsesammanträde 1 Oktober 1966,' p. 6, and 'Protokoll fört vid sammanträde vid den Socialdemokratiska riksdagsgruppen tisdagen den 8 november 1966 Kl 18 i Stora partilokalen,' p. 4.
4. Tage Erlander, ARAB, 'Protokoll partiytrelsesammanträde 3 November 1966 §26 Författningsfrågan,' pp. 1–2, 9.
5. Gunnar Lange, 3 November 1966, ibid. p. 11.

Bibliography

Primary sources

ARAB (Arbetarrörelsens Arkiv och Bibliothek). Protokoll, Social Demokatiska Partisytrelsen, 1948–1968. (Minutes of the Meetings of the Executive Council of the Swedish Social Democratic Party).

ARAB (Arbetarrörelsens Arkiv och Bibliothek). Protokoll, Social Demokatiska Riksdagsgruppen, 1948–1968. (Minutes of the Meetings of the Parliamentary Group of the Swedish Social Democratic Party).

Secondary sources

Abbott, A. (2001) *Event Sequence Analysis: Methods. International Encyclopedia of the Social and Behavioral Sciences*. Amsterdam and New York: Elsevier Science 4966–9.
Abbott, Andrew (1990) 'Conceptions of time and events in social science methods: causal and narrative approaches'. *Historical Methods* 23 4 (Autumn) 140–50.
Abbott, Andrew (1991) 'History and sociology: the lost synthesis'. *Social Science History* 2 (Summer) 201–38.
Abbott, Andrew (1992) 'From causes to events: notes on narrative positivism'. *Sociological Methods & Research* 20 4 (May) 428–55.
Almond, Gabriel A. & Coleman, James S. eds. (1960) *The Politics of the Developing Areas*. Princeton: Princeton University Press.
Anderson, Perry (1974) *Lineages of the Absolutist State*. London: Verson Editions.
Bendix, Reinhard (1969) *Nation-Building and Citizenship. Studies of Our Changing Social Order*. Garden City, New York: Anchor Books.
Berger, Suzanne (1981a) 'Introduction'. In: Berger, Suzanne ed. *Organizing Interests in Western Europe: Pluralism, Corporatism and the Transformation of Politics*. Cambridge: Cambridge University Press 1–23.
Berger, Suzanne ed. (1981b) *Organizing Interests in Western Europe: Pluralism, Corporatism and the Transformation of Politics*. Cambridge: Cambridge University Press.
Boudon, Raymond & Bourricaud, Francois (1994) 'Historizismus'. In: Boudon, Raymond & Bourricaud, Francois eds. *Soziologische Stichworte: Ein Handbuch*. Opladen: Westdeutscher Verlag 205–12.
Bridges, Amy (2000) 'Path dependence, sequence, history, theory'. *Studies in American Political Development* 14 (Spring) 109–12.

Buchanan, James M. & Vanberg, Viktor (1989) 'A Theory of leadership and deference in constitutional construction'. *Public Choice* 61 15–27.

Calhoun, Craig (1991) 'The problem of identity in collective action'. In: Huber, Joan ed. *Macro–Micro Linkages in Sociology*. Newbury Park (NJ)/ London/ New Delhi: SAGE Publications 51–75.

Calhoun, Craig (1994) 'Social theory and the politics of identity'. In: Calhoun, Craig ed. *Social Theory and the Politics of Identity*. Oxford: Blackwell Publishers 9–36.

Calhoun, Craig (1996) 'The rise and domestication of historical sociology'. In: McDonald, Terrence J. ed. *The Historic Turn in Social Sciences*. Ann Arbor: University of Michigan Press 305–37.

Collier, Ruth Berins & Collier, David (1991) *Shaping the Political Arena: Critical Junctures, the Labor Movement, and Regime Dynamics in Latin America*. Princeton: Princeton University Press.

Colomer, Joseph M. (2001) 'Disequilibrium institutions and pluralist democracy'. *Journal of Theoretical Politics* 13 3 (July) 235–47.

Cutler, David M. & Johnson, Richard (2001) 'The birth and growth of the social-insurance state: explaining old-age and medical insurance across countries'. Working Paper RWP 01-13. Kansas City: FRB.

Dahl, Robert A. (1961) *Who Governs?: Democracy and Power in an American City*. New Haven, London: Yale University Press.

Easton, David (1965) *A Systems Analysis of Political Life*. New York: John Wiley.

Edgington, Dorothy (1995) 'On conditionals'. *Mind* 104 (April) 414 235–329.

Elwert, G. (2001) Boas, Franz (1858–1942). *International Encyclopedia of the Social and Behavioral Sciences*. Amsterdam and New York: Elsevier Science 1266–70.

Frank, Andre Gunder (1978) *Dependent Accumulation and Underdevelopment*. London: Macmillan.

Genschel, Philipp (2001) 'Pfadabhängigkeit: Metapher oder Erklärungsansatz'. Unpublished Paper. Cologne: Max Planck Institute for the Study of Societies.

Hall, Peter (2003) 'Aligning ontology and methodology in comparative politics'. In: Mahoney, James & Rueschemeyer, Dietrich eds. *Comparative Historical Analysis in the Social Sciences*. Cambridge: Cambridge University Press 373–406.

Hall, Peter A. & Taylor, Rosemary C. R. (1996) 'Political science and the three new institutionalisms'. *Political Studies* 5 936–57.

Hattam, Victoria C. (2000) 'History, agency, and political change'. *Polity* 32 3 (Spring) 333–8.

Hedström, Peter & Swedberg, Richard eds. (1998) *Social Mechanisms: An Analytical Approach to Social Theory*. Cambridge: Cambridge University Press.

Ikenberry, John G. (1988) 'Conclusion: an institutional approach to American foreign economic policy'. *International Organization* 1 (Winter) 219–43.

Ikenberry, John G. (1994) 'History's heavy hand: institutions and the politics of the state'. Prepared for The New Institutionalism, University of Maryland 14–15 October, 1994.

Immergut, Ellen M. (1998) 'The theoretical core of the new institutionalism'. *Politics and Society* 26 1 (March) 5–34.

Immergut, Ellen M. (2002) 'The Swedish constitution and social democratic power: measuring the mechanical effect of a political institution'. *Scandinavian Political Studies* 25 3 231–57.

Jupille, Joseph, Caporaso, James A. & Checkel, Jeffrey T. (2003) 'Integrating institutions: theory, method, and the study of the European Union'. *Comparative Political Studies* 36 1/2 (February/March) 7–41.

Katznelson, Ira (1992) 'The State to the rescue?: political science and history reconnect'. *Social Research* 4 (Winter) 719–37.
Katznelson, Ira (1997) 'Structure and configuration in comparative politics'. In: Lichbach, Mark Irving & Zuckerman, Alan S. eds *Comparative Politics: Rationality, Culture, and Structure*. Cambridge: Cambridge University Press 81–112.
King, Gary, Keohane, Robert O. & Verba, Sidney (1994) *Designing Social Inquiry: Scientific Inference in Qualitative Research*. Princeton: Princeton University Press.
Kitschelt, Herbert (1986) 'Political opportunity structures and political protest: antinuclear movements in four democracies'. *British Journal of Political Science* 16 57–85.
Krasner, Stephen D. (1989) 'Sovereignty: an institutional perspective'. In: Caporaso, James A. ed. *The Elusive State: International and Comparative Perspective*. Newsbury Park: Sage Publications 69–96.
Kuran, Timur (1998) 'Ethnic norms and their transformation through reputational cascades'. *Journal of Legal Studies* 27 623–59.
LaPalombara, Joseph ed. (1966) *Political Parties and Political Development*. Princeton, NJ: Princeton University Press.
Lebow, Richard Ned (2000) 'What's so different about a counterfactual?' *World Politics* 52 (July) 550–85.
Lehmbruch, G. (2001) *Corporatism: International Encyclopedia of the Social and Behavioral Sciences*. Amsterdam and New York: Elsevier Science 2812–16.
Lewis, David (1993) 'Causation'. In: Sosa, Ernest & Tooley, Michael eds. *Causation*. Oxford: Oxford University Press 193–04.
Mahoney, James (2000) 'Path dependence in historical sociology'. *Theory and Society* 29 507–48.
Mahoney, James & Rueschemeyer, Dietrich (2003) 'Comparative historical analysis: achievements and agendas'. In: Mahoney, James & Rueschemeyer, Dietrich eds. *Comparative Historical Analysis in the Social Sciences*. Cambridge: Cambridge University Press 3–40.
Maier, Charles S. ed. (1987) *Changing Boundaries of the Political: Essays on the Evolving Balance between the State and Society, Public and Private in Europe*. Cambridge: Cambridge University Press.
March, James G. & Olsen, Johan P. (1984) 'The new institutionalism: organizational factors in political life'. *American Political Science Review* 78 3 (September) 734–49.
March, James G. and Olsen Johan P. (1986) 'Popular sovereignty and the search for appropriate institutions'. *Journal of Public Policy* 4 (Oct.–Dec.) 341–70.
March, James G. & Olson, Johan P. (1989) *Rediscovering Institutions: The Organizational Basis of Politics*. New York: Free Press.
Moore, Barrington Jr (1966) *Social Origins of Dictatorship and Democracy: Lord and Peasant in the Making of the Modern World*. Boston: Beacon Press.
Offe, Claus (1996) 'Designing institutions in East European transitions'. In: Goodin, Robert E. ed. *The Theory of Institutional Design*. Cambridge: Cambridge University Press 199–226.
Orren, Karen & Skowronek, Stephen (1994) 'Beyond the iconography of order: notes for a "new" institutionalism'. In: Dodd, Lawrence C. & Jillson, Calvin eds. *The Dynamics of American Politics: Approaches and Interpretations*. Boulder: Westview Press.
Parsons, Talcott (1951) *The Social System*. New York: The Free Press.
Peters, Guy B. (2001) *Institutional Theory in Political Science: The 'New Institutionalism'*. London: Continuum.
Pierson, Paul (2000a) 'Dr. Seuss and Dr. Stinchcombe: a reply to the commentaries'. *Studies in American Political Development* 14 (Spring) 113–19.

Pierson, Paul (2000b) 'Increasing returns, path dependence, and the study of politics'. *American Political Science Review* 94 2 251–67.

Pierson, Paul (2000c) 'The limits of design: explaining institutional origins and change: governance' *An International Journal of Policy and Administration* 13 4 (October) 475–99.

Pierson, Paul (2000d) 'Not just what, but when: timing and sequence in political processes'. *Studies in American Political Development* 14 1 (Spring) 72–92.

Poggi, Gianfranco (1978) *The Development of the Modern State: A Sociological Introduction*. Stanford University Press.

Powell, Walter W & DiMaggio, Paul J. eds. (1991) *The New Institutionalism in Organizational Analysis*. Chicago and London: The University of Chicago Press.

Pye, Lucian W. ed. (1965) *Political Culture and Political Development*. Princeton, NJ: Princeton University Press.

Riker, William H. (1980) 'Implications from the disequilibrium of majority rule for the study of institutions'. *American Political Science Review* 74 432–47.

Roberts, Clayton (1996) *The Logic of Historical Explanation*. Pennsylvania: Pennsylvania State University Press.

Rokkan, Stein ed. (1970) *Citizens Elections Parties*. Oslo: Universitetsforlaget.

Rothstein, Bo (1998) 'Political institutions – an overview'. In: Goodin, Robert E. & Klingemann, Hans-Dieter eds. *A New Handbook of Political Science*. Oxford University Press.

Ruin, Olof (1990) *Tage Erlander: Serving The Welfare State, 1946–1969*. Pittsburgh: University of Pittsburgh Press.

Sabel, Charles F. & Zeitlin, Jonathan (1997) 'Stories, strategies, structures: rethinking historical alternatives to mass production'. In: Sabel, Charles F. & Zeitlin, Jonathan eds. *Worlds of Possibility: Flexibility and Mass Production in Western Industrialization*. Cambridge: Cambridge University Press 1–36.

Schmitter, Philippe & Lehmbruch, Gerhard eds. (1979) *Trends Towards Corporatist Interest Intermediation*. Beverly Hills, CA: Sage.

Sewell, William H. Jr (1985) 'Ideologies and social revolutions: reflections on the french case'. *Journal of Modern History* 57 1 (March) 57–85.

Sewell, William H. Jr. (1996) 'Three temporalities: toward an eventful sociology'. In: McDonald, Terrence J. ed. *The Historic Turn in the Human Sciences*. Ann Arbor: University of Michigan Press 245–80.

Shepsle, Kenneth A. (1986) 'Instituitonal equilibrium and equilibrium institutions'. In: Weisberg, Herbert F. ed. *The Science of Politics*. New York: Agathon Press 51–81.

Shepsle, Kenneth A. (1989) 'Studying institutions: some lessons from the rational choice approach'. *Journal of Theoretical Politics* 1 2 131–47.

Shepsle, Kenneth A. (2001) 'A comment on institutional change'. *Journal of Theoretical Politics* 13 3 (July) 321–5.

Shepsle, Kenneth A. & Weingast, Barry R. (1981) 'Structure-induced equilibrium and legislative choice'. *Public Choice* 37 503–19.

Skocpol, Theda (1979) *States and Social Revolutions: A Comparative Analysis of France, Russia and China*. Cambridge: Cambridge University Press.

Somers, Margaret R. and Gloria D. Gibson (1994) 'Reclaiming the epistemological "other": narrative and the social constitution of identity'. In: Calhoun, Craig ed. *Social Theory and the Politics of Identity*. Oxford/ Cambridge, MA: Blackwell Publishers 37–99.

Stjernquist, Nils (1996) *Tvåkammartiden: Sveriges riksdag 1867–1970*. Stockholm: Sveriges Riksdag.

Tarrow, Sidney (1994) *Power in Movement: Collective Action, Social Movements and Politics*. Cambridge, NY: Cambridge University Press.
Tetlock, Philip B. & Lebow, Richard Ned (2001) 'Poking counterfactual holes in covering laws: cognitive styles and historical reasoning'. *American Political Science Review* 95 4 (December) 829–43.
Thelen, Kathleen (1999) 'Historical institutionalism in comparative politics'. *Annual Review of Political Science* 2 369–404.
Thelen, Kathleen (2000) 'Timing and temporality in the analysis of institutional evolution and change'. *Studies in American Political Development* 14 (Summer) 101–8.
Thelen, Kathleen (2003) 'How institutions evolve: insights from comparative historical analysis'. In: Mahoney, James & Rueschemeyer, Dietrich eds. *Comparative Historical Analysis in the Social Sciences*. Cambridge: Cambridge University Press 208–40.
Thelen, Kathleen & Steinmo, Sven eds. (1992) *Historical Institutionalism in Comparative Politics*. Cambridge: Cambridge University Press.
Tilly, Charles ed. (1975) *The Formation of National States in Western Europe*. Princeton: Princeton University Press.
Tilly, Charles (1978) *From Mobilization to Revolution*. London: Addison-Wesley Publishing Company.
Truman, David B. (1971 [1951]) *The Governmental Process: Political Interests and Public Opinion*. New York: Alfred A. Knopf, Inc.
van Parijs, Philippe (1982) 'Perverse effects and social contradictions: analytical vindication of dialectics?' *The British Journal of Sociology* 33 4 (December) 589–603.
Vanberg, Viktor & Buchanan, James M. (1989) 'Interests and theories in constitutional choice'. *Journal of Theoretical Politics* 1 1 49–62.
Vanberg, Viktor J. & Buchanan, James M. (1996) 'Constitutional choice, rational ignorance, and the limits of reason.' In: Soltan, Karol Edward & Elkin, Stephen L. eds. *The Constitution of Good Society*. University Park, PA: Pennsylvania State University 39–56.
Voigt, Stefan (1997) 'Positive constitutional economics: a survey'. *Public Choice* 90 11–53.
von Sydow, Björn (1989) *Vägen till enkammarriksdagen: Demokratisk författningspolitik i Sverige 1944–1968*. Stockholm: Tidens förlag.
Wallerstein, Immanuel (1974) *The Modern World-System I: Capitalist Agriculture and the Origins of the European World-Economy in the Sixteenth Century*. New York/London: Academic Press, Inc.
Weber, Max (1978) 'Objective possibility and adequate causation in historical explanation'. In: Shils, Edward A. ed. *The Methodology of the Social Sciences*. New York: The Free Press 165–88.
Weingast, Barry R. (1996) 'Political institutions: rational choice perspectives'. In: Klingemann, Hans-Dieter and Goodin, Robert eds. *A New Handbook of Political Science*. London/Oxford/New York: Oxford University Press 1–27.
Zald, Mayer N. & McCarthy, John D. (1987) *Social Movements in an Organizational Society*. New Brunswick, NJ: Transaction Books.
Zucker, Lynne G. (1991) 'The role of institutionalization in cultural persistence'. In: Powell, Walter W. & DiMaggio, Paul J. eds. *The New Institutionalism in Organizational Analysis*. Chicago and London: The University of Chicago Press 83–107.

18
Social Science and History: How Predictable Is Political Behavior?

Roger D. Congleton

The chapter by Ellen Immergut provides a very nice survey of the methodological issues faced by historians who search for the proper lens through which to understand historical events and by other social scientists who use history as data to test the limits of alternative theories from social science. Toward the end of the essay, she poses a methodological puzzle from the history of Swedish constitutional reform.

Both her methodological survey and her overview of Swedish constitutional history are very well written, thorough, and interesting; however, Professor Immergut neglects two significant points in her analysis. The first is methodological: a difference exists between the aims of social science and history. From the vantage point of social science *much is inherently unpredictable* insofar as patterns of causality may be so complex as to defy systematic analysis, or truly stochastic events exist. From the vantage point of history, every historical event is open to explanation, because every event is a direct consequence of particular decisions and circumstances. The second point is an implication of the first. If the future is not entirely predictable, then much about the future is necessarily unknown to decision makers at the moment of choice. Consequently, rational decisions reflect both uncertainty and ignorance, and mistakes will be made. This may well have been the case for the 1970 reforms of the Swedish Riksdag, as suggested by Immergut's analysis. However, mistakes do not imply irrationality.

Determinism and uncertainty in social science and history

To understand why social science is willing to accept uncertainty, perhaps even more so than modern physics, which has increasingly come to be erected on statistical foundations, consider the following example. Suppose that a leading government official is rolling two six-sided dice and desperately wants the numbers to add up to seven at the moment the dice come to rest. For a physicist, the solution to this problem is entirely within the realm of calculation. A sufficiently precise analysis of initial conditions: shape of the hand, weight and size of the dice, the coefficients of friction, gravity field, and inclination of the surface on which the dice will be rolled will imply that a wide range of forces and vectors that could, potentially, cause the dice to stop rolling at a particular place and with a particular numerical configuration. There are many perfect solutions; there are many ways to roll a seven on a particular surface!

The problem faced by an engineer who wishes to implement the physicist's theory is a bit more difficult than ordinary physics implies, because physicists tend to focus on general rather than specific cases. To design a machine that causes two dice to land at a particular spot and in a specific configuration involves other factors, which make the problem more demanding than implied by a physicist's precise and sophisticated computations of Newtonian forces and inertia. For example, the material of the dice and machines, themselves, absorb and release energy through time, and also slightly change shape as these processes take place. This does not mean that the physicist's conclusions are incorrect, but it does imply that other neglected factors may affect the final design of a dice-throwing machine.

A talented engineer might well be able to design a machine that would cause a pair of dice to stop at more or less the intended place with exactly the 'correct' number of spots on the top, given specific characteristics of the dice, gravity, wind, temperature, and the surface upon which the dice are to be thrown. However, people are not machines. Historical experience has shown that no person can exercise sufficient control over his or her hand to achieve such predictable results if significant rolling of the dice is required. It is for this reason that casinos have long been profitable and that many commercial board games use dice to induce a bit of playful uncertainty. It is entirely because of the limited precision of human coordination and calculation that games of chance remain entertaining and profitable.

Consequently, the extent to which a social scientist can predict the outcome of a particular roll of the dice by a top government official is limited. We can predict with absolute certainty that the numbers will add up to no less than two, nor to more than twelve, but we cannot predict much else about any single roll of the dice.

Fortunately, statistical theory allows us to go a bit beyond such well-informed statements of ignorance. Statistics implies that little can be said about a single roll of the dice, but that a variety of predictions can be made about a series of dice throws – the outcomes of the case in which our government official rolls the dice repeatedly. These predictions are testable, insofar as a series of rolls may refute a number of hypotheses about dice rolling – for example that 'dice can be hot' if they are fair. Social scientists can, thus, provide explanations of particular 'histories' of governmental dice rolling in more or less similar circumstances and can make predictions about as yet unrealized 'histories' that would emerge in the future. A government official will roll a seven about 1/6 of the time using unweighted dice in ordinary circumstances.

For a historian the question is a bit different and in many ways more interesting. Having observed a particular roll of the dice, the historian wants to understand exactly why the values observed arose. Here, there are clearly proximate causes – more or less the same ones used by our physicist – and also more indirect causes: the government official rolling the dice was upset, was under pressure, had been exposed to different theories of rolling dice, was affected by beliefs about divine causality, was left handed, near sighted, weak from age, lived north of the equator, and so on. All these factors might affect the manner in which the dice were thrown and, therefore, would largely determine the flight of the dice actually observed. It is entirely possible that this partial list of factors might have 'determined' the exact trajectory of the dice imposed by the official who 'controlled' the dice and the numbers that appeared on top.

Such completely accurate histories may, thus, fully account for what happened without shedding light on what will happen on the next roll. Although 'history will repeat itself,' about 1/6th of the time in this case; little of the detail that applies to a particular instance of dice rolling will be relevant for explaining the next similar event (rolling a seven). Either the underlying chain of causality is too complex to be fully understood or truly stochastic phenomena occur.

This is not to say that social science is only about prediction or that history only analyzes particular historical events, because the persons who engage in these enterprises are often themselves interested in both questions to varying degrees, and properly so. Social science provides a lens through which particular historical events can be understood, and historical research often produces new hypotheses to be tested as well as facts that may be used to test existing hypotheses. Such 'convex combinations' of research interests produce a more useful and compact body of knowledge for fellow travelers, teachers, readers, and practitioners than would have been produced by methodological 'purists.'

Moreover, in areas where there are few determining factors, the analysis of historians and social scientists tend to be very similar. The light went on because a person flipped the wall switch. The building survived a direct lightening strike unharmed because it was protected by Ben Franklin's invention (the lightning rod). The battle was lost because the losers were greatly outnumbered, outgunned, and caught by surprise. Prices rose in 17th century Spain because of the influx of gold from South America. In cases where causal relationships are simple, even a single instance may generalize perfectly to a wide variety of settings.

In other cases where causality is more complex, there are often many plausible claims and counter claims. Here disagreements are commonplace both across disciplines and within disciplines.

The scope of uncertainty in social science

Controversy, however, is not always caused by differences in research interests, as might be said about differences between social scientists and historians. Disagreements within social science exist, at least in part, because there is disagreement about the extent to which human behavior is predictable, in general or in particular circumstances, and therefore on the extent to which particular empirical results can be generalized.

To appreciate this point, consider the time series of data points depicted below in Figure 18.1 For those who believe that the world is completely explainable, the 'finely nuanced' dashed fitted line, g(x), will be the sort of theory they aspire to. For those who believe that the world is not so readily explained, the 'essential' dotted linear line, f(x), is all that they believe can be accounted for. Disagreements of this sort may cause social scientists to disagree for reasons that are similar to those discussed above, but that are subtly different. Some social scientists would insist that 'we' can, or will be able to, *predict* each successive dice roll; others would regard such precision to be very unlikely.

It seems clear that we know a good deal about social phenomena that can be generalized and a good deal that cannot be generalized. Yet, there is little systematic evidence on the 'meta-questions' that might allow us to assess the extent to which long-standing theories will explain new cases or the extent

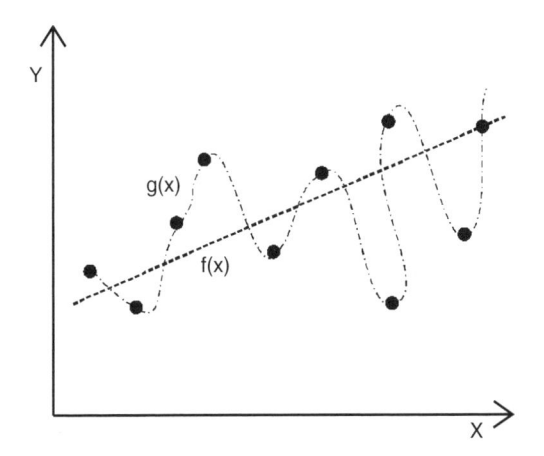

Figure 18.1 How predictable?

to which special factors or new theories will be necessary to understand the cases not yet analyzed. Indeed, each 'side' can point to scientific episodes in which 'they' have been proven correct.

Rational choice and Swedish constitutional reform

To make this point a bit more concrete, consider the case of Swedish constitutional history. There is clearly a sense in which it represents a time series of events analogous to the data points in Figure 18.1. The constitution of 1809 underwent three major reforms over the course of a century and a half. In 1866 the four-chamber unelected parliament was replaced with an elected bicameral parliament, with various wealth restrictions on voting and qualification for office. Between 1909 and 1920, universal suffrage was adopted and proportional rule replaced the weighted first-past-the-post electoral system as the method of counting the votes of the new much broader electorate. In 1970, as noted by Immergut (2002), the bicameral legislature was replaced with a unicameral legislature. These three major episodes of reform led to the core features of the modern Swedish constitution formally adopted in 1976.

To a political historian, it is obvious that the main results of these reforms can be accounted for. Particular people wrote and accepted each of the constitutional reforms in particular political circumstances. For example, Baron de Geer is credited with the ingenious constitutional reform of 1866 that used bicameralism and wealth-weighted voting to secure the required approval by the four chambers of the old Riksdag. Wealth weighted voting in the new first chamber secured majority approval by the noble and burger chambers. The new directly elected second chamber secured approval of the farmer's chamber, and a new church council helped obtain the consent of the clerical chamber (Verny, 1957). Credit for engineering the electoral reforms of 1909 is attributed to Arvid Lindman, who combined proportional representation, universal suffrage and bicameralism to secure supermajorities in the first and second chambers for radical reforms of election law (Verny, 1957). Similarly, Tage Erlander is credited with engineering the end of bicameralism that took place in 1970 (Ruin, 1990).

How much of this can be attributed to general features of the political and historical setting and how much is peculiar to the men and circumstances that confronted constitutional reformers is not immediately obvious, and well-informed individuals may disagree about what is causal and what is the result of chance in given circumstances. Although there were just three major episodes of constitutional reform in Sweden during the past two hundred years, proposals for major and minor constitutional reforms were continuously advanced during the entire period. It seems clear that at least some of the reforms adopted were particular to Swedish personalities and circumstances. Nowhere else in Europe was an explicit wealth-weighted voting system adopted. None the less, broadly similar patterns of reform were adopted in several other northern kingdoms during the same time period. Denmark, the Netherlands, the United Kingdom, and Norway also adopted constitutional reforms in the nineteenth century that produced broad increases in suffrage and a gradual transfer of power from their kings to their parliaments.

How much of this pattern of reform is explainable by general economic, social, and political forces might be debated by serious and well-informed scholars for a variety of reasons. For example, a good deal of the controversy within social science reflects differences in hypotheses about human behavior. Is human behavior driven by narrow self-interest – wealth and power – or more generalized political and economic interests. Is individual behavior largely determined by social pressures and genetic influences that are beyond the individual's control; a consequence of rational decisions to make effective use of what is available in his or her historical circumstances; the result of impulse, whimsy, and creativity – or some combination of all three?

Moreover, as noted above, even social scientists who agree about the aim of science and share a common vision of human behavior may reach different conclusions, because they disagree about how predictable a particular historical event is, or series of such events, can be. A rational self-interested individual cannot know the future any better than a well-informed social scientist can and, therefore, is bound to make mistakes both in assessing his or her interests and in predicting the consequences of the range of actions that may be taken, at least on occasion. Such mistakes produce an irreducible residual of uncertainty in rational choice models and imply that predictions based on those models are better able to describe families of similar events than particular case histories.[1]

This residual of uncertainty is bound to exist even if the rational choice model is perfectly true – as long as individual actors cannot be perfectly informed. To predict human behavior in such cases requires social scientists to know what individual interests were, what they believed about the connection between their actions and consequences, and the limits of both types of knowledge.[2]

In the Swedish case, it seems clear that the main political decisionmakers were very aware of some of the effects that constitutional reforms would have on their own future interests, on their own parties, and on the average Swede, all of which are interconnected. A current member of parliament is more likely to be reelected if the consequences of its policies are good for his or her party, and that party is more likely to be successful if the policies are good for its country. Unfortunately, these are complex relationships that are difficult to fully model and estimate. Consequently, even very well informed legislators may differ in their predictions about the consequences of particular public policies or institutional reforms, and mistakes will be made.

The rational choice hypothesis predicts that political decisionmakers 'get it right' on average. The self-interest hypothesis implies that parliamentary decisions will generally advance member political and economic interests. The very high incumbent success rates in parliamentary systems suggest that members of parliament do get it right on average. Indeed, it is sometimes argued that a member of parliament or congress is more likely to lose office because of death than electoral failure, barring truly outrageous behavior. The observed advantage of incumbency suggests that elected officials do understand and promote their long-term electoral interests, which requires doing a good job of anticipating the consequences of public policies.

The same logic applies to constitutional reforms. For example, the 1970 Swedish reform of parliament did not literally eliminate the first chamber, but merged the two chambers together in a manner that was likely to yield a 'new' parliament with essentially the same membership as the old. The increase in proportionality also tended to increase the power of party leaders. Similarly, in the other two periods of major constitutional reform, the members favoring reform generally continued in office after the reforms were adopted, although many who voted against the reforms did not (Verney, 1957). Yet, it certainly is possible that parties make mistakes on constitutional matters. For example, the liberal party evidently did not do as well in the long run under popular suffrage as they had anticipated after the 1909–20 reforms. The consequences of constitutional reforms are often more difficult to predict than are the effects of ordinary policies.

Immergut (2002) makes a convincing case that the unicameral reforms of 1970 were mistakes as far as the partisan interests of the Swedish Social Democrats were concerned. I have argued elsewhere (Congleton, 2003) that the reforms were also a mistake for the country as a whole, insofar as unicameralism made Swedish public policies less transparent and less predictable. Thus, there clearly is evidence that mistakes were made in 1970. However, these *ex post* analyses do not necessarily shed light on the thoughts of the members at the time of the reforms, because our research does not directly address the knowledge question.

What members believed would happen following the reforms cannot be directly inferred from what did happen, and neither can alternative futures that did not happen – for roughly the same reason. The members of the Swedish parliament in 1967 clearly could not have read the Immergut or Congleton pieces, because they were not available at the time the constitutional negotiations were underway. And, to the extent that those pieces of research meet current professional standards, their analyses are nontrivial and not intuitively obvious. Thus, it is unlikely that the members of parliament during the late 1960s would have had these exact consequences in mind when they voted, more or less along party lines, to approve the new Riksdag act by an overwhelming majority.[3] Such work is more capable of uncovering political mistakes than irrational calculation.

It bears noting, however, that the votes cast by the nonsocialist MPs are clearly explainable in terms of narrow partisan and self interest. The shift to unicameralism made it more likely that the nonsocialists would gain control of the government at some point in the near future, and they did gain control within ten years of the reform. It is the votes cast by the Social Democrats that are difficult to explain from a rational choice perspective, as indicated by Immergut's analysis.

There are several possible rational choice explanations. For a good bit of their history, the Social Democrats favored unicameralism for partisan reasons. The indirectly elected first chamber remained in the control of the non-socialists for nearly three decades after the adoption of universal male suffrage in 1909. During this period, Social Democrats often argued that the bicameral system was 'undemocratic' and should be eliminated to make Swedish politics more democratic. This normative case for unicameralism provides a rational choice explanation for some of its ongoing support among Social Democrats. However, the ideological or norm-driven explanation cannot be the only source of support for the unicameral reform, because the Social Democrats could easily have eliminated the first chamber when they finally gained control of it in 1937 (as some Social Democrats proposed at the time), but they did not.

A second and complementary explanation is that after the unicameral proposal was clearly on the table, it became more difficult for the Social Democrats to hold on to the moral high ground, which they had successfully defended for decades, and which may have accounted for a significant fraction of their electoral support. Fear of future losses in the absence of reform is mentioned by several scholars as an explanation for the Social Democrat vote in favor of unicameralism (Holmberg and Stjernquist, 1996). The non-socialists had won control of the second directly elected chamber briefly during 1957, partly by running against bicameralism, and it was this result that again focused attention on the 'non democratic' aspect of the old bicameral system.

Party leaders may have been believed that by voting in favor of unicameralism the party would do better in subsequent elections than they would have by appearing 'too' partisan. This motivation is entirely compatible with a rational choice model, even if subsequent evidence demonstrated that those fears were unfounded. Mistakes are possible in rational choice models. It bears noting that even correct decisions in uncertain circumstances can look like mistakes, *ex post*. The leaders of the Social Democrats might have been entirely correct in their assessment of the full range of possible outcomes that might follow their constitutional choice, but *appear* to have voted 'incorrectly' given the particular events that transpired. There are losers as well as winners in every fair bet.

Immergut's counterfactual history only examines what happened given that the Social Democrats *did* vote for unicameralism. It does not examine what would have happened among their supporters – at the margin – had the party behaved in an extremely pragmatic fashion and rejected unicameralism simply because it temporarily protected their control of the Swedish parliament. That is to say, Immergut's counter-factual history does not analyze the political consequences of *blocking* constitutional reform. There clearly is a puzzle here, and Immergut's analysis sheds important light on it, but her work is not sufficient to challenge the rational choice explanation of the event. The Social Democrats might well have done worse under bicameralism in the long run, if they had become an 'undemocratic' party.

Moreover, even if Professor Immergut is entirely correct about the effects of the constitutional reform, her results only allow one to reject the perfectly informed model of rational decisionmaking. This is a very limited critique, although not an unimportant one given the widespread use of rational expectations theories by many economists and political scientists for the past two decades.

Mistakes, however, are predictions of rational choice models in settings where causal connections are difficult to untangle and information is incomplete or unavailable (Congleton, 2001a, 2001b). Nowhere is this more likely to be the case than on constitutional issues. Acknowledgement of this fact and the risk associated with mistakes is evidently one of the reasons why major constitutional reforms are infrequent, and subject to more scrutiny and review than are more narrow forms of legislation.

Notes

1. Such 'meta' disagreements can lead to differences in methodology as well. Social scientists will be more or less interested in historical detail according to their beliefs about the underlying predictability of the events being analyzed, because this affects priors about what is likely to be learned from different kinds of data. If not much is truly predictable, a good deal of historical data is simply random noise, rather than part of the generalizable causal chain.

    For example, scholars would clearly be more inclined to carefully review the deliberations of the Riksdag if they believed that the same constitutional reforms would have been adopted if 'clock were reset' and the relevant sessions of the Swedish parliament were repeated, than if they believed many of the details of constitutional reform to be the result of unrepeatable chance thoughts or conversations between members.
2. Failure does not necessarily conflict with rationality. Purchasing a losing lottery ticket is not necessarily a mistake as far as the individual is concerned. He or she may freely purchase another on the hope of winning next time. In the case of lotteries, a series of such purchases may be mistaken, in that it reflects a poor understanding of probability theory, if lottery tickets are purchased to generate increases in income, but may be entirely rational given what is believed at the time the decisions were made.
3. On 17 May 1968, a series of decisions were made regarding the proposed reorganization of the Riksdag. First, a decision was made regarding the transition regulations, which were approved by visual inspection of opinion, but a vote count was

demanded by member of parliament (MP) George Pettersson; 105 voted for, 18 against, and 8 abstained. Next, the remainder of the constitutional amendment was considered. Again, the vote was visually determined to be overwhelmingly pro. The Speaker noted that these reforms were the most important for Swedish democracy in a long time and was pleased that so little opposition existed. (Only the very small Swedish communist party opposed the reform.)

Bibliography

Congleton, R. D. 2001a. 'Rational ignorance and rationally biased expectations: the discrete informational foundations of fiscal illusion. '*Public Choice* 107: 35–64.
Congleton, R. D. 2001b. 'In Defense of Ignorance.' *Eastern Economic Journal* 27: 391–408.
Congleton, R. D. 2003. *Improving Democracy Through Constitutional Reform: Some Swedish Lessons*. Dordrecht: Kluwer Academic Publishers.
Hadenius, S. 1999. *Swedish Politics During the 20th Century: Conflict and Consensus*. Stockholm: The Swedish Institute.
Heclo, H. and H. Madsen, 1987. *Policy and Politics in Sweden: Principled Pragmatism*. Philadelphia: Temple University Press.
Herlitz, N. 1939. *Sweden: A Modern Democracy on Ancient Foundations*. Minneapolis: University of Minnesota Press.
Holmberg, E. and N. Stjernquist, 1996. *The Constitution of Sweden: Constitutional Documents of Sweden*. Translated by U. K. Nordenson, F. O. Finney, and K. Bradfield. Stockholm: the Swedish Riksdag.
Immergut, E. M. 2002. 'The Swedish constitution and social democratic power: measuring the mechanical effect of a political institution.' *Scandinavian Political Studies* 25: 231–57.
Ruin, O. 1990. *Serving the Welfare State, 1946–1969*. Pittsburgh, PA: University of Pittsburgh Press.
Verney, D. V. 1957. *Parliamentary Reform in Sweden 1866–1921*. London: Oxford University Press.

19
Reconstructing Change in Historical Systems: Are There Commonalties Between Evolutionary Biology and the Humanities?

Joel Cracraft

How do we reconstruct history? And how do we explain it? These two broad questions cut across virtually all sciences because most of them are concerned with historical systems. Biological, astronomical, chemical, geological, and anthropological phenomena, for example, are parts of evolutionary trajectories – they are imbedded in systems that have changed over time. Although many scientists may not think of themselves primarily as 'historians' – they may be studying problems that can be assumed to be atemporal, or they may use 'explanatory' equations that are time-independent (such as much of traditional physics and chemistry) – nevertheless the systems they study are the result of previous evolutionary change and, to some degree, however imperceptible, are still changing.

Among the social sciences, historians, obviously, are directly concerned with reconstructing past events and attempting to understand why they happened the way they think they did. So are social scientists, philosophers, political scientists, literary scholars, and linguists, among others. Understanding and explaining change permeates human thought, thus one might assume there would be some epistemological commonality in the mechanics of reconstructing history and some commonality in how we approach causation.

Despite this expectation, there are a variety of opinions on what it means to reconstruct history, how to do it, and the theory and methods one uses to explain how and why specific historical events might have happened. This is certainly true for the quintessential historical science, evolutionary biology. For over 150 years the history of systematics – that part of evolutionary biology directly concerned with reconstructing life's evolutionary past – has encompassed opinions that evolutionary history is only tractable if one can link together a sequence of fossils – the 'right fossils at the right time' – or if one can quantify overall similarity (relationships) among organisms ('phenetics'), or if one can use contemporary 'cladistic' methods. Many of the giants of the so-called evolutionary synthesis in the mid-20th century thought the study of phylogeny was largely speculation and barely rigorous science because history was to be found in the rocks, and if fossil evidence were absent, history was elusive. Today, however, because of new methods, mostly cladistic in approach, and new data, the field is rich with results (Cracraft and Donoghue, 2004).

Thinking about how to reconstruct the history of life on Earth, like that of any historical system, is partly a problem of scale, meaning that a sequence of historical events can be described at a variety of spatiotemporal scales, and thus detail. To some, history is writ large – thinking about relationships among the major groups of organisms, for example – whereas to others it is an issue of piecing together genetic change within a set of populations. Although working at different temporal and spatial scales need not imply differences in understanding what history is and how it might be reconstructed, in fact it generally entails radically different sets of questions and worldviews. This is the case with population biologists and population geneticists who focus their domain of interest at the level of evolving populations, on the one hand, and those systematic biologists who seek to understand the history of species and groups of species, on the other. These different approaches to evolutionary biology not only arise from different sets of questions being asked but also from the 'canalization' of thought that has developed within the various disciplines over the past century and a half (Eldredge, 1985 describes this in some detail). Examples of these conceptual conflicts abound (for definitions and general overviews of these conceptual positions, see Levin, 2001 and Pagel, 2002): 'adaptationists' versus 'nonadaptationists' or 'epigeneticists' (e.g. Williams, 1966; Gould and Lewontin, 1979; Goodwin and Saunders, 1989), 'selectionists' versus 'nonselectionists' and 'macroevolutionists' versus 'microevolutionists' (Williams, 1966; Stanley, 1979; Stebbins and Ayala, 1981; Charlesworth *et al.*, 1982; Vrba and Eldredge, 1984), equilibrial diversification versus nonequilibrial diversification (Sepkoski, 1978, 1979; Cracraft, 1985; Brooks and Wiley, 1986; Benton, 1997), and so on.

Can one really describe history and what does it mean to do so?

Those who study the history of any system understand that one cannot describe its history in any complete sense. Because the description of historical trajectories can be undertaken at different spatiotemporal scales, any historical sequence is far too complicated for an investigator to capture all those elements that might comprise a 'complete' description, and as time passes the observer knows that more and more of those elements will be lost to recovery. This difficulty, however, does not stand in our way, and as a generality, most would agree that history is recoverable at some level of accuracy and acceptability.

It may seem trivial to say that the description of history is limited by the available evidence. Yet, what constitutes 'evidence' is itself nontrivial for a variety of reasons. First, the relevance of evidence can depend on a particular 'worldview.' For many years, for example, the predominant paradigm for reconstructing life's history was to 'connect' fossils from older to successively younger stratigraphic layers using overall similarity and stratigraphic position. As biologists increasingly adopted 'tree thinking,' evolutionary history was depicted more and more as a branching hierarchy, even as an assessment of overall similarity was used to cluster species (including fossils) together. Gradually, systematists began to posit relationships on the basis of shared novelties in structure (derived characters) rather than simple overall similarity, and that method became codified as the cladistic method of Hennig (1966; see also Eldredge and Cracraft, 1980; Wiley, 1981; Nelson and Platnick, 1981).

Although most systematic biologists concerned with reconstructing life's history use cladistic methods, including paleontologists (e.g. Smith, 1994), the conceptual idea of seeing evolution as a gradual change in form over time exerts a powerful influence on the work of many paleontologists and of evolutionary biologists in general. To cite one example: traditional ideas of 'adaptation by natural selection' to explain change in form necessarily imply a continuous, if not gradual, process of genetic change within evolving lineages.

A second reason for the notion of 'evidence' not being straight-forward is that observations can be ambiguous. Systematists who use morphological or behavioral data, for example, argue frequently about whether two features being examined are the 'same' (what is usually called 'homologous'). Indeed, how characteristics are to be identified (individuated) and coded so that they can be used in a quantitative cladistic analysis is a highly contentious issue within systematics (see Scotland and Pennington, 2000). How one delineates the evidence, therefore, has significant effects on how history is reconstructed.

Finally, what constitutes evidence is frequently a function of how data are analyzed. Cladistic, or other kinds of algorithms, build phylogenetic trees on the basis of having taxa (species or groups of species) scored for numerous characters. Depending on the input parameters, and the clustering algorithm, different relationships can result. There is intense debate, for example, over how to analyze DNA sequence data to produce trees (Hillis *et al.*, 1996), and different trees imply that our interpretations of the evidence itself can vary. 'Evidence' supportive of one hypothesis of relationship among three or more taxa can be interpreted as contradictory evidence on another tree in which the taxa are not seen as close relatives. The standard criterion for choosing among alternatives is the notion of philosophical parsimony: the best-fit tree, under some criterion such as minimizing the amount of change on the tree, is the one that is accepted, given the data.

Entities of change

Explanation within the sciences is most commonly described as involving causal statements about things (entities, events) that make reference to universal causal 'laws' (Nagel, 1961; Hempel, 1965a). Thus, an explanation about a specific event in space and time (called the explanandum, what is being explained) is said to follow the logical structure of the 'covering-law' or 'hypothetical–deductive' model, which includes a statement about various initial conditions and one or more universal laws (together termed the 'explanans'), and the explanatory sentence links the explanandum to the explanans. Although this view of explanation arose out of philosophical analyses of the physical sciences, it has also been seen as applying to the biological sciences (Ruse, 1973; Hull, 1974), as well as to explanation within the humanities, including human history (Hempel, 1965b; Roberts, 1996). The critical point to be made here is that universal laws, statements specifying initial conditions, as well as the explanandum itself all include reference to entities. Thus, explanatory structures must take a philosophical stance about ontology – the things said to be changing over time.

More often than not major debates in biology center around ontology, not methodology or differences in theory *per se*. Within systematic and evolutionary biology there is a huge literature debating entities such as species, homologues, characters, among others. Whereas numerous scientists are dismissive about such discussions and characterize them as being pedantic and boring – the 'I know what a species is so why all this argument' attitude – these confrontations over entities persist and frequently become intemperate precisely because so much is at stake. If theories are about entities – and not just any entity but a specific one – then different views of entities might imply that particular theories about those entities, tests of predictions from those theories, or observations themselves may be wrong:

> Depending on how we construe 'reality', some [explanatory] discourses will not explain 'reality', or will offer explanations in terms of entities which are not 'real'. It is therefore of some importance that the question of the relation between ontologies and reality, and the question of the criteria by which something is established as being 'real' in a discourse are dealt with. (Gaukroger, 1978: 41)

The *raison d'être* of the historical sciences is to describe and explain change, which raises a number of fundamental questions about the nature of those entities said to be changing. Indeed, how we conceive of change and describe it depends on how we see the entities of interest: Are the entities real? Are they observable, or if not, is their existence necessitated by theory? Are they discrete and what is the nature of their boundaries?

Debates over species concepts exemplify the importance and role of entities in scientific progress (e.g. Cracraft, 1987, 1989a,b, 2000; Wheeler and Meier, 2000; Hey et al., 2003). Prior to the Darwinian revolution, species were seen as discrete and immutable entities, specially created supernaturally. The Darwinian worldview challenged this notion and imposed a radically alternative ontology: if species transformed into other species – ancestors gave rise to descendants – then such entities must neither be discrete nor immutable, rather they must blend together over space and time. Neontologists of the time began looking more closely at patterns of geographic variation to see if species that were seemingly discrete at one point in space might intergrade (interbreed) with closely related species elsewhere, and paleontologists sought to document the slow, gradual change of one species into another up through the stratigraphic column.

The Darwinian paradigm of slow gradual change of one species into another has provided the dominant worldview of much of paleontology for the past 150 years, yet its implications for ontology are infrequently discussed. How can species actually be discrete real entities if evolution works through slow changes in individuals and populations over vast geological time? Are species therefore 'fictitious' entities? Some have suggested such. Most of evolutionary biology, of course, never abandoned the concept of species, and neontology in particular has been uncomfortable with a gradualist worldview because species appear to be discrete units in the real world (even though species have been defined differently by different investigators). Species, moreover, are said to function in a host of processes, most notably 'speciation' (see also Cracraft, 1989a), and it would be nonsensical to think of a process producing fictitious entities.

One answer to this conundrum over species (previously termed 'the paradox of discrete entities' by Cracraft, 1987) lies both in ontology and in empirical science. From an ontological perspective, whether entities are 'seen' as being discrete or not is dependant not only on the theoretical mileu but also on the spatio-temporal context of the observer. Thus, many entities in empirical science are 'theoretical' in the sense their existence is required by current theory even though they may not (yet) be directly observable, or they are ambiguously observable (some entities of particle physics come to mind). At the same time, if the spatio-temporal scale of observation is shifted, an entity might be considered to be discrete, or its boundaries might be seen as 'fuzzy.' If the time-frame of change from one entity to another is short relative to the longevity of the entities themselves, then entities will likely be interpreted as being both discrete and 'immutable,' but at the same time as transmutable (e.g. an element that decays to another element, or a species that begets another). This ontological view is intimately related to theories about change. The long-held paleontological notion of gradual change over time was challenged by the idea that most species show long periods of morphological stasis in the stratigraphic record and are characterized by a sudden 'punctuated' change while in allopatry (isolation) that gives rise to new species (Eldredge and Gould, 1972).

Describing and understanding historical change is thus crucially dependent upon individuating the entities of change. Often, that is an issue of how those entities are defined. Species, even in the here and now, are not directly observable, thus our ideas about what species are depends upon the interplay of observation (pattern recognition of organismal variation in space and time) and our understanding of how those things we call species function in theories and whether a particular notion of species is consistent with those theories. Species have been defined in numerous ways, and each of those definitions generally implies a different way of thinking about what is being observed in nature as well as how those entities function in theories about them (Cracraft, 2000).

Entities beyond species

The above discussion hints at the complexity of describing 'evolutionary change.' This complexity increases further when we add thinking about the many other entities besides species involved in a domain as broad in scope as 'evolutionary change in biological systems.' Genes, environmental descriptors such as niches, habitats, or ecosystems, or elements of form and function all come to mind. Likewise, understanding something as complex as 'institutional change' within political-economies must surely need to confront what it means for institutions to change. What exactly is changing – that is, what are the relevant entities of such system descriptions and explanations? It appears from Immergut's essay (this volume) that many historians and political scientists may not see the necessity of addressing these questions, or their philosophical worldview toward history may preclude seeing them as being relevant:

> It may not be possible or desirable to make historical analysis more scientific and exact, if the very purpose of the historical approach is precisely to capture the unpredictable, contingent nature of human actions. (Immergut, this volume)

Leaving the idea of contingency aside for the moment, Immergut's later discussion of 'thick description' and the idea that institutional analysis requires correct specification of institutional variables, as well as the analysis of discrete events, parallels remarks made above about evolutionary analysis. Thus, descriptions of 'biological evolution' or of 'institutional change' only make sense if we individuate the elements of each that are changing.

Perhaps more importantly, if this ontology is general, it will permit a more nomothetic, less idiographic, description of change. If the entity 'species' means one thing in beetles, another in earthworms, and another in birds or mammals, then there is little hope one will find something general about the origin of 'species.' Likewise, in analyzing institutional change, the more the 'same' entities of change can be precisely individuated across various case studies, the more likely it will be that common patterns and process can be identified (or denied).

A note on causation in historical analysis

In common parlance, 'to explain' connotes an answer to a 'why' or 'how' question, or to specify a causal process of some event, action, or observed pattern. A causal explanation is generally cast within the context of an applicable hypothetico-deductive covering-law model for the thing being explained. Roberts (1996) and Immergut (this volume) draw attention to the fact that many historians and social scientists reject this approach to causation in historical analysis, claiming instead that a detailed description of change, whereby all sequential events are captured in the description, is sufficient in and of itself and is 'explanatory' as a consequence of such detail. Roberts (1996) and Hempel (1965b) before him, among others, argue persuasively against this approach to explanation. In contrast to 'why' or 'how', description tends to establish a 'what' question. Immergut notes, correctly I believe, that such an approach also lacks generality and results in a history that is merely a collection of descriptive stories. Moreover, it relies on the unrealistic proposition that history can indeed be described in sufficient detail such that 'no lacuna is tolerated' in Oakeschott's words (quoted in Roberts, 1996: 39). The 'no lacuna left behind' approach to reconstructing history has perhaps been as widespread in evolutionary biology and paleontology as it has been in the social sciences: fill the gaps, and we'll be able to 'see' history. Neontologists can also become effusive, after studying a problem in so much detail, that they believe they can 'see' the process of evolution in action. The typical basis for a claim such as this is that between two points in time (say, T1 and T2, usually taken to be generations) populations are observed to be different in the mean of some measurement (e.g. body size, bill shape), and that metric can be correlated with change in some environmental variable. For years, a not infrequent claim of this 'microevolutionary' approach to evolutionary change has been that those who study the genes and demographics of populational change over 'ecological time' are studying the 'processes' of evolution, whereas those concerned with the history of species or groups of species over 'evolutionary time' are merely describing evolutionary pattern. Similarily, in the social sciences, 'big picture' modernization theory (see Eisenstadt, this volume) has been opposed to the micro-analysis of specific 'mechanisms' of change (Hedström and Swedberg).

Reconstructing Change in Historical Systems 277

This caricature over 'pattern' versus 'process' (that which explains pattern) is, however, misguided. Consider those patterns and processes involved in explaining the origin of species, or speciation (Figure 19.1). The 'process' of differentiation, at least for most organisms, is generally assumed to entail geographic isolation of populations, the appearance of genetic/phenotypic novelties in those populations, and the fixation of these new novelies to characterize a new taxon. But, if one adheres to a biological species concept (e.g. Mayr, 1963), those novelties must result in the population being reproductively isolated from other populations before we can speak of a new species being born; other species concepts, on the other hand, might accept any diagnosably distinct variant as evidence for a new species-level taxon. The point of Figure 19.1 is two-fold: (a) a description of any time-slice of the 'causal nexus' of the origin of a species will be inordinately complex, and it would therefore be unreasonable to think that many time-slices could be described at all completely, and (b) the distinction between pattern and process is problematic. If one takes a descriptive/idiographic approach to history, then a process is merely pattern differentiated with respect to time (two time-slices of patterns conceptually linked together). The escape from this, perhaps, is to assume a covering-law model of explanation and deduce the existence of one or more processes involved in the particular instance one is studying (many philosophers have invoked covering-law explanation in processes such as those shown in Figure 19.1).

Contingency in historical analysis: is there something nontrivial here?

The claim that history is contingent is generally taken to mean that singular events, had they not occurred, would have sent the trajectory of history in a different direction. But in what sense is this statement not trivial? Surely an answer lies more in the realm of philosophy than in empirical science. If the history of natural systems or human culture is indeed irreversible and the unfolding of events must follow time's arrow, then excising any singular event would, by definition, lead to an alternative history at some spatiotemporal scale (I leave aside the logical inconsistency of time travel to retrodictively describe that history). The flipside of this view of contingency could be as follows: given certain deterministic laws that explain the behavior of a large-scale system, micro-scale 'contingent events' might be said to have no measurable effect on large-scale behavior (e.g. even if a particular moon had not been formed by cometary impact, its presence or absence would have had no measurable effect on the large-scale expansion of the universe). Yet surely, at the small-scale, if Earth's moon had not been formed we would have written our history differently (the importance of tides, howls, and spooning would have all been rewritten). So contingency could be construed as a

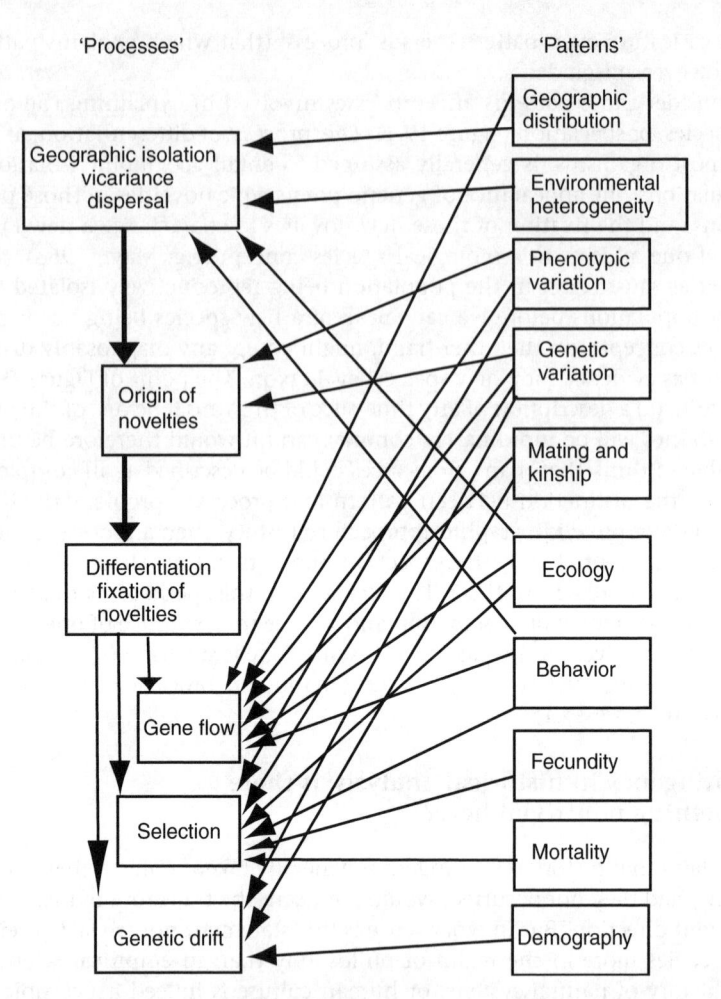

Figure 19.1 A simplified causal nexus for the origin of a species. Three primary processes leading to the origin of new species are often identified: the geographic isolation of populations, the origin of new genetic/phenotypic variants, and their fixation. Each of these 'higher-order' processes are, in turn, manifestations of one or more other processes that often have very different time-frames. Thus, geographic isolation could come about via long-distance dispersal or vicariance (splitting apart of an ancestral range by a barrier); the origin of new novelties might arise as point mutations of DNA, morphogenetic reorganization, or both), fixation of novelties could arise through gene flow, selection, drift, or some combination of these 'processes.' At each time slice many observations ('patterns') contribute to inferring each identified process. But 'processes' are inferences about change in pattern over time; in effect processes are little more than patterns (observations) that are conceptually linked across time intervals (we do not see the change directly). The spatio-temporal scale of the observer (including instrumentation) influences the smoothness of the perceived change

matter of micro-level description where singular events matter, versus explanation on the large-scale where they don't. Similar remarks could be made with regard to human history and modernization. Immergut's article, for example, is definitively not situated at the level of large-scale processes that might determine the evolution of democracy, but rather focuses on one particular event in the development of a system of voting. There a 'contingent event' might be of paramount importance.

But what is the rationale for picking one chunk of time, and an event that happened in it, as being contingently important? We can easily *imagine* how a particular event, say Lincoln's assassination or the Cretaceous-Tertiary asteroid impact, *if it had not occurred*, might have changed history (at the very least Lincoln and all those ancient critters would have lived a bit longer), but we cannot actually reconstruct and interpret a history that never happened, therefore what is the point of such an exercise? One person's scenario would be as plausible as another's. Such speculation lies outside the domain of naturalistic explanation. What is not outside the realm of naturalistic explanation, however, are descriptions and interpretations (hypotheses) about how those singular events causally influenced subsequent history. Those conjectures can be examined and debated though evidence, but the scope of any questions about contingent events must be circumscribed in a way that is empirically meaningful. That this is an important issue is illustrated by some of the writings of the most well-known proponent for 'the science of contingency' in evolutionary biology, S. J. Gould (1989, 1998, 2002):

> ŠI began to explore the role of history's great theoretical theme in my empirical work as well-contingency, or the tendency of complex systems with substantial stochastic components, and intricate nonlinear interactions among components, to be unpredictable in principle from full knowledge of antecedent conditions, but fully explainable after time's actual unfoldings. (2002: 46)

Gould's most frequently cited example invokes the so-called Cambrian explosion as recorded in the 520 million year old Burgess Shale in the Canadian Rocky Mountains (see also Conway Morris 1998). Paleontologists have found in those reocks a large diversity of complex, and often strange, creatures that have been assigned taxonomically to most of the modern groups of animals. Prior to that interval of time, in contrast, comparatively little morphological and taxonomic diversity has been recorded. Gould (1989, 1998) uses the Burgess Shale fauna as the centerpiece of his discourse on the importance of contingency in history:

> *Wonderful Life* argues for greater disparity during the explosion [than afterward], with subsequent trimming on the 'lottery model'-thus raising the interesting implication (and central theme of my book) that if we

could perform the great undoable thought experiment of 'rewinding the tape of life' back to the Cambrian and 'distributing the lottery tickets' at random a second time, the history of animals would follow an entirely different but equally 'sensible' course that would almost surely not generate a humanoid creature with self-conscious intelligence. (Gould 1998: 53)

In *Wonderful Life* Gould makes the chordate fossil *Pikaia* the focus of his argument on contingency. After a disclaimer that 'I do not, of course, claim that *Pikaia* itself is the actual ancestor of vertebrates, nor would I be foolish enough to state that all opportunity for a chordate future resided with *Pikaia* in the Middle CambrianŠ' he then goes on:

> *Pikaia* is the missing and final link in our story of contingency-the direct connection between Burgess decimation and eventual human evolutionŠ Wind the tape of life back to Burgess times, and let it play again. If *Pikaia* does not survive the replay, we are wiped out of future history-all of us, from shark to robin to orangutanŠ why do humans exist? – major part of the answerŠ must be: because *Pikaia* survived the Burgess decimation. (Gould 1989: 322–3)

A deeper consideration of Gould's example lays bare the problems with the notion of contingency and its implications for historical analysis. It is certainly possible that a sequential chain of events might be so tightly linked causally that one could reasonably envision that the interruption of one event meant subsequent events in the chain would be lost. As noted above, by definition this is trivially true: if George Washington had been stillborn, to be sure he would not have become first President of the United States. And all we can say is that either there would have been a first president or there would not have been. Just as a contingent event does not necessarily predict the future, neither does that contingent event necessarily specify a causal chain. In the case of George Washington being stillborn, it merely predicts he is not a participant in further history.

Gould's view of contingency can be questioned, not only scientifically but also for what it says about historical analysis. First, despite his disclaimer quoted above, his argument does in fact rest on the assumption that *Pikaia* is the ancestor of all that followed: the chain is linear (ancestor → descendant), yet those who have paid attention to the issue of identifying ancestors have argued that it is an extremely difficult analytical problem (e.g. Eldredge and Cracraft 1980; Smith 1994). Ontologically, ancestors are not species but parts (populations) of species that become isolated and differentiate to 'begat' a new species. Most historical analysis of candidate fossil ancestors finds them not to be direct ancestors but off on a side branch (see Smith 1994). But if we accept, for the moment, Gould's line of argumentation, that these ancestors can be specified, then it is surely true that between *Pikaia gracilens* and

Homo sapiens there must have been hundreds of thousands, if not millions, of other ancestors, the demise of any of them would have terminated the possibility of us having this discussion. What makes *Pikaia* so special then? It is only special if someone picks it out, claims it is an ancestor, and sees contingency as being something more than trivially true. The contingency argument is another example of the genre of untestable historical narrative. Another example includes those who have argued that 'evolution must have proceeded through this sequence of events.' Narratives of this type have been most common among functionalists who argue that biomechanical principles can be used to specify which evolutionary pathway must have been followed among different body plans. Virtually all of these 'just-so stories' have ignored standard historical analytical approaches (cladistic analysis) that are available to reconstruct species interrelationships as well as character change through time. That some of these stories have involved the use of fictitious organisms as presumed intermediate ancestors only amplifies the problems with this form of historical reconstruction (let alone thinking one can identify something as a true ancestor).

What might historical analysis and explanation in evolutionary biology have to say about change in the social sciences?

It is arguable whether there are 'natural laws' applicable to the study of human social history and institutions (although some 'sociobiologists' seek to explain human behavior using Darwinian law-like statements). But if there are statistical regularities, as noted by Roberts (1996), it might be argued that a covering-law-like explanation could be constructed. As in much of evolutionary biology, the social sciences also seek generality through the comparative historical method (see Mahoney and Rueschemeyer 2003). Observing (reconstructing) similar historical patterns across cases implies common causation, and confidence in common cause becomes stronger the more common patterns we observe. This mode of reasoning is seemingly not that different from the idea of statistical regularity mentioned by Roberts. A crucial part of this form of causal analysis is to ensure that one is comparing *the same thing* across groups (be those groups species, institutions, or historical events).

Because the evolution of life has been largely divergent and branching, and therefore hierarchical – although this is not to deny reticulation (lineages at the population or species level do sometimes merge together) – biologists use cladistic methods to reconstruct (formulate hypotheses about) that history. That methodology has likewise provided insight into other disciplines of historical inquiry where one might expect a significant amount of reticulation, including the history of language (Hoenigswald and Wiener 1987; Hurles *et al.* 2003), texts (Platnick and Cameron 1977; Cameron 1987), and cultural change (O'Brien and Lyman 2003). These types of studies – biological and cultural – suggest that all historical systems will have both hierarchical divergence and interchange (e.g. gene or cultural exchange), followed by more divergence. If such is the case, then methods designed to retrieve hierarchy can provide a framework for identifying reticulation. Cladistic methods, by their nature, attempt to find the most parsimonious distribution of shared similarity across a hierarchy (tree). Conflicts in shared similarity will often yield conflicting hierarchies, which can point to reticulation. These studies, then, suggest that the methods of evolutionary biology may prove useful in any historical discipline in which historical trajectories have a strong hierarchical (divergent) component.

Culture as a whole clearly has both divergence and reticulation, in part because across human history populations have expanded, become isolated, diverged in language and material culture, and then come back together again and exchanged genes and culture, and in some instances diverged again. But it is not clear how this 'model' of change applies to Immergut's example of the Swedish Constitution, which is a tiny snapshot of political history that seemingly examines the 'motives' of politicians – the causes or reasons – behind a change in political structure rather than the temporal 'microanatomy' of the structure itself. Perhaps what is being analyzed here is not so much the historical change in political structure as the nature of political decision making (why they made the changes they did). Immergut's analysis therefore appears to have something in common with Hull's (1988) evolutionary (selectionist) approach to the history of systematic thinking in which some ideas propagate via selection (because of more citations, better empirical results, and so forth). In other words, scientists act on behalf of their own ideas (and their friends' ideas), much like politicians do. If scientific change might be conceived of as a selection process (Hull 1988), so too might political change.

Acknowledgements

I want to thank Andreas Wimmer for inviting me to participate in this project, even though my schedule did not allow me to participate in the conference. I am indeed grateful for his encouragement.

Bibliography

Benton, M. J. 1997. 'Models for the diversification of life'. *Trends in Ecology and Evolution* 12: 490–5.

Brooks, D. R. and E. O. Wiley. 1986. *Evolution as entropy*. Chicago: University Chicago Press.

Cameron, H. D. 1987. 'The upside-down cladogram: problems in manuscript affiliation'. In *Biological Metaphor and Cladistic Classification* (H. M. Hoenigswald and L. F. Wiener, eds). Philadelphia: University Pennsylvania Press, pp. 227–42.

Charlesworth, B., R. Lande, and M. Slatkin. 1982. 'A neo-Darwinian commentary on macroevolution'. *Evolution* 36: 74–98.

Conway Morris, S. 1998. *The Crucible of Creation*. New York: Oxford University Press.
Cracraft, J. 1985. 'Biological diversification and its causes'. *Ann. Missouri Bot. Gard.* 72: 794–822.
Cracraft, J. 1987. 'Species concepts and the ontology of evolution'. *Biology and Philosophy*, 2: 29–46.
Cracraft, J. 1989a. 'Species as entities of biological theory'. In *What the Philosophy of Biology Is: the Philosophy of David Hull* (M. Ruse, ed.). Dordrecht: Kluwer Academic Publishers, pp. 31–52.
Cracraft, J. 1989b. 'Speciation and its ontology: the empirical consequences of alternative species concepts for understanding patterns and processes of differentiation'. In *Speciation and Its Consequences* (D. Otte and J. Endler, eds). Sunderland, MA: Sinauer Assoc., pp. 28–59.
Cracraft, J. 2000. 'Species concepts in theoretical and applied biology: a systematic debate with consequences'. In *Species Concepts: A Debate* (Q. Wheeler and R. Meier, eds). New York: Columbia University Press.
Cracraft, J. and M. J. Donoghue (eds). 2004. 'Assembling the tree of life'. New York: Oxford University Press.
Eldredge, N. 1985. *Unfinished Synthesis*. New York: Oxford University Press.
Eldredge, N. and J. Cracraft. 1980. *Phylogenetic Patterns and the Evolutionary Process*. New York: Columbia University Press.
Eldredge, N. and S. J. Gould. 1972. 'Punctuated equilibria: an alternative to phyletic gradualism'. In *Models in Paleobiology* (T. J. M. Schopf, ed.). San Francisco: Freeman, Cooper, pp. 82–115.
Gaukroger, S. 1978. *Explanatory Structures*. Atlantic Highlands, NJ: Humanities Press.
Goodwin, B. and P. Saunders (eds). 1989. *Theoretical Biology: Epigenetic and Evolutionary Order from Complex Systems*. Edinburgh: Edinburgh University Press.
Gould, S. J. 1989. *Wonderful life*. New York: W.W. Norton & Co.
Gould, S. J. 1998. 'The reply in showdown' on the Burgess shale' (S. Conway Morris and S. J. Gould). *Natural History* 107(10): 48–55.
Gould, S. J. 2002. *The Structure of Evolutionary Theory*. Cambridge: Harvard University Press.
Gould, S. J. and R. C. Lewontin. 1979. 'The spandrels of San Marco and the Panglossian paradigm: a critique of the adaptationist programme'. *Proc. Roy. Soc. Lond.* 205B: 81–98.
Hedström, Peter, and Richard Swedberg (eds). 1998. *Social Mechanisms: An Analytical Approach to Social Theory*. Cambridge: Cambridge University Press.
Hempel, C. G. 1965a. *Aspects of Scientific Explanation and Other Essays in the Philosophy of Science*. New York: The Free Press.
Hempel, C. G. 1965b. 'The function of general laws in history'. *Aspects of Scientific Explanation and Other Essays in the Philosophy of Science*. New York: The Free Press, pp. 231–43.
Hennig, W. 1966. *Phylogenetic Systematics*. Urbana, IL: University Illinois Press.
Hey, J., R. S. Waples, M. L. Arnold, R. K. Butlin, and R. G. Harrison. 2003. 'Understanding and confronting species uncertainty in biology and conservation'. *Trends Ecol. Evol.* 18: 597–603.
Hillis, D. M., C. Moritz, and B. K. Mable (eds). 1996. *Molecular Systematics*. Sunderland, MA: Sinauer Associates.
Hoenigswald, H. M. and L. F. Wiener (eds). 1987. *Biological Metaphor and Cladistic Classification*. Philadelphia: University of Pennsylvania Press.
Hull, D. L. 1974. *Philosophy of Biological Science*. Englewood Cliffs, NJ: Prentice-Hall Inc.
Hull, D. L. 1988. *Science as a Process*. Chicago, University of Chicago Press.

Hurles, M. E., E. Matisoo-Smith, R. D. Gray, and D. Penny. 2003. 'Untangling oceanic settlement: the edge of the knowable'. *Trends in Ecology and Evolution* 18: 531–40.

Levin, S. A. (ed.). 2001. *Encyclopedia of Biodiversity*. 5 vol. San Diego, Academic Press.

Mahoney, James, and Dirk Rueschemeyer. 2003. *Comparative Historical Analysis in the Social Sciences*. Cambridge: Cambridge University Press.

Mayr, E. 1963. *Animal Species and Evolution*. Cambridge: Belknap Press, Harvard University Press.

Nagel, E. 1961. *The Structure of Science*. New York, Harcourt: Brace & World.

Nelson, G. J. and N. I. Platnick. 1981. *Systematics and Biogeography: Cladistics and Vicariance*. New York: Columbia University Press.

O'Brien, M. J. and R. L. Lyman. *Cladistics and Archaeology*. Salt Lake City, University of Utah Press.

Pagel, M. (ed.). 2002. *Encyclopedia of Evolution*. 2 vols. Oxford: Oxford University Press.

Platnick, N. I. And H. D. Cameron. 1977. 'Cladistic methods in textual, linguistic, and phylogenetic analysis'. *Systematic Zoology* 26: 380–5.

Roberts, C. 1996. *The Logic of Historical Explanation*. University Park, PA: Pennsylvania State University Press.

Ruse, M. 1973. *The Philosophy of Biology*. London, Hutchinson.

Scotland, R. and T. Pennington. 2000. *Homology and Systematics*. London: Taylor & Francis.

Sepkoski, J. J. Jr. 1978. 'A kinetic model of Phanerozoic taxonomic diversity. Analysis of marine orders'. *Paleobiology* 4: 23–51.

Sepkoski, J. J. Jr. 1979. 'A kinetic model of Phanerozoic taxonomic diversity. II. Early Phanerozoic families and multiple equilibria'. *Paleobiology* 5: 22–51.

Smith, A. B. 1994. *Systematics and the Fossil Record*. Oxford: Blackwell Scientific Publications.

Stanley, S. M. 1979. *Macroevolution: Patterns and Process*. San Francisco: W. H. Freeman.

Stebbins, G. L., and F. J. Ayala. 1981. 'Is a new evolutionary synthesis necessary?' *Science* 213: 67–71.

Vrba, E. S. and N. Eldredge. 1984. 'Individuals, hierarchies and processes: towards a more complete evolutionary theory'. *Paleobiology* 10: 146–71.

Wheeler, Q. D. and R. Meier (eds.). 2000. *Species Concepts: A Debate*. New York: Columbia University Press.

Williams, G. C. 1966. *Adaptation and Natural Selection*. Princeton: Princeton University Press.

Wiley, E. O. 1981. *Phylogenetics*. New York: John Wiley.

20
History, Uncertainty, and Disciplinary Difference: Concluding Observations by a Social Scientist

Reinhart Kössler

Change involves risk since it contains uncertainty. Putting together first a workshop and then a volume which draws together authors and contributions from a great diversity of scholarly disciplines has also been a somewhat risky undertaking. The fact that there have been fruitful discussions is certainly also due to the readiness of most participants in the original exercise and of the contributors to this volume to look for common strands, even though these may be attained only at some difficulty in communication and through controversial debate.

Significantly, such difficulty and controversy seems to increase as we move from concepts or paradigms drawn from natural science disciplines to those stemming from social science. This increasing difficulty in communication may be due to the fact that social science conceptions also include idiographic approaches, as opposed to nomothetic ones. This shall be addressed in a little more detail further below. In addition to the observations made in the introduction on the asymmetry of concept migration between the various disciplinary fields, it may also be noted that natural scientists seem to be much more prepared to offer transfer insights on questions related to society than social scientists are heard to argue in the opposite direction. While concepts are imported from natural into social science, social scientists themselves will comment only rarely on the subject matter of natural science. In this context, Raghavendra Gadagkar has pointed out a crucial general concern: to be precise about what we are doing, particularly in cases when we are borrowing from each other. Certainly, Gadagkar's particular research on social insects beckons cross-references with social science research and thus sensitises the practitioner also for possible pitfalls. This is an apt warning to be mindful of our language which both on the level of everyday talk and in scholarly discourse is replete with various hidden meanings and metaphors. When we import concepts from other fields, we should do so in a controlled and that means, above all, in a reflected manner.

To try and sketch out the main strands of the discussion contained in this book is a daunting task. It can hardly be accomplished by doing justice to all the arguments advanced and subjects touched upon. Not only would this require near encyclopaedic knowledge. It would also need something approaching one-to-one representation of the complexity present in these discussions. Such a feat is on principle not possible. When it comes to theory building, complexity can be dealt with only in one way – by reducing it. As I shall point out later, this is not quite as trivial as it may seem to some. Above all, this will not, and indeed, cannot be done from some quasi-Archimedian vantage point which might afford an 'objective', decontextualised view to judge all the problems we have been dealing with. We shall return later to the futility of any such pursuit. *A fortiori*, this position obliges the author to be explicit and specific about the point of view or otherwise, the point of observation, which vitally informs his argument. My remarks will be tinged by the perspective of a social scientist chiefly concerned with process, novelty and alterity, while working from long-term, comparative perspectives. Still, I hope to give some indications about how the confrontation with very diverse disciplinary concerns and methodological approaches can shed light on issues that at first sight may appear to be fairly widely apart from each other. Furthermore, the sociology of knowledge may help to understand in how far ideas, concepts, metaphors, models and methodologies are conveyed by people as members of society whose mindset is also informed by their daily social experience and practice.

In dealing with the overall issues of processes of change, of their conceptualising and representation by formal models or discursive accounts, practically all the contributors in one way or the other have addressed the issue of complexity. Here, I do not want so much to discuss complexity as such but rather, I want to look at the analytical strategies and concepts employed in understanding the behaviour and dynamics of complex systems. I would like to first recapitulate some salient points about the conceptualisation of change in the present contributions (1). This will lead up to a brief reflection about the different dimensions addressed in the contributions, both in relation to time and to scale (2). After that I shall look at the ways in which the issue of uncertainty emerges, after the demise of teleological concepts, as a central theme, not only for understanding the subject matters dealt with in science, but for the meaning of science for and its role in society as well (3). The problem of uncertainty is linked closely to the issue of human agency, which again plays very different parts within the various processes discussed in the papers, and also within the analytical and methodological approaches that inform them. This will lead us to the much debated divide between the broad field of social or hermeneutic sciences on the one hand and natural sciences on the other, which has played an important part also in the debates contained in this book (4). In closing, I shall remind of the underlying and fundamental problems that confront, albeit in diverse forms, the quest for objectivity in our enquiries, be it in the field of human action or in that of natural processes (5).

Concepts of change: structure and event

Walter Fontana's title is indicative for one of the chief problems of conceptualising change. To the social scientist, the 'topology of the possible' evokes the significant subtitle of the opening volume of Fernand Braudel's (1979) last *opus magnum*: *'le possible et l'impossible'*. Mapping patterns of change means, among other things, delineating a broader or narrower range of 'objective possibilities' that are inherent in any given situation. This implies that within any given complex system – which may indeed represent a climate system, a RNA chain or a social formation – change will occur *not* in a perfectly predictable fashion, *but* within the broader or narrower structural bounds set by systemic features. Whether change will occur, and which amongst the range of possible paths, trajectories or cases within this horizon of possibilities will actually be realised, is up to concrete constellations and events. In Fontana's phenotype space, the possibilities of change, nearness, as well as the probability of specific changes, accessibility, are given by the structures and relative positions of the neutral networks. Yet all this apparently does not determine or explain the singular case of one amongst the enormous number of *possible* though not equally *probable* phenotype changes actually to take place. Moreover, the neutral networks and their relationships to each other define the range of possible change even though this potential may not be apparent in the phenotype. In this way, we arrive not only at a juxtaposition of structure and event (potential vs. actual change), but also at one between underlying structure and appearance, since from the latter, the potential for change can obviously not readily be inferred. For all the differences, Paul Higgins at one point articulates a similar problem when discussing the long-term accumulation of rather small and unnoticeable change in forces shaping climate that then might result in rather sudden, palpable change, such as the possible effects of a break-down in the thermohalyne Gulf Stream system. Higgins also notes the role of human agency both in effecting change and possibly also in preventing projected change and disaster. Fontana's model seems particularly stimulating since it allows to conceptually link processes of change on a macro and micro level, as well as to evaluate the relationship between structure and event, inherent determinism and chance. These are themes that run through most other contributions to this volume. A given complex structure or system does not map out any definite, determinate trajectory of change. Rather, more than one or even multiple possible outcomes are present in a situation of bifurcation, and the actual trajectory the system will pursue from there is determined by rather fortuitous or contingent conditions and constellations. It should be noted, however, that this does not imply mere fluctuation. As indicated by the trope of the *possible*, initial conditions set out a corridor of change, and further, bifurcation, or multiple change, is not understood as a constantly present condition. Rather, concepts such as punctuated equilibrium or path dependence accentuate both the directionality of the processes under discussion and the intermittent phases of stability of the complex systems in question.

Overall, the problems of structure and event, determinism and probabilism derive from the shared emphasis, common to all authors of this volume, on the non-linear character of processes of change: This means that in certain situations, the rules of causality do not guarantee that a process of change will yield any one particular result; rather, two or more outcomes are possible such as in a non-equilibrium situation, and this will also qualify the predictive power of a model. As shown in the introduction, in many cases prediction is reduced to various degrees of probability. This becomes quite clear from Paul Higgins's discussion of climate change. Complexity, which here means the interaction of 'multiple sub-systems', may give rise to 'multiple equilibria' situations, where it is uncertain which among several possibilities the resultant new equilibrium will be. Such ambiguous situations are of importance not only for understanding the difficulties in predicting climate change, but also as a conceptual model for change in general. Jeffrey Nugent, in his exposition of New Institutional Economics (NIE), addresses a broadly similar problem within a totally different disciplinary framework when he takes up the reasons for differential outcomes of elite formation in various coffee exporting countries. To be sure, in these cases uncertainty or ambiguity will be reduced with a higher degree of concretisation, in other words in looking at each country separately. As I shall discuss in a little more detail below, the specific ways such contingency is dealt with in the various fields vary widely.

These ambiguous situations are reminiscent of the specific 'far-from-equilibrium situations' discussed by Prigogine and Stengers (Prigogine & Stengers, 1984: 176–7; Prigogine, 1997: 61–71) where in a non-equilibrium state, the outcome of change can also not be predetermined from the original set-up. From a social science point of view, fascination with complex systems that may yield two or more (equally or differentially) possible outcomes is motivated, in particular, by the potential of generating novelty. Complex system analysis thus allows to understand not only crisis but, even more importantly, structural innovation.

Time structure and scale

We can discern clear differences in the time structures or timescapes (cf. Adam, 2000) that are characteristic of the processes of change considered in this volume. Some are cases of slow and gradual change, such as certain forms of climate change and institutional development that may be traced over several centuries. Even there, however, the speed of change is by no means even. Rather, as noted, there are phases of acceleration and deceleration. Thus, Higgins refers both to long-term stability and to rather abrupt changes such as those linked to thermohalyne circulation which may result in an incivise climate change in the North Atlantic merely within a couple of years. In looking at institutions, we also observe both long-term stability and surreptitious change. Despite these similarities, however, we should also take into account the differential timescapes involved in the various types of systems and processes discussed in the different chapters.

Take Joel Cracraft's cladistic analysis of species formation. When juxtaposed to the constitutional change analysed by Ellen Immergut, the grossly different time-frames are striking. While not in every single case of the formation of a species, yet certainly in terms of the overall evolutionary process, the processes analysed are extremely long-term, which also makes for protracted phases of equilibrium. The whole story of constitutional reform in Sweden, on the other hand, was played out within merely a few years. Taken together with the over-all development of parliamentarism in Sweden that Immergut mentions, it still covers only a few decennia. Furthermore, there seems to have been no equilibrium or stability over any more or less prolonged period of time. The change is rather set within an ongoing political process. It may be considered within a trajectory of establishing and evolving parliamentary institutions and a periodisation of that trajectory may be considered. Yet it may be inappropriate to consider specifically the change from a bicameral to a unicameral system along with the change in the system of representation as a turning point, given the insertion of these changes into an ongoing process of political debate and strategy, where institutional change forms only one, even though important aspect. An intriguing question would be whether the importance of non-ergodicity that is stressed repeatedly in contributions to this volume as an important feature in complex systems may be differentiated in relationship to different timescapes. Obviously, over very long periods of time, numerically many more cases of unstable constellations may occur than may be expected within the comparatively extremely limited timeframe addressed by Immergut.

'Turning points' form a characteristic of more or less all the other patterns of change that have been presented in this volume. This feature dovetails with the concept of punctuated equilibrium that is adduced above all as a characteristic of the evolutionary pattern. In particular, this applies to Fontana's account of RNA change and also to the concept of path dependence and lock-in. As mentioned, in Fontana's research strategy, change is differentiated into potential for change and actual change. This in itself is an extremely useful tool also for conceptualising other processes of change, where such potential builds up in various forms, even though, in contradistinction to Fontana's example, often going unnoticed for some time, until momentary and visible change occurs. The trajectories of path dependence analysed by Castaldi and Dosi may be said to pinpoint the starting point of the equilibrium phase within a larger trajectory of punctuated equilibrium. Since seemingly unimportant and inconspicuous events or decisions may show unexpected and unforeseeable consequences, path dependency also introduces forcefully the theme of contingency or chance. To refer to the celebrated examples, there is no conceivable specific reason why it was precisely the choice of the QWERTY set-up of the typewriter keyboard or the DOS operating system that set into motion a mechanism of increasing returns and thereby technological as well as economic lock-in. Initially, other solutions might have been equally likely candidates. Considering timescape, the notable feature here is the close linkage that exists between short-term or even momentary events or decisions and very long-term, unforeseen or unforeseeable consequences.[1]

From Mahoney's discussion emerges a further difference between the approaches: scale. Along with his insistence that we should distinguish more rigorously between 'types' and 'sources' of path dependence, he also points to the different scales of cases discussed ranging from single – even though pervasive – technological arrangements to the encompassing social structure envisaged in Immanuel Wallerstein's 'world system'. Szathmáry, in discussing the evolution of different species and their organic structure, explicitly points out the difference between path dependence on the level of the 'individual' and a 'population'. On the basis of the present volume, we may add the conceptually adjacent cases of punctuated equilibrium whose scale also varies hugely, from RNA chains to the climate system. Inevitably, all this has consequences for the systemic environment in which path dependent structures or complex systems evolve. Sticking to the path dependence cases adduced by Mahoney, at least from the viewpoint of social science, the forces acting on the system or structure said to be path dependent vary widely and are basically conditioned by the economic, social or natural context.

Non-teleological trajectories and uncertainty

The issue of the starting point in path dependent processes provides an adequate background to now address the interrelated topics of uncertainty and agency. As Mahoney brings out very clearly, social science approaches, possibly by their very 'eclecticism' or otherwise, by their inherent methodological pluralism, introduce a much wider range of perspective into our understanding of the mechanisms producing path dependency than the reference to the rationality of the *homo oeconomicus*. I would like to take this one step further. Contingency in the selection of a technology or an institutional set-up implies that such decisions do not necessarily yield the 'optimal' result in terms of the goals set by current decision makers and much less in terms of considerations of efficiency, coherence etc. from a later perspective.

In some cases, decision makers may not even be aware of a decision taking place, much less of its consequences. However, even if there exists such awareness, actors can still not determine the outcomes of their decisions adequately in an *ex ante* fashion. There are three different dimensions to these limitations: the systemic limits that exist for prediction and prognosis; incomplete information available to actors; and the multiplicity of actors' considerations and motives and their interrelationships – in other words, the complexity of agency driven decision making.

Still within the framework of path dependence, systematic incongruence between goal-setting and goal-attainment forms one important aspect of uncertainty. This links up to broader considerations, casting doubt on the prognostic potential of the social sciences, at least as long as that potential was understood as eliminating the risk of decision making with the help of assured scientific predictions. Such a view is clearly at variance with older conceptions of science and more commonly with mechanistic world-views that held out exactly such promises.

All the concepts discussed in this volume clearly and almost without question eschew teleology. This may even appear almost trivial to the contributors, but it is certainly not so when we consider broader and also more everyday views and conceptions of what science is about. The absence of teleology means that there is no necessity that conditions the preliminary results of any of these processes of change that we see before our eyes. Causal connections may certainly be traced, and previous stages may be identified as *conditions* in the absence of which later ones would not be possible or conceivable. However, while this doubtlessly gives direction to such processes and introduces irreversibility, direction in this sense does definitely not imply that the process is evolving or striving towards any specific or even knowable goal. Even though it is possible to observe – in retrospect – an almost unrelenting build-up of complexity in the evolution of biological as well as social systems, we may infer from this little more than to suspect a continuance to even more complexity for the future, while it is much more questionable to be specific about any of the forms this may take. Consequently, prediction and prognosis are approached much more modestly than has formerly been the case. At the level of complex systems and processes, prognostic statements can only be conditional at best, and moreover, they are subject to the consequences of complexity reduction. In contradistinction to earlier conceptions of scientific progress which were pinned to the discovery of universal laws and inexorable, unilinear necessity, this opens space for uncertainty. In a bifurcation situation, the outcome is not predetermined, the likelihood of phenotype change is far removed from certainty, and the adoption of a certain technical set-up may be evaluated and revised at least initially, while however the importance of doing so may again emerge only in retrospect. Some of the implications may emerge more concretely when looking at Eisenstadt's paper.

It is only retrospectively that we can understand the consequences of different ways of dealing with the challenges of modernity and of different societal routines mobilised for this task. The consequences of the Meji Restauration of Japan in 1871 were hardly transparent to the Imperial coterie who engineered the toppling of the Shogun and initiated the adaptation of Western techniques in a wide array of fields. They undertook an extremely risky venture which might have failed easily, as did similar modernisation drives at the time, such as the mid-century reforms in Russia and even more so, the Tongzhi Restoration in China. Still, it is significant that the risks envisaged by the actors at the time probably were of rather different from the reasons Eisenstadt makes responsible for different trajectories of modernity. While 19th century latecomers and modernisers generally were fearful of the emergence of a sedentary urban proletariat and the associated social disruption, Eisenstadt points precisely to the different potentials and ways that existed or were open in various societies to articulate and absorb such dissent and social contradiction as the decisive factor in the concrete shaping 'multiple modernities'. In this way, Eisenstadt maps out one possible explanation for variations within the broad process of continuous transformations dubbed as modernisation. This variation is linked to systemic givens such as the different 'basic cosmologies', i.e. fundamental traits of intellectual and spiritual culture, and 'institutional formation' in the course of 'historical experience'. To be sure, Eisenstadt does not address the inherent uncertainty in the decision making that is implicit in his account. Still, if only by the fact that the actors could not foresee all and even the main consequences of their actions and strategies, such uncertainty was certainly present in the trajectories he considers as examples for multiple modernities.

In a way that has some similarity with Eisenstadt's account, Jeffrey Nugent demonstrates the potential of NIE to explain differences in long-term trajectories of otherwise similar national economies, such as coffee exporting countries, by constellations of interests that result in differential institutional arrangements. While clearly much more short-term than Eisenstadt's grand approach, here also a break – the introduction of coffee production for export – is viewed as overarched, as it were, by more fundamental arrangements and dispositions. Significantly, Somdatta Sinha explores some of the possibilities to conceptualise the actual rise of what she, from the point of view of biology and physics, terms 'differentiation'. Not only does she underscore once again the crucial interplay between stability and change as distinct stages in observed system behaviour, but she also looks at factors that induce change and may explain different modalities of change. Sinha's account of models of different multivariable systems also highlights the complex interplay between internal dynamics and external factors, termed here as perturbations, to bring about changes with specific results and consequences. If we project this back onto Eisenstadt's problematic, we would be beckoned to look for the specific responses – and the societal and cultural *potential* for such responses – to the kind of external 'disturbance' that was indeed experienced with the onset of modernisation, e.g. by Commodore Perry's man of war appearing in Edo (Tokyo) Bay, by the Opium Wars that 'opened up' China or by the Russian disaster in the Crimean War. In contradistinction to much of earlier modernisation theory, however, this would not mean to look for either 'external' or 'internal' factors conditioning the further trajectories of those societies, but rather, for the specific constellation of structural givens, cultural dispositions and contingent settings. In this way, we are called upon not only 'to look deeply into the differences in similar systems', as Sinha rightly stresses, but also into the differences that exist between seemingly similar constellations of systems.

Here, we need moreover to be careful about the exact place of contingency in the whole process. If we stick to the paradigm of world historical encounters in the course of the spread of modernity, we can see that perturbance as such must not be conflated with contingency. Rather, the expansion of the modern world market with all its consequences and its clash with other societal systems, along with the *possibilities* that arose from that clash can be understood as systemic givens, on account of the different, counteracting dynamics of the various societal systems involved. Apart from the uncertainty mentioned, which was conditioned by a sort of systematically incomplete information on the side of the actors, contingency enters this picture on two levels. On the one hand, in the clash of these systems, each system is contingent to the other and on the hand, it may be surmised that the realisation of *possible* trajectories actually is connected with contingent features, such as the crisis of the Shogunate and even the mindset of the Meiji Emperor in Japan, the decay of the Qing dynasty in China or the retardation of socio-economic reform in Russia, which again is sometimes is seen as connected with certain personality traits of Nicholas I, besides other, more structural factors.

This leads us up to the mode of change presented by Ellen Immergut and to the specific methodological problems that come out when discussing ways to understand that mode. Her argument revolves around one particular decision of institutional reform: changing the Swedish parliament from a bicameral to a monocameral system and shifting the form of parliamentary representation from majoritarianism to proportionalism. As Immergut demonstrates, this basic set-up opens up two avenues of inquiry: On the one hand, we can look for the political, and more specifically, partisan design behind the decision to opt for constitutional change; we can analyse the opportunity structure of this strategy which includes the general public opinion pressing for the change to a one-cameral system. On the other hand, an exercise in counterfactual analysis can shed some light on the question whether the presuppositions and projections that informed the decision to go ahead with parliamentary reform have in fact been borne out by later developments, or indeed, by those developments that did in fact not materialise because of the reform. Both forms of reconstructing the process of constitutional change and of assessing its consequences have to deal with considerable imponderabilities. These imponderabilities were hardly limited to the consequences of limited or incomplete information.

All in all, when considering change in complex systems, we find two forms of uncertainty and two basic constellations whose basic set-ups differ strongly from each other: On the one hand, we are faced with the dynamic of a system that is observed within a certain situation that, on account of the intrinsic systemic forces at work, may yield at least two possible outcomes in the sense of the bifurcation of a trajectory; on the other hand, the lack of knowledge due to the very complexity of the system also results in uncertainty. To be sure, this lack of knowledge is not identical with the kind of incomplete information referred to in rational choice arguments. In cases such as this, such patchy knowledge is conditioned by the very complexity of the issues involved. The difficulties in the way of prognosis stem from the probabilistic character of the basic set-up. The knowledge available for planning and prediction may therefore be improved upon in a kind of asymptotic process, but this will not eliminate the fundamental inherent uncertainty.

Under both perspectives, uncertainty is a consequence of complexity, but in the first instance, it is mediated through structural processes that lead up to bifurcation, while in the second, uncertainty is due to the difficulty and even the impossibility to mentally deal with a large complex system, to assemble all the relevant factors and forces such that we would be in the position to predict the system's future behaviour with a degree of assurance that would substantially alleviate or even dispel uncertainty.

It is noteworthy that this dual notion of uncertainty in models of change refers us to the more general role uncertainty plays for modern societies. Eisenstadt underlines the enormous expansion of uncertainty in all walks of life that is one of the main markers of modernity. This is linked both to the massively increased potential of technology and the much narrower scope for people on the ground to control their lives, and also to societal structure in a more rigorous sense: In modern societies, people's lives are determined by and dependent on technologies that are well beyond their comprehension and control. Moreover, modern life places people into manifold relationships, most of them of an anonymous nature, that make their very existence dependent on preconditions, processes and changes far beyond their control and even their understanding.

As Anthony Giddens has argued, one main antidote to this systemic uncertainty inherent in the modern condition is trust in expert systems (1990: 83–92). While not the only one, science and technology derived from science arguably make up the pivotal expert system in modernity. However, this position rests on the assurance that science can deliver not only knowledge that is applicable technologically, but knowledge that eliminates uncertainty.

Indeed, this may be considered as the great promise of modern science since the days of Francis Bacon and Isaac Newton. During the 19th and well into the 20th centuries, received opinion had it that this promise had actually been realized, and that acting according to scientifically deduced laws would yield foreseeable and even assured results in all fields, including society. However, the discovery of complexity that underlies many of the approaches discussed in this volume, as of course many other branches of the sciences, also implied acknowledging uncertainty, both in the sense of mere statistical probability, and in the sense that the processes studied involve human action, as will be discussed in the following section. In this we may even see a surprising *rapprochement* between natural sciences on the one hand and social and historical sciences on the other – united as one might say, in their disillusionment with objective, eternally iron laws (cf. also McNeill, 1998). One question that warrants further thought and investigation concerns the implications this development carries for the modern quest for certainty.

Agency and the specifics of social science

In all the paradigms represented in this volume, except in discussions of micro-biological evolution, human action as yet another source of complexity (and uncertainty) is an important common theme, albeit in very divergent ways. To be sure, it makes a huge difference if by human action we refer to the anthropogenic forces considered by Higgins in their potential consequences on the climate, or if we address the strategic considerations of Swedish politicians discussed by Immergut. Higgins may conceptualise human action as just another – even though in important ways, a decisive – force to be reckoned with when constructing and refining his model of climate change. In Immergut's chapter, human action and associated strategies take centre stage.

Significantly and unsurprisingly, it is the social science concepts that introduce the problem of agency as an integral component of the complex systems under discussion. This differs markedly from conceptualising agency as an outside force acting on the system as Higgins suggests with respect to climate change. Within and amongst the social science concepts, agency is treated in rather diverse fashion. I would first like to comment briefly on some of the implications for change in the contributions by Eisenstadt and Immergut. Samuel Eisenstadt accounts for the multiplicity of responses to the provocations and challenges of modernity, and consequently for the different forms of modern societies, chiefly by pointing out differential *chances* which various cultures offered for dissent and protest and thereby, for a specific form of agency, one that digresses from received norms. Ellen Immergut, on the other hand, dissects the consequences of one single collective act, the change in the Swedish parliamentary system. Here, strategies and their success or failure can be assessed, also by constructing a counterfactual for ascertaining what might have happened had the bicameral system of parliamentary representation been preserved.

In Eisenstadt's set of comparative cases, then, agency, or rather the potential for and possibility of the specific kind of agency considered as protest, is seen as a consequence of cultural conditions and also as a force making for specific forms of modernity in Eisenstadt's understanding. This still implies responding to certain exigencies which the spread of modernity across the globe implies for regionally bounded societies. Immergut addresses the issue of unforeseen and even unforeseeable consequences of strategic action within one such regionally bounded modern society defined as a nation state. Her concern is not so much the long-term effect of structural or cultural properties of Swedish society, but the modalities and consequences of a specific decision and its conceivable alternative.

It is in Immergut's case study that the consequences of the integration of agency come out with particular clarity. This is something fundamentally different from the treatment of human agency as conditioning the starting point of a lock-in situation, where the process leading to the starting point can be treated rather as a black box as long as we focus on the problem of path dependence as such and are not interested in the actions that led to the adoption of the QWERTY keyboard in the first place. As has already been indicated, a fundamental change occurs once the underlying strategy for a particular choice or decision is addressed. This leads us into the middle of central problems of social science. There are several levels to this problematic, which is basically concerned with conceptualising and understanding the motives of actors.

Congleton maintains that, in analysing the same empirical case as Immergut does, a 'rational choice' explanation of the covering law (or nomothetic) type *may* still hold under the premise of incomplete information. Whether this applies can only be ascertained by an examination of the historical record. This certainly brings us back to the micro-level of the single event, albeit on the rather large scale of decision making in constitutional matters of a nation state. It is precisely on this level, in the analysis of one concrete decision making process, that Immergut stresses the importance and even necessity of idiographic approaches, besides nomothetic ones. What is necessary here may be termed an 'analytical narrative' (Bates) or indeed also a 'thick description' (Geertz); it introduces, besides privileging the single case, the need to delve into actors' motives – if only to ascertain whether these may be rightly be considered as 'rational'. As I shall argue later, this inevitably leads towards a hermeneutical approach and its concomitant problems and exigencies.

Cracraft pinpoints vital problems involved with idiographic arguments: Such an approach is bound to refer to a very small time-scale and thereby to miss long-term, large-scale processes. Cladistic analysis, as applied to the origins or evolution of species, certainly appears as a superb instrument in conceptualising such (very) long-term processes. Yet again, we may question the validity of applying such concepts to society, politics or culture. This is the case not least because of the high degree of reflexivity of human society when we regard social systems in comparison to other complex systems. Here, feedback loops are enhanced and complicated by the intervention of mechanisms of diffusion, i.e. by imitation, learning, adapting and also rejecting features observed in different societies besides mere innovation out of local resources. Even though Gadagkar refers to cases of 'horizontal transfer' of fungus cultivars by communities of leafcutter ants, we still may maintain that this feature of systematic diffusion by observing others and by learning from them is unique to human society.

To refer to a larger-scale example, we may therefore question whether analysis along the lines of a cladistic pattern would be appropriate to account for instance for the multiple forms of modernity Eisenstadt has sketched out. At first sight, such an approach may appear an intriguing proposition: It might be seen as helpful in conceptualising the spread of a type of social formation and indeed, of a process of constant social transformation dubbed modernisation from North-western England to the entire globe. Above all, it would ease the taking on board of such a conception the continuous differentiation in the forms of modernisation that Eisenstadt has called 'multiple modernities'. However, once we consider that Eisenstadt claims precisely that longer-term, regionally rooted 'cultural' factors can account for the diversity he observes, cladistic analysis would at least have to be refined by taking into account the processes of conscious selection, of locally informed adaptation strategies and not least, of the struggles waged over such strategies. We may recall here that Stichweh, in his response to Fontana, even questions the validity of distinguishing, along the lines current in biology, between evolution (phylogenesis) and development (ontogenesis) when referring to what is generally called social evolution. This leads up to a twofold conclusion: In looking at social evolution in the sense of directional change, we cannot evade to address the problem of conscious human action and the variously grounded strategies that inform it. A hermeneutical approach, among others, is therefore indispensable. At the same time, this is closely connected, though not identical, with the aspect of reflexivity constantly present in social evolution, and particularly so in the modern epoch. Since reflexivity articulates in diffusion and selection, we are also talking here about communication and strategic action, which reinforces the argument for the need of hermeneutical approaches.

To be sure, the problem raised by Immergut can be analysed on yet a different level. Obviously, if we are to understand a specific institutional change, we are referred back to the small-scale, short-term variations Immergut has addressed. A different argument may perhaps be made with respect to longer-term processes of change in social structure or culture, which Cracraft also notes in referring back to Eisenstadt's paper. This means we are challenged to interlink these different time-scales or timescapes in our account if we wish to arrive at a picture of the overall process as well as its concrete local and regional variations. We may make some headway here by remembering Max Weber's employment of the term 'elective affinity' in his quest to account for the onset of modern capitalism in a tiny corner of North-western Europe (1963b: 77): As is often overlooked, Weber not only explicitly shunned any causal explanation, but he stressed the singularity of the chain of events and developments that eventually led to the transformation of all social life on this planet and which today is termed modernisation. In retrospect, we may account today for the reasons why factors such as changes in the agricultural system, the European price revolution induced by American precious metals, Calvinist Puritanism, the technological quest of early modern science, an effective state apparatus, and commercial expansion in an emerging world market all acted together to produce the result that today is before our eyes. However, it is hard to imagine both that this constellation came about of necessity or that it might be repeated at another time and place. The latter is open to pure speculation, since in relation to England after the Industrial Revolution, all other regional societies were latecomers who were deeply influenced and informed by that prior event and the developments it had sparked off. The constellation that brought about the Industrial Revolution was itself highly contingent and, as again Weber stresses, quite transitory. As has been hinted, similar arguments, albeit on a smaller structural scale, might also be made up for the cases of path dependence cited by Castaldi and Dosi, where lock-in stems from specific, in many cases rather inconspicuous decisions. In order to understand not only how lock-in or path dependence functions on a structural level once such – often indeed minute – turns have actually occurred, but also to have an idea of how they came about, we have to address the specific, often rather unlikely circumstances of their emergence.

On account of this, Cracraft's criticism of Immergut's approach is very helpful and instructive in yet another way. In closing, he notes that not only does Immergut address a 'tiny snapshot of political history', but that she also examines what motivated politicians to reach a particular decision. In analysing this conjuncture, Cracraft posits that such motives must be separated from 'change in political structure' and thus might be seen as functioning somewhat like selection in evolutionary concepts. However, this analogy appears questionable once we integrate the perspective of actors. Their motives are indeed part of the process of change itself, and they cannot always be subsumed under formal concepts of rationality – not least since criteria to determine rationalisation vary so widely, as once again Max Weber has stressed impressively. After all, such criteria can encompass such unlikely fields as a yogi's introspection just the same as the capitalist enterprise (cf. Weber 1963a: 11–12).

For these reasons, social science, including historical inquiry, has to address the meaning of actors' motives in their own terms. In a celebrated article,

Theodore Abel (1948) pointed to the triviality of seeing my neighbour assembling a pile of wood and 'understanding' that he is about to make a fire. Over and above Abel's behaviourist argument, what is *not* trivial in motives and meaning has to be ascertained by means other than mere observation. This is precisely why hermeneutical procedures, the business of 'translating' meaning (Gadamer, 1965: 361–7), are indispensable here. Since we cannot dispense with the endeavour to capture meaning, we also cannot do away with 'soft' methods such as thick description, participant observation, qualitative interviewing or interpreting textual source material. In terms of analysing social processes, anything else would quite unreasonably narrow down our vision. This is not to deny difficulties and pitfalls which are connected with the inherent circularity of 'understanding'. Because of this, hermeneutical interpretation has to be performed in a controlled way, and its results are rightly and indispensably subject to controversy.

The observer's viewpoint or the 'blind spot'

The problems of hermeneutic understanding of action may not be as specific to the social sciences as it may seem at first sight. Rather, it may be a special instance of a more general epistemological problem. In discussing climate change, Higgins points specifically to the difference that always exists between models, however complex they may be, and 'the more complicated conditions of the natural world'. This difference implies an epistemological uncertainty that cannot be overcome. To be sure, such uncertainty differs fundamentally from the kind that is rooted in bifurcation situations or incomplete information. In this case, we are faced with a fundamental feature of human apperception, a systematic consequence of the fact that to grasp, much more to understand the 'natural world' (or the social world, for that matter), we inevitably are constrained to reduce that world's actual complexity. Commonly, such operations may be perceived as the exclusion of 'noise', with regard to the vital workings of the system in question, yet what is involved in the definition and elimination of 'noise' is precisely an essential part in the process. Reduction of complexity is not only an exigency when building a scientific concept; rather, it represents one of the basic features of the human condition. Humans could hardly survive were they not conditioned to filtering out that minute part of possible apperceptions that is required and useful in any given moment and situation (Gehlen, 1986: 39–46). Put in terms of a general systems theory, this represents a special case of the general need of any system to define the difference between itself and its environment and thereby eliminate non-systemic features as noise and disorder (cf. Luhmann, 1984: 361f).

However, recognition of this fact and its consequences also entails far-reaching implications for what we may expect from the scientific endeavour. Doing science means not just only to excise from reality a rather limited, even tiny portion and to observe or manipulate it under certain rules and controlled conditions, e.g. in the laboratory or by simulation models, thus contributing to the gradual accumulation of knowledge. Rather, science, in particular modern science, has passed through a whole series of 'revolutions' in the systems of scientific representations which transformed noise into meaning. Ptolemaian was superseded by Copernican cosmology, Newtonian by quantum physics. In such scientific revolutions, fundamentally novel constructions of reality disrupt heretofore conventional 'normal science', only to be channelled back into a new 'normal' routine of enquiry. It can be shown that the former concepts that have been discarded by such revolutions were not simply false, but proved inefficient defences against a slow accumulation of countervailing evidence (Kuhn 1962). Scientific revolutions involve radical revisions in cosmologies, each of which implies a transformation of the observer's point of view or in other words, a rather radical change in perspective.

Since any observation of reality can be performed only from a certain, specific perspective, it is impossible to evade the problem of the 'blind spot' that is defined by the observer's position. More dramatically, this makes it difficult, as a long tradition of research in the philosophy of science after Kuhn has shown, to clearly establish whether a specific system of observation is suited for capturing reality 'better' than another. Meta-levels of observation, while useful, will result in infinite regression (cf. e.g. Luhmann 1997: 187). In other words, our conceptual constructs can in no way be expected to give a full picture of reality. The sociology of knowledge takes its point of departure from here and analyses which kinds of constructs are adopted by which particular kinds of people under which circumstances and which kinds of questions and answers result from these basic constructs – for obviously, the answers we can hope for are vitally conditioned by the questions we ask.

Take our discussion on NIE. Jeffrey Nugent mentions the fact that this approach is not easily or very frequently applied to issues of change. This impression has been confirmed by a perusal of the relevant literature. By its underlying cognitive quest, NIE is concerned with conditions that ensure the viability and stability above all of markets. NIE deals with this basic question in looking at the institutional environment of markets. This is different from the 'functionalist trap' against which Nugent cautions us with much justification, since such a perspective does not presuppose in any way that what exists is actually also efficient, let alone optimally efficient, for the purpose ascribed to it. However, if stability and equilibria are considered the main problems to be explained, it is not easy to address change, let alone fundamental transformations. Nor is that all. Significantly, when discussing differential developments starting from seemingly similar pre-conditions, Nugent pointedly mentions that this 'NIE explanation draws upon political economy' when addressing diverse sets of power relations in various coffee growing countries. In other words, Nugent here performs a shift in his perspective that opens up themes that cannot be discussed so easily, or not at all, from a conventional NIE point of view.

In his discussion of the broad NIE approach, John Harriss makes another important point that should be considered here. Harriss insists on the need to take in the broader societal context if we want to understand the history of institutional arrangements. Not only power relations are important here, which Nugent mentions in the way just addressed, but also class relations and class struggle, which of course may easily be translated back into the relative bargaining power of certain social groups and thereby integrated more easily into a NIE framework. Again, this can be read not only as a plea for further and vital contextualisation, but also as an invitation to change the observer's point of observation. In this way, a discussion between representatives of various paradigms may help to shed light on their respective blind spots –an aim which is difficult to achieve through pure introspections and best arrived at in a dialogic process.

It is in this vein that the authors of this book embarked on the adventure of confronting their own practice as scientists with those of others active in so many, so diverse fields. Multiple shifts in perspective enable us to reach a better understanding of what we are doing when carrying on also with our routine research tasks. This includes also recognition of the limits to commonality that exist between disciplinary fields, let alone any idea of the unity of science. It is not a new or surprising insight that such a limit has again emerged, in my view, in the different roles played by human action and in the different understandings of action and rationality. Despite these dividing lines, it is remarkable that similar concepts of change have appeared across the various discussions of this volume and that there seems to be at least a partial convergence of views regarding the role contingency and uncertainty. In other words, we seem to have returned to the grand old scheme for understanding a changing natural and human world: history.

Note

1. Significantly, Adam (2000) broaches this general problematic with reference to the various short and long range timescapes implied by the introduction and spread of gene modified organisms.

Bibliography

Abel, Theodore (1964): 'The Operation called Verstehen.' In: Hans Albert (ed.), *Theorie und Realität*. Tübingen: J.C.B. Mohr (Paul Siebeck), pp. 177–88 (*The American Journal of Sociology 53*, 1948).

Adam, Barbara (2000): 'The temporal gaze: the challenge for social theory in the context of GM food'. In: *The British Journal of Sociology* 51(1): 125–42.

Braudel, Fernand (1979): *Civilisation matérielle, économie et capitalisme, XVe–XVIIIe siècle. Les structure du quotidien: Le possible et l'impossible*. Paris: Armand Colin.

Gadamer, Hans-Georg (1965): *Wahrheit und Methode*. 2nd enl. edn. Tübingen: J.C.B. Mohr (Paul Siebeck) (English ed. Truth and method. New York: Crossroad 1989).

Gehlen, Arnold (1986): *Der Mensch. Seine Natur und seine Stellung in der Welt.* 13th edn, Wiesbaden: Aula (1940).

Giddens, Anthony (1990): *The Consequences of Modernity.* Stanford, CA.: Stanford University Press.

Kuhn, Thomas S. (1962): *The Structure of Scientific Revolutions.* Chicago: University of Chicago Press.

Luhmann, Niklas (1984): *Soziale Systeme. Grundriss einer allgemeinen Theorie.* Frankfurt am Main: Suhrkamp (English ed.: Social systems. Stanford, CA: Stanford University Press, 1995).

Luhmann, Niklas (1997): *Die Gesellschaft der Gesellschaft.* Frankfurt am Main: Suhrkamp.

McNeill, William H. (1998): 'History and the scientific world view.' In: *History and Theory* 37, pp. 1–13.

Prigogine, Ilya (1997): *The End of Certainty. Time, Chaos, and the Laws of Nature.* New York *et al.*: The Free Press.

Prigogine, Ilya and Isabelle Stengers (1984): *Order Out of Chaos.* London: Flamingo.

Weber, Max (1963a): 'Vorbemerkung.' In: Id., *Gesammelte Aufsätze zur Religionssoziologie.* Vol. 1., 5th printing, Tübingen: J.C.B. Mohr (Paul Siebeck) (1920), pp. 1–16.

Weber, Max (1963b): 'Die protestantische Ethik und der Geist des Kapitalismus.' In: Id., *Gesammelte Aufsätze zur Religionssoziologie.* Vol. 1., 5th printing, Tübingen: J.C.B. Mohr (Paul Siebeck) (1905/1920), pp. 17–206 (English edn. *The Protestant Ethic and the Spirit of Capitalism.* Los Angeles, CA: Roxbury 2002).

Index

Abbott, A., 6, 10, 15, 19, 137, 242, 244
Abel, T., 299
Abell, P., 19
Acemoglu, D., S. Johnson and J.A. Robinson, 173
Adam, B., 288
adaptive landscape, 142
agency, and social science, 295–9
agent-based modeling, 54
agglomeration economies, 102
aggregation, 108–9
agriculture, 188–90
Akerlof, G.A. and W.T. Dickens, 118
'algorithmic chemistry', 89, 91, 93–4
Allen, P.M., 56, 118
Alley, R.B. et al., 42
Almond, G.A. and J.S. Coleman, 238
altruism, 220
Aminzade, R., 137
analogy
  in the history of thought, 24
  requirements for effective, 89
'analytical narratives', 6, 296
Anderson, P., 239
Anderson, P.W. and D. Pines, 51
anthropic principle, 141
ants, fungus cultivation, 188–90, 297
Aoki, M., 107, 115
Archibugi, D. et al., 107
Arrow, K., 102
Arthur, W.B., 4, 103, 116, 130, 131
Arthur, W.B. et al., 111
Aunger, R., 147

Bailer-Jones, D. M., 15, 20
Bardhan, P., 179
Barrow, J.D. and F.J. Tippler, 141
Barth, F., 7
Bassanini, A.P. and G. Dosi, 100, 104, 140, 141
Bates, R.H. et al., 6
Bauer, H.H., 18
Bayesian logics, 15

bees, 191
  dance language, 187–8
  behaviour, 80–2, 81, 241
  non-average ('deviant'), 118
  political, 260–9
  and system, 86
Bendix, R., 239
Benner, S.A., 149
Benton, M.J., 271
Berger, S., 239
Berger, S. and R. Dore, 214
Berry, B.J.L. et al., 54
Bickerton, D., 146, 151
'bifurcation', 224
  Lorenz system, 226–7
  one dimensional maps, 225–6
  as a process of change, 225–7
Binswanger, H.P. and M. Rozensweig, 171
biology, 3
  individual and population levels, 141
  and ontology, 273, 274
  path dependence and historical contingency in, 140–57
biota, 19–20
Birchandani, S. et al., 174
'blind spot' problem, 300
Bond, G. et al., 42
Boudon, R. and F. Bourricord, 243
'boundary concepts', 21
Bowers, J. et al., 190
branching effects, 8
  in the social system, 53
Braudel, F., 287
breakpoints, in history, 136
Brennan, G. and J. Buchanan, 167
Bridges, A., 243
Brock, W.A. et al., 114
Brodbeck, M., 89
Broecker, W.S., 41
Brooks, D.R. and E.O. Wiley, 271
Brüning, R. and G. Lohmann, 16, 21
Buchanan, J.M. and V. Vanberg, 249
bureaucracy, 92, 93

Butlerow, 144
'butterfly effect', 40, 55, 59, 227

Calhoun, C., 244
Camazine, S. et al., 191
Cambrian explosion, 279–80
Cameron, H.D., 281
canalization, 73, 271
Canguilhem, G., 21
capitalism, 132
Carr, E.H., 179
Carroll, S.B. et al., 69
Castaldi, C. and G. Dosi, 129, 130, 131, 134, 135, 136
Casti, J., 54
causality, 5–6, 12, 15, 90, 136
causation, in historical analysis, 276–7
Cavalier-Smith, T., 151
Cavalli Sforza, L.L. and M.W. Feldman, 232
Central America, coffee, 172, 292
Central and Eastern Europe, 174
certainty, markers of, 203, 204
chance and determinacy, in evolution theory, 3
change
  continuous and discontinuous, 77
  entities of, 273–5
  neutrality as a paradigm of, 85–8
  as a reversible process, 7
  and robustness, 74–6, 79–80
  structure and event, 287–8
Changeux, J.-P., 146
chaos
  and climate, 39–40
  and economic models, 114
  in social systems, 51–63
chaos models, 60–1
chaos theory, 3, 135
  and climate change, 22–3
  and economics, 23
  patterns in, 12
  and social science, 23, 52
  and social system change, 52–3
chaotic dynamics, 110–11
Charlesworth, B. et al., 271
Chavance, B., 118
chemical systems, in heredity, 144–5
'chronocentrism', 1
civilization, evolution of, 225
civil society, Japan, 209–10

cladistic method, 15, 271, 272, 281, 289, 296–7
Clarke, C.W., 61
class
  conflict, 239
  structure, 93–4
Claussen, M., 45
Claussen, M. et al., 46
Clement, A.C. et al., 41
climate, and weather, 40
climate change, 3, 28, 287, 288, 289, 299
  anthropogenic, 37, 38–9, 47
  and chaos theory, 22–3
  chaotic dynamics, 39–40
  and complexity, 39–41
  controlled experiments, 38
  and economic behaviour, 59, 62
  and economics, 62–3
  'emergent properties', 37, 41
  mathematical models, 38–9
  and multiple equilibria, 40, 41, 43, 47
  reversible, 47–8
  thermohaline circulation, 41–3, 48
  use of characteristics in other fields, 56
  vegetation cover and climate dynamics, 43–8
Coase, R.H., 4–5
coffee, Central America, 172, 292
'cognitive dissonance', 118
coherence, 117
collective identity, 206–7, 215
  Europe, 215–16
Collier, R.B. and D. Collier, 5, 243
Colomer, J.M., 241
colonisation, 173
Communism, 216
communities, expansion of, 220–1
complexity, 54–5, 229, 286, 288
  and climate change, 39–41
  and social systems change, 55
complex systems, differentiation and change in, 223–5
concept migration, 2, 14–20, 285
  and innovation, 20–2
  metaphor move, 16–17, 21
  methodological analogies, 15–16
  misapplication, 17–18
  misunderstanding, 17
  model migration, 15, 20
  problems of misfit, 18
  tool transfer, 15, 20–1

conflict, and modernity, 214–15
Congleton, R.D., 266, 268, 296
conjuncture, 136
constancy, 230
contextual causality, 245–6
contingency, 10–11, 28–9, 293
   and conjunctures, in reactive sequences, 136
   and determinism, 132–5
   in historical analysis, 277–81
   in historical research and institutional analysis, 246
   in a self-reinforcing sequence, 134
contingent irreversibility, 150–1
contracts, 164–5
contra-factual analysis, 6
convergent evolution, 153–5
Conway Morris, S., 279
cooperation, 220
Coriat, B. and G. Dosi, 114, 118
corporate organizations, 105–6
Costanza, R., 62
counterfactual reasoning, 137, 249
Cracraft, J., 271, 274, 275, 289
Cracraft, J. and M.J. Donoghue, 271
cross-disciplinary borrowing, 26
Csete, M.E. and J.C. Doyle, 232
cultural inheritance, 146–9
cultural programmes, 208
culture, 132, 181–2, 184, 220–1, 282
   concept of, 17
   and social system, 87
Currie, C.R. et al., 190
Cutler, D.M. and R. Johnson, 246
cyclical chains, 10

Dahl, R.A., 237
Daily, G.C., 47
Dansgaard, W. et al., 41
Darden, L., 6
Darwinism, 274
David, P.A., 4, 5, 99, 103, 111, 112, 113, 116, 130, 132, 133, 134
Dawkins, R., 144, 146, 154
decision making, 291
   and irreversibilities, 100–1
De Duve, C., 141, 151–2
de-locking, 117–18, 135
   and technological innovation, 118
democracy, 167–8
Dendrinos, D., 53

de Soto, H., 171
determinism
   and contingency, 132–5
   in history and social science, 260–3
Devaney, R.L., 225
development, 68, 79, 86, 89
Diamond, J., 117
Diaz, H.G. and V. Markgraf, 38
Dietrich, M.R., 223
differentiation
   in biology, 223–4, 292
   Eisenstadt's use of term, 223
   in societies, 200
Dirks, N., 182
disciplinary boundaries, 'migration', 2
discontinuity, 85, 88
dissipative structures, theory of, 52
diversity, 230
   in social systems, 91
division of labour, 190–2, 213, 219
   in insect societies, 190–1
DNA, 70–1
   sequence data, 272
Dollo's law, 152–3
Donoghue, M.J., 231
Dosi, G., 103, 113, 118
Dosi, G. et al., 100
Dosi, G. and G. Fagiolo, 115
Dosi, G. and J.S. Metcalf, 100, 109
Dosi, G. and S.G. Winter, 91, 112, 114, 115
Dunbar, K., 21
dynamic increasing returns, 103–4, 130

Easter Island, 117
ecological–economic dynamics, 62
economic behaviour, and climate change, 59
economic evolution, path dependence in, 112–15
economic history
   and path dependence, 133
   reproduction of inefficient outcomes, 133
economic processes, path dependence in, 99–128
economics
   chaos models in, 60–1
   and chaos theory, 23
   and climate change, 62–3

economics – *continued*
  and environment, 61–2
  and nonlinearity, 60
  and path dependence, 129, 131–2
  and social sciences, 27
Edgington, D., 249
Eigen, M., 73, 142
Eisenstadt, S.N., 5, 199, 202, 203, 204, 205, 206, 209, 210, 212, 215, 222, 232
Eisenstadt, S.N. and B. Giesen, 206
Eisenstadt, S.N. *et al.*, 206
Eldredge, N., 271
Eldredge, N. and J. Cracraft, 272, 279
Eldredge, N. and S.J. Gould, 20, 68, 79, 114, 275
'elective affinity', 298
elites, 211, 213, 215
El Niño/La Niña, 38
Elwert, G., 241
Engerman, S.L. and K. Sokoloff, 172
entities
  beyond species, 275–6
  of change, 273–5
environment, and economics, 61
environmental Kuznets curves, 62
epigenetic inheritance, 145–6
epistasis, 72
'epistatic correlations', 105, 106
Epstein, J., 51
Epstein, J. and R. Axtell, 54
ergodicity, 111
  in stochastic processes, 119
Ertman, T., 131
eukaryotic cells, 151
Europe, 207
  collective identity, 215–16
'event', concept of, 6
event chains, 11–12, 13, 19, 130
  analysing with narrative analysis, 136–7
event sequencing programmes, 244
Eve, R. *et al.*, 51, 52
evidence, 272
evolution
  biological, 230–1
  of civilization or society, 225
  contingent irreversibility, 150–1
  convergent, 153–5
  of the eye, 142, 153–5
  and history dependence, 116
  irreversibility in, 152–3
  major transitions in, 149–52
  and path dependence, 141
  repeated, 151–2
  unique transitions, 151
evolutionary analogies
  benefits in social sciences, 90–2
  in social science, case studies, 89–95
evolutionary biology, and the humanities, 270–84
evolutionary change, 67
  change and robustness, 74–6
  mutations, 74
  neighborhood concept, 76
  neutral networks, 73–4, 75–6, 79, 80, 92–3, 287
  phenotype and genotype, 67, 68–9, 79, 80–1
  phenotype space, 76–7
  phenotypic innovation, 67–8, 74
  RNA shape, 70–3, 78, 82
  sequence space, 74
'evolutionary games', 115
evolutionary processes, hierarchically nested, 115
evolutionary units, 142
evolution theory, 3–4, 23–4, 27–8
  chance and determinacy, 3
  'development', 3
  genetic variation, 3–4
  teleological perspective, 3
eye, evolution of, 142, 153–5

Faubion, J.D., 202
Fearon, J.D., 6
Ferguson, N., 6
Firth, R., 7
fitness landscapes, 104–5, 107
Fontana, W., 115, 225, 287
Fontana, W. and L.W. Buss, 93
Fontana, W. and P. Schuster, 69, 74, 76, 77
Forni, M. and M. Lippi, 108
Forrester, J.W., 51, 53
Fowles, J., 1
Freedman, D., 193
Freeman, C., 118
free riding, 167
Frisch, K. von, 187
functional differentiation, 86

Index 307

Gadagkar, R., 191, 194, 285
Gadamer, H.-G., 299
Galison, P., 21
game theory, 18, 20
Ganopolski, A., 46
Gaukroger, S., 273
Gehlen, A., 299
general systems theory, 15
genes, 89, 90
  equivalents in social systems, 90
genetic drift, 11
Genschel, P., 243
geographical boundaries, 77
geographic conditions, 173
globalization, 213–14
Goldstone, J.A., 130, 134
Goodwin, B. and P. Saunders, 271
Gould, J.L. and C.G. Gould, 187
Gould, S.J., 3, 16, 69, 82, 279–80
Gould, S.J. and E. Vrba, 88
Gould, S.J. and R.C. Lewontin, 271
greenhouse gas (GHG) emissions, 38, 39, 42
Griesemer, J.R., 80
Griffin, L.J., 137
Griffin, L.J. and C.C. Ragin, 137
Guastello, S., 56
Gunder Frank, A., 239

Hacker, J., 131
Haldane, J.B.S., 224
Hall, P.A., 246
Hall, P.A. and D. Soskice, 107, 132
Hall, P.A. and R.C.R. Taylor, 240, 242, 243
Hammerstein, P., 22
Hanson, L.P. et al., 53
Harriss, J., 79, 301
Harriss, J., J. Hunter and C.M. Lewis, 181, 182
Hartwell, L.H. et al., 232
Hartz, L., 201
harvesting models, 61–2
Hattam, V.C., 244
Hawthorn, G., 6
Hayami, Y. and V.W. Ruttan, 172
Hedström, P. and R. Swedberg, 244, 276
Heise, David R., 19
He, L. et al., 71
Hempel, C.G., 273, 276
Hennig, W., 272

heredity
  chemical systems, 144–5, 155
  epigenetic inheritance, 145–6
  limited and unlimited, 143–9
  memes and cultural inheritance, 146–9
Hesse, M., 89
Hey, J. et al., 274
hierarchically nested evolutionary processes, 115
Higgins, P.A.T. et al., 28, 41, 43, 47, 56, 287, 289, 299
Hilborn, R.C., 227
Hildenbrand, W., 109
Hillis, D.M. et al., 272
Hirschman, A.O., 167
historical analysis
  causation in, 276–7
  contingency in, 277–81
  historical contingency, and path dependence in biology, 140–57
  historical explanation, logic of, 244–6
  historical institutionalism, 240, 241–4
    change and continuity, 246–8
    and history, 244–5
    in political science, 237–59
    Swedish constitutional reform, 248–51
  historical narrative, 136–7
  historical systems, change in, 270
  historical variation, 182
history, 129, 229
  approaches to, 6
  breakpoints in, 136
  description of, 271–3
  determinism in, 260–3
  evidence, 272
  and historical institutionalism, 244–5
  institutions as carriers of, 106–7
  and path dependency, 13
  uncertainty in, 260–3
history of thought, analogy in, 24
history-dependence, 99–100, 108, 115, 134
  degrees of, and detection, 112–15
  and evolution, 116
Hodgson, G., 177
Hoenigswald, H.M. and L.F. Wiener, 281
Hogeweg, P., 224
Holland, J., 54, 55, 225
Hölldobler, B. and E.O. Wilson, 189
Hollingsworth, J.R. and R. Boyer, 107

Index

Holocaust, 214
Host–Parasite Model, 228
Huberman, B., 92
Hull, D.L., 273, 282
human action, 295–9
　expansion, 212
human evolution, 231
humanities, and evolutionary biology, 270–84
Hurles, M.E. et al., 281
Hurrell, J.W., 38
Huynen, M.A., 74
Huynen, M.A. et al., 74
hysteresis, 41, 242

Ikenberry, J.G., 242
Imbrie, J. et al., 38
Immergut, E.M., 26, 244, 250, 264, 275
increasing returns, 111, 130
　dynamic, 103–4, 130
　non-utilitarian rationales for, 131–2
　path dependence as, 130–1
　properties of sequences, 130–1
India
　agrarian institutions, 178, 179
　caste, 182–3
　divergence of states, 180
　history, 209
industrial diversification, 93–4
Industrial Revolution, 178
information, 163
　properties of, 102
inherent sequentiality, 137
inheritance, 143–9, 155
Inkeles, A. and D.H. Smith, 203
innovation
　and concept migration, 20–2
　and invention, 80
insect societies, division of labour, 190–1
institutional arrangements, 162, 163–4, 301
　states as, 167
institutional change, 240–1, 275–6
institutional inertia, 4–5
institutions, 106–8, 131, 132, 161, 177–86
　as carriers of history, 106–7
　'collective action' problem, 166
　complementarities, 107
　definitions of, 162
　demand for, 162, 164–6, 177
　functions of, 162–3

holdup problem, 166
　and social sciences, 177
　supply of, 162, 166–8, 177
interdisciplinary co-operation, 61–3
interest-groups, 238–9
international systems, 208–9, 211
international trade, 107
invasions, 117–18
invention, and innovation, 80
IPCC, 38, 39, 42
irreversibility, 8–9, 99, 129
　contingent, 150–1
　and decision making, 100–1
　in evolution, 152–3
　of increasing returns, 130
　and path dependency, 101
Isaac–L.W. et al., 135

Jablonka, E. and M.J. Lamb, 145, 146
Jaeger, J.A. et al., 71
Janssen, M.A., 62
Jantisch, E., 51
Japan, 209, 209–11, 292
　civil society, 209–10
　political dynamics, 210
Jastrow, R., 144
Johnson, P.A. et al., 142
Jouzel, J., 41
Jupille, J. et al., 244

Kaneko, K. and T. Yomo, 224
Katznelson, I., 238, 242, 244
Kauffman, S.A., 224
Kaviraj, S., 233
Keeling, C.D., 38
Kellert, S.H., 16, 17
Kiel, L.D., 52, 53, 56
Kiel, L.D. and E. Elliott, 51, 52
Kimura, M., 76
King, G. et al., 249
King, Martin Luther, 135–6
Kirschner, M. and J. Gerhart, 69
Kitschelt, H., 239
Kleidon, A. et al., 45
Klein, J.T., 15, 20
knowledge, organizational, 105–6
Kolakowski, L., 216
Krasner, S.D., 243
Kuhn, T.S., 300
Kuran, T., 118, 174, 244

labour, division and organization, 190–2
Lamarckism, 90, 147, 148
Landes, D., 177, 183, 184
Landes, D.S., 118
Langton, C.G., 225
language, 151, 155, 281
LaPolombara, J., 238
La Porta, R. *et al.*, 173
Latin America, growth rate, 171–2
learning process, 100–1, 146, 155
Lebow, R.N., 249
Lefort, C., 203
legal traditions, 173–4
Lehmbruch, G., 239
Leigh Star, S. and J. R. Griesemer, 21
Lenski, R.E. and M. Travisano, 142
Lerner, D., 203
Levin, S.A., 271
Levin, S.A. and S.W. Pacala, 224
Levinthal, D., 105
Levitus, S., 41
Lewis, D., 249
Libecap, G., 170
Liebowitz, S.J. and S.E. Margolis, 113
life course, 'turning points' in, 6
Lincoln's assassination, 279
Lipset, S.M. and S. Rokkan, 131
Lister, A.M. *et al.*, 190
local interactions, 101
lock-in, 104, 107, 117–18, 131, 133, 135, 151
Lorenz, E.N., 40, 226
Lorenz system, 226–7
Löwry, I., 21
Loye, D. and R. Eisler, 52, 53
Luhmann, N., 56, 86, 299, 300
Lundvall, B.A., 107

McAdams, H. and A. Arkin, 223, 232
McDonald, T.J., 6
McNeill, W.H., 295
Mahoney, J., 5, 130, 132, 134, 135, 243
Mahoney, J. and D. Rueschemeyer, 244, 281
Maier, C.S., 239
Manabe, S. and R.J. Stouffer, 42
March, J.G., 102, 241
March, J.G. and J.P. Olsen, 240
Marée, A.F.M. and P. Hogeweg, 225
Marengo, L., 105

Marion, R., 53, 56
markers of certainty, 203, 204
Markov chains, 9
Markov processes, 111, 119, 140
Marshall, C.R. *et al.*, 153
Marxism, 239
mass production, 133
mathematics, 19
Mathews, D.H. *et al.*, 71
Maturana, H. and F.J. Varela, 86
Maynard Smith, J., 68, 140, 142, 149, 154
Maynard Smith, J. and E. Szathmáry, 143, 147, 149
Mayntz, R., 15, 16
Mayr, E., 277
May, R.M., 225
mechanistic models of change, 7
Med, M., 193
Meheus, J. and T. Nickles, 20
memes, 231
  and cultural inheritance, 146–9
metaphors, 16–17, 21
Metcalfe, J.S., 120
methodological analogies, 15–16
Mirowski, P., 55
Mittenthal, J.E. and A.R. Baskin, 230
model migration, 15, 20
modernity, 292
  and conflict, 214–15
  as a distinct civilization, 202–7
  Weber's conception of, 202–3
  and wellbeing, 219–21
  see also multiple modernities
modernization, 213, 276, 298
  fifties studies, 212
  theory, 5
Mohan Rao, J.M., 179
molecular phylogeny, 149
Moore, B., 239
Moore, B.J., 118
Morgan, M.S. and M. Morrison, 15
Mueller, U.G. *et al.*, 189
Müller, G.B. and G.P. Wagner, 69
'multilinear evolution', 5
multilinearity, 26–7
multiple equilibria, 7
  and climate change, 40, 41, 59
'multiple modernities', 27, 199–200, 222, 231–2, 297
  autonomy of man, 203–4

'multiple modernities' – *continued*
  collective identities, 206–7
  internal tensions, 208
  political order, 204–6
  and power, 208
  and protest, 200–1
  roots of, 207–11
multiple steady-state equilibria, 219
multiplicity
  bifurcation, 225–7
  in non-linear systems, 222–34
  perturbation and system variables, 228–9
  reactions to external perturbations, 227–8
  sources in nonlinear dynamic systems theory, 225–9
Murray, J.D., 225
mutations, 142–3, 244

Nabli, M.K. and J.B. Nugent, 162
Nagel, E., 273
narrative analysis, 136–7, 243
national growth patterns, 107
nationalism, 207
natural sciences
  export of metaphors, concepts and methods to social sciences, 187–95
  and social sciences, 285
Nelson, G.J. and N.I. Platnick, 272
Nelson, R.R., 116
Nelson, R.R. and B. Sampat, 106
Nelson, R.R. and S.G. Winter, 112
neoclassical theory, 133, 161, 181
network, 74, 76
  social, 92, 93
'network technologies', 104
neutral drifts, 116
neutrality
  in biological systems, 85
  concept of, 24, 88
  as a paradigm of change, 85–8
  and social systems, 85–6
neutral networks, 287
  in RNA, 73–4, 75–6, 79, 80
  and social structure, 92–3
New Institutional Economics, 4, 25–6, 161–76, 288, 292, 300–1
  explanations of institutional changes, 169–74, 178

and private property rights, 170–3, 178–9
  shortcomings as a paradigm of change, 168–9
'new' institutionalism, 240
  historical institutionalism, 240
  rational choice (or positive political theory), 240
  sociological institutionalism (or Organization theory), 240
Nimwegen, E. *et al.*, 74
Nisbet, R.A., 16
'noise', 299
non-ergodicity, and unpredictability, 130, 131
nonlinear dynamics and chaos, 109–10
nonlinear dynamic systems theory, 225–9
nonlinearity, 2, 8
  detecting, 114
  and economics, 60
  social systems, 51
nonlinear systems, multiplicity in, 222–34
nonlinear transaction function, 109–10
non-teleological trajectories, and uncertainty, 290–5
North America, and Latin America, growth rates, 171–2
North, D.C., 5, 107, 169, 175, 179–80, 181
novelty, 91–2
Nugent, J., 177, 288, 300
Nugent, J.B. and J.A. Robinson, 172
Nussinov, R. and A.B. Jacobson, 71

Oakeshott, 247
O'Brien, M.J. and R.L. Lyman, 281
Odling-Smee, F.J. *et al.*, 232
Offe, C., 243
'old' institutionalism, 182–3
operation, and system, 86
opportunism, 165
organization, and structure, 86
organizational change, 56, 92–3
organizational knowledge, 105–6
organizations, corporate, 105–6
Orgel, L.E., 144
Orren, K. and S. Skowronek, 244
Ortony, A., 16
Oster, G.F. and E.O. Wilson, 191
Ottoman Empire, 209

Pagel, M., 271
paleomolecular resurrection, 149
Pantin, C.F.A., 19
Parekh, N. et al., 229
Parekh, N. and S. Sinha, 229, 230
Parsons, T., 87, 212
Parthasarathy, S. and S. Sinha, 229
partial determination, 8
Pascal, B., 6
path dependence, 4, 24–5, 288, 290–1, 298
  contingency and determinism, 132–5
  definitions of, 129–30
  in economic evolution, 112–15
  and economic history, 133
  in economic processes, 99–128
  and economics, 129, 131–2
  and evolution, 25, 141
  and historical contingency in biology, 140–57
  and history, 13
  and irreversibility, 101
  and local interactions, 101
  and new institutionalism, 5
  and properties of selection, 104–5
  and reactive sequences, 10, 135–6
  in social learning, 146
  and social sciences, 26–7, 129–39
  sources of, 100–9, 129–30, 131
  strong and weak forms, 140
  types of, 130
  see also increasing returns
path dependency models, 242–4
path dependent processes
  negative and positive definitions, 111
  nonlinear dynamics and chaos, 109–10
  theoretical representations of, 109–12
Peters, G.B., 240
phenotypic plasticity, 224
'physics envy', 55
PICC, 38
Pichaud, F. et al., 154
Pierson, P., 107, 130, 131, 242, 244, 246
Piore, M.J. and C.F. Sabel, 133
Platnick, N.I. and H.D. Cameron, 281
Platteau, 183
'pluralism', 238
Poggi, G., 239
political behaviour, predictability, 260–9

political history, 5–6
political science, 5
  'behavioural' revolution, 237–8
  change and continuity in, 237–8
  historical-institutionalism in, 237–59
  institutional approach, 238–9
  new institutionalism, 240
  path dependence, 130
  political development, 238
  qualitative research in, 244
politics, 107
Polya urn processes, 111–12, 116, 119–20
post-mechanistic models of change, 6–7
Powell, W. and P.J. DiMaggio, 241
power, and 'multiple modernities', 208
preference formation, 244–5
Prigogine, I., 1, 7, 52
Prigogine, I. and I. Stengers, 53, 288
processes, 1
'process-tracing', 246
property laws, 4
property rights, 170–3, 178–9, 182
protest, 200–1, 205, 221, 238–9, 296
punctuated equilibria, 20, 79, 114, 243, 288, 290
Pye, L.W., 238

QWERTY typewriter keyboard, 103–4, 121n14, 130, 132, 133–4, 296

race-based poor relief, 135–6
Ragin, C., 15
Rahmstorf, S., 41, 42
'random walks', 111
Rao, M.G., 180
rational choice
  and structuralism, 91
  and Swedish constitutional reform, 264–8, 296
reactive sequences, 10
  contingency and conjunctures in, 136
  and path dependence, 135–6
Rebaglia, A., 20
Reish, G., 137
replicator dynamics, 120
research methods, transfer, 90
research networks, 22
Riker, W.H., 241
RNA sequences, 69

RNA shape, and a systemic phenotype, 70–3
Roberts, C., 247, 249, 273, 276, 281
robustness, and change, 74–6, 79–80
Rokkan, S., 239
Roniger, L. and C.H. Waisman, 199
Roniger, L. and M. Sznajder, 206
Rosenberg, N., 118
Rosser, J., 62
Rostow, W.W., 7
Rothstein, B., 240
Roy, W.G., 131, 132, 134
Ruin, O., 264
Ruse, M., 273
Ruttan, V.W., 162, 172

Sabel, C. and J. Zeitlin, 244
Sahara, 46
salamanders, 153
'Santa Fe *Zeitgeist*', 2
Scheinkman, J. and B. LeBaron, 53
Schmitter, P. and G. Lehmbruch, 239
Schneider, S.H. and S.L. Thompson, 42
Schultes, E.A. and D.P. Bartel, 75
Schuster, P. *et al.*, 74
Scotland, R. and T. Pennington, 272
Segerstrale, U., 17
segregation phenomena, 101
selection, properties of, 104–5
selection mechanisms, 117
selections models, 115
self, 97
semantics, and social structure, 87
Sepkowski, J.J., 271
sequence/sequencing, 69, 70–5, 78, 79, 135–6
Sewell, W.H., 6, 136, 244
Shepsie, K.A., 241
Shepsie, K.A. and B.R. Weingast, 241
Shils, E., 206, 212
Simon, H., 51
Sinha, S. and S. Parthasarathy, 229
Skocpol, T., 131, 239
Slack, J.M.W., 223
Smith, A.B., 272, 279
Smith, T.F. and H.J. Morowitz, 140
social adaptation, 117, 118
social learning, and path dependence, 146
social networks, 92

social perturbation, 55
social science models, 53–4
social sciences
  and agency, 295–9
  benefits of evolutionary analogies, 90–2
  and chaos theory, 23, 52
  determinism in, 260–3
  eclecticism, 132
  and economics, 27
  evolutionary analogies, case studies, 89–95
  evolutionary perspective, revaluation, 211–16
  export of metaphors, concepts and methods to natural sciences, 187–95
  and institutions, 177
  and mathematical models, 15
  and natural science, 285
  and path dependence, 26–7, 129–39
  structural differentiation, 212–13, 222
  uncertainty in, 260–4
social structure
  and neutral networks, 92–3
  and semantics, 87
social systems
  bipolarity, 87
  chaos in, 51–63
  and culture, 87
  diversity in, 91
  and neutrality, 85–6
  nonlinearity, 51
  stratified social order, 86–7
social systems change, 52
  agent based simulation, 53–5
  and chaos theory, 52
  and complexity, 55, 56
  evolutionary perspectives, 53
society
  differentiation in, 200
  evolution of, 225
'socio-biology', 115
socio-economic lock ins, 118
sociological differentiation theory, 86
sociology, 5
Sole, R.V. *et al.*, 228
Sombart, W., 201
Somers, M.R., 244
speciation, 223–4, 277, 278, 289

species, entities beyond, 275–6
species concepts, 274–5
Stacey, R., 53
Stadler, B.M.R. et al., 69, 76, 77
'standard operating procedures', 241
Stanley, S.M., 271
Stark, O. and Y.Q. Wang, 221
states, as institutional arrangements, 167
statistical theory, 261–2
steady states, 110, 114, 219
Stebbins, G.L. and F.J. Ayala, 271
Steward, J.H., 5
Stichweh, R., 86
stochastic matrices, 9
Stocker, T.F. and O. Marchal, 41, 42
Streeck, B.M.R. and K. Yamamura, 107
Strogatz, S.H., 224
structuralism, and rational choice, 91
structure, and organization, 86
Subrahmanyam, S., 233
'sunspot equilibria', 108
'survival of the fittest', 9, 90
Swain, P.S. et al., 232
Swedish constitutional reform, 248–51, 282, 293, 295–6
and rational choice, 264–8, 296
'symmetry breaking', 118
system, and behaviour and operation, 86
system dynamics, 108
systemic stability, 7
system level teleology, 91
'Systems Biology', 231
Szathmáry, E., 147, 150, 151

Tagore, R., 232
Tarrow, S., 239
technical knowledge, 103
technological innovation, as a de-locking force, 118
technology
  adoption of, 103
  reversal, 133
Tetlock, P.B. and A. Belkin, 6
Tetlock, P.B. and R.N. Lebow, 249
Thelen, K., 5, 134, 242, 243, 246
Thelen, K. and S. Steinmo, 238, 242
Thom, R., 225
Tilly, C., 239
time, 7–8
  as a succession of instances, 9

time irreversibility, 99
time structure and scale, 288–90
Tiryakian, E., 207
tool transfer, 15, 20–1
Toye, J., 179
'trading zone' metaphor, 21
transformation, 10–11, 12
transition probability matrices, 9–12
Truman, D.B., 237, 238
trust, 219–20, 294
Turing, A.M., 225
Turner, D.H. et al., 71
'turning points', 289–90
  in the life course, 6
Tyson, J.J. et al., 225

uncertainty, 57, 194, 221, 229, 265, 285, 286
  in history, 260–3
  and non-teleological trajectories, 290–5
  in social science, 260–4
unpredictability, and non-ergodicity, 130, 131
Urry, J., 16

Vanberg, V. and J.M. Buchanan, 249
van Parijs, P., 243
variability, 73
variation, 80
Verney, D.V., 264, 266
Via, S. and R. Lande, 224
Vila, C. et al., 190
Vogel, M.P., 62
Voigt, S., 250
voter model, 112
Vrba, E.S. and N. Eldredge, 271

Wächtershäuser, G., 145
Waddington, C.H., 73, 82
Wagner, A., 73
Wagner, G.P. and L. Altenberg, 73
Wahl, L.M. and D.C. Krakauer, 142
Wake, D.B., 230
Wallerstein, I., 2, 131, 132, 239
Walter, A.E. et al., 71
Wang, G.L. and E.A.B. Eltahir, 44
war, 214
Washington, W.M. and C.L. Parkinson, 39
Waterman, M.S. and T.F. Smith, 71

weather
  and climate, 40
  long-term predictions, 59
Weber, M., 92, 249, 298
Weingast, B.R., 241
wellbeing, and modernity, 219–21
West Africa, Sahel region, 44
West Rudner, J., 193
Wheeler, Q.D. and R. Meier, 274
Wiley, E.O., 272
Williams, G.C., 271
Wilson, D.S., 87
Wilson, E.O., 190
Wimmer, A., 17, 214

Winter, S.G. *et al.*, 115
Wolpert, L., 152
Wood, R.A. *et al.*, 42
world systems, 207

Yedid, G. and G. Bell, 142

Zald, M.N. and J.D. McCarthy, 239
Zapf, W., 5
Zhang, J. and H.F. Rosenberg, 149
Zucker, L.G., 241
Zuckerman, A.S., 136
Zuker, M. and D. Sankoff, 71
Zuker, M. and P. Stiegler, 71